Atomic Spectroscopy in Elemental Analysis

Analytical Chemistry Series

Series Editors: John M. Chalmers and Alan J. Handley

A series which presents the current state of the art in chosen sectors of analytical chemistry. Written at professional and reference level, it is directed at analytical chemists, environmental scientists, food scientists, pharmaceutical scientists, earth scientists, petrochemists and polymer chemists. Each volume in the series provides an accessible source of information on the essential principles, instrumentation, methodology and applications of a particular analytical technique.

Titles in the series:

Inductively Coupled Plasma Spectrometry and its Applications
Edited by S.J. Hill

Extraction Methods in Organic Analysis
Edited by A.J. Handley

Design and Analysis in Chemical Research
Edited by R.L. Tranter

Spectroscopy in Process Analysis
Edited by J.M. Chalmers

Gas Chromatographic Techniques and Applications
Edited by A.J. Handley and E.R. Adlard

Atomic Spectroscopy in Elemental Analysis
Edited by M. Cullen

Pharmaceutical Analysis
Edited by D.C. Lee and M. Webb

Chemical Analysis of Contaminated Land
Edited by K.C. Thompson and C.P. Nathanail

Atomic Spectroscopy in Elemental Analysis

Edited by

MICHAEL CULLEN
LGC
Runcorn, UK

Blackwell Publishing

CRC Press

Editorial Offices:
9600 Garsington Road, Oxford OX4 2DQ, UK
Tel: +44 (0) 1865 776868
108 Cowley Road, Oxford OX4 1JF, UK
Tel: +44 (0)1865 791100
Blackwell Munksgaard, 1 Rosenørns Allé, P.O. Box 227, DK-1502 Copenhagen V, Denmark
Tel: +45 77 33 33 33
Blackwell Publishing Asia Pty Ltd, 550 Swanston Street, Carlton South, Victoria 3053, Australia
Tel: +61 (0)3 9347 0300
Blackwell Verlag, Kurfürstendamm 57, 10707 Berlin, Germany
Tel: +49 (0)30 32 79 060
Blackwell Publishing, 10 rue Casimir Delavigne, 75006 Paris, France
Tel: +33 1 53 10 33 10

Published in the USA and Canada (only) by CRC Press LLC 2000 Corporate Blvd., N.W. Boca Raton, FL 33431, USA
Orders from the USA and Canada (only) to CRC Press LLC

USA and Canada only:
ISBN 0–8493–2817–9

First published 2004

British Library Cataloging-in-Publication Data

A catalogue record for this title
is available from the British Library

ISBN 1–84127–333–3
Originated as Sheffield Academic Press

Library of Congress
Cataloging-in-Publication Data
A catalog record for this title is available from the Library of Congress

Atomic spectroscopy in elemental analysis / edited by Michael Cullen.
p. cm.
Includes bibliographical references and index.
ISBN 1–84127–333–3 (hbk.: alk. paper)
1. Atomic spectroscopy. I. Cullen, Michael, 1956–

QD96.A8A865 2003
539.7′028′7—dc21 2003005150

Set in 10.5/12 pt Times
by Integra Software Services Pvt Ltd, Pondicherry, India
Printed and bound in Great Britain by MPG Books Ltd, Bodmin, Cornwall

For further information on Blackwell Publishing, visit our website:
www.blackwellpublishing.com

Contents

Contributors

Professor Michael J. Adams	Department of Applied Chemistry, RMIT University, City Campus, GPO Box 2476V, Melbourne, Victoria 3001, Australia
Mr Michael Cullen	LGC, The Heath, Runcorn WA7 4QD, UK
Miss Vicki Barwick	LGC, The Heath, Runcorn WA7 4QD, UK
Professor Annemie Bogaerts	University of Antwerp, Department of Chemistry, Universiteitsplein 1, B-2610 Wilrijk-Antwerp, Belgium
Dr Andrew Fisher	Department of Environmental Sciences, University of Plymouth, Drake Circus, Plymouth, Devon PL4 8AA, UK
Professor Albert Kh. Gilmutdinov	Department of Physics, Kazan State University, 18 Kremlevskaja Street, Kazan 420-008, Russia
Dr Volker Hoffmann	Institute for Solid State Research, Helholtzstr. 20, D-1069 Dresden, Germany
Professor Stephen J. Hill	Department of Environmental Science, University of Plymouth, Drake Circus, Plymouth PL4 8AA, UK
Dr Norbert Jakubowski	Institute of Spectrochemistry and Applied Spectroscopy, Postfach 10 13 52, Bunsen-Kirchhoff Strasse 11, D-44013 Dortmund, Germany
Mr Ed McCurdy	Agilent Technologies UK, Lakeside, Cheadle Royal Business Park, Stockport, Cheshire SK8 3GR, UK

Mr Don Potter — Agilent Technologies UK, Lakeside, Cheadle Royal Business Park, Stockport, Cheshire SK8 3GR, UK

Dr Melissa J. Romeo — Department of Applied Chemistry, RMIT University, City Campus, GPO Box 2476V, Melbourne, Victoria 3001, Australia

Professor Julian F. Tyson — Department of Chemistry, University of Massachusetts 701, Lederle Graduate Research Tower, 710 North Pleasant St, Amherst, MA 01003-9336, USA

Professor Peter C. Uden — Department of Chemistry, 701 Lederle Graduate Research Tower, University of Massachusetts, 710 North Pleasant St, Amherst, MA 01003-9336, USA

Dr Bruno Vrebos — Philips Analytical, Lelyweg 1, NL 7602 EA Almelo, The Netherlands

Dr Emily Yourd — Department of Chemistry, University of Massachusetts, 701 Lederle Graduate Research Tower, 710 North Pleasant St, Amherst, MA 01003-9336, USA

Preface

Atomic spectroscopy techniques form an essential part of any elemental analysis laboratory. This book gathers together under one cover the most important techniques of atomic spectroscopy currently available to the elemental analyst. Each chapter presents the key principles of a technique, descriptions of the instrumentation and representative applications. The book is intended as an up-to-date source of reference for analytical chemists, environmental chemists, earth scientists, food scientists and petrochemists working in the field of elemental analysis, in both academic and industrial laboratories.

The book begins with a chapter on method validation, written in such a way that it is of general relevance to any of the techniques covered in later chapters. Chapter 2 addresses inductively coupled plasma mass spectrometry (ICP-MS) instrumentation and operation, providing an insight into the ability of collision cell technology to control interference. Chapter 3 on inductively coupled plasma atomic emission spectroscopy (ICP-AES) begins by explaining plasma theory, before moving on to discuss interferences and applications. The chapter includes a useful section on pre-concentration methods. Chapter 4 on analytical glow discharges provides a thorough description of the basic principles and theory of glow discharges. The techniques of glow discharge optical emission spectroscopy (GD-OES) and glow discharge mass spectrometry (GD-MS) are explained, and applications for both conducting and non-conducting samples are included. Chapter 5 on microwave plasma atomic emission spectroscopy describes the application of microwave-induced plasmas (MIP) as detectors – mainly in the area of gas chromatography, but liquid and supercritical fluid applications are also considered. Chapter 6 covers the technique of X-ray fluorescence spectroscopy (XRF), discussing the principal components of wavelength dispersive (WD) and energy dispersive (ED) methods. In Chapter 7, electro thermal atomic absorption spectroscopy (ETAAS) is examined, beginning with the background to furnace operation and presenting the principles behind the electrothermal atomisation process. There is a description of potential interferences and how they may be overcome. Chapter 8 on flame atomic absorption spectroscopy brings a classic technique up to date, with references to new applications in a variety of sectors, including hydride and cold vapour methodologies. Chapter 9 on chemometrics considers experimental design, multivariate statistics and data analysis, as applied to atomic spectroscopy.

This book would not have been possible without the cooperation of the chapter authors, and I would like to thank them for their hard work and the high quality of their manuscripts. I am grateful to Professor John Marshall for his initial suggestions of chapter authors, and to my colleagues at LGC Teddington for their kind permission to use some material from their method validation course.

I would like to thank the publishers for their guidance, and my colleague Alan Handley for his helpful advice. I am also grateful to my family – Liz, Rebecca and William – for the encouragement and support they gave to me throughout the preparation of this book.

Michael Cullen

1 Method validation for atomic spectroscopy

Michael Cullen and Vicki Barwick

1.1 Introduction

In analytical chemistry, methods are regularly developed for a specific purpose and then placed into use in the laboratory. A part of the method development process is to validate the method to prove that the results it gives are reliable. To be able to do this correctly, the intended use of any results must be considered. It might be that the method is required just to give some indication of the elemental composition of a sample, such as that provided by means of a qualitative scan. This type of method can be especially useful for the analysis of unknown materials, where the aim is to identify their composition quickly. Alternatively, the method may have been developed to measure a component in a sample of a pharmaceutical material intended for human application. In this instance, any significant sources of error in the methodology ought to have been identified during the development stage and eliminated. Once the method is placed into routine use in the laboratory, the analytical process must be tightly controlled and monitored to ensure that the results produced continue to be fit for purpose. Although these examples are very different in their purpose, method validation is needed in each case. In both cases, the laboratory should ideally work to a quality standard such as ISO/IEC 17025 [1]. There is a need to demonstrate that methods are fit for purpose and that any results can be relied upon for each specific application. The phrase *fit for purpose* exemplifies the whole idea behind method validation and binds the analytical method to its end use.

Depending on the type of work carried out in a laboratory, there may be a regulatory requirement to use validated methods. It is always worth following any guidance available for your sector of work in order that your validation procedure is sufficiently thorough and focused. This will ensure that it is compatible with other laboratories in your sector. With fully validated methods such as an ISO or ASTM standard, the method validation has generally been thoroughly investigated involving many experiments and *round robin* analysis. Therefore, the amount of validation subsequently required by laboratories working to the standard is only a verification to show that the Repeatability achieved by the laboratory is compliant with the published value and that its results are unbiased.

The EURACHEM publication on the Fitness for Purpose of Analytical methods [2] and the joint EURACHEM/CITAC Working Group guide for Research and Development and Non-routine Analysis [3] are useful guides to method validation.

1.2 The process

In the process of method validation, the very first step is to consider why the method is required and how any results will be used. This will then help to decide how much effort needs to be given to the validation process. If an existing validated method is being modified, such as the addition of another element to a multi-element inductively coupled plasma mass spectrometry (ICP-MS) procedure, then this should not be too large a task to re-validate. On the other hand, developing a new method, possibly in a matrix that the measuring laboratory has had little or no experience of handling, is likely to require a significant amount of effort to ensure a reliable method is produced. Even where the analysis may be considered to be a *one-off* or infrequent test, validation is still needed and must be sufficient to demonstrate that the method is fit for purpose. The requirements specification is a very important part of the initial validation process and care must be taken to get this right. The final performance of the developed method must be shown to meet the needs set out at the beginning. If this is not the case then the method is unlikely to be fit for its intended use and must be developed further before implementation is allowed. Failure to meet the original specification, assuming that it was correct in the first place, will lead to the results being unreliable and could cost the analytical laboratory dearly in time, money and reputation.

Having chosen the candidate method for evaluation, and assuming that the staff carrying out any tests are suitably experienced and qualified, the main stages in a typical validation study are:

- Define requirements specification and create a method validation plan.
- Carry out precision experiments.
- Measure bias.
- Test the method for ruggedness.
- Evaluate the selectivity, limit of detection (LoD) and limit of quantification (LoQ).
- Define the method working range and linearity.
- Carry out an assessment of uncertainty.
- Check for successful outcome compared to the specification in the validation plan.

- Finally, document the validation procedure and write the validated method as a standard operating procedure (SOP), including any performance criteria and quality control (QC) protocols for the method.

If due consideration is given to the above points then the analytical laboratory should have a reliable method fit for the intended purpose at the end of the method development/validation process. In addition, documentation of the validation study and the preparation of an SOP, together with suitable QC protocols, ensures that the method will continue to produce valid results in the future.

1.3 Create the validation plan

After agreeing the requirements specification with the end user or customer, the next stage is to plan a set of carefully designed experiments to evaluate the performance of the method. This plan will include details such as the number of replicate tests, a definition of the parameters to be tested, the types of matrix to be tested, the range of analyte concentrations to be studied and appropriate reference standards. The target values that the method must achieve in order to be considered fit for purpose should be defined in the plan. Depending on the application, some performance parameters may be more important than others. Accuracy will be crucial for calculating absolute values of analytes and for this certified reference materials (CRM) will be required. Precision is important in comparative level studies. The working range of the method will be significant in most applications with the LoD being most relevant for trace work. The use of another validated method as a comparison to the method under test is sometimes beneficial and worth considering as part of the plan. Details of who will carry out the work, the equipment to be used and a time schedule for the work should be included. If the Reproducibility of the method is to be evaluated this will require collaboration with other laboratories and the purchase (or production) of homogeneous and stable reference samples for inter-laboratory testing. To validate a complex method, considerable effort may be required to do the experimental work and funding may need to be sought to finance the project. Therefore, an assessment of the cost associated with the activities in the plan has to be made. However, cost must not be the primary factor in mind when developing the plan. The requirement for the method to be fit for purpose must be the key. It is not good practice to automatically cut down the validation work to fit the budget. There is a danger that without sufficient validation the method may prove unreliable and a risk assessment of the consequences of reporting poorly validated results should be carried out.

Having defined the plan, the next stage is to carry out the various activities taking care to evaluate the results at regular intervals to ensure the method is continuing to meet the target values set in the specification.

1.4 Precision

Precision is a measure of the spread of results, that is, how varied the individual results are from each other. It is not, in itself, related to how close the result is to the true value, and it is dependent on the conditions under which it is measured. It may also be dependent on the concentration at which it is measured and this should be tested by carrying out experiments over a range of concentrations appropriate to the method. There is a likelihood that lower levels of analyte will have poorer relative precision than concentrations at the top of the range of the method, but this may not be significant. Another factor likely to affect precision is the sample matrix and this too should be tested by experiments on a range of typical matrix compositions covering the scope of the method.

The determination of precision is carried out by repeating the analysis under a standard set of conditions. Then an analysis of the data is made using statistical methods. Only a brief discussion of statistical techniques will be given in this chapter, a more detailed coverage of statistical methods is available [4] and the subject of chemometrics in atomic spectroscopy is covered elsewhere in this book.

The precision may be stated in a variety of ways, e.g.:

- Standard deviation (s)
- Relative standard deviation (rsd, CV)
- Standard deviation of the mean ($s/\sqrt{n}$)

Initially, the standard deviation s is calculated, and assuming a normal distribution, approximately 95% of the results will lie between $\pm 2\ s$ of the mean. The relative standard deviation (rsd) is calculated by dividing the standard deviation by the mean. This may then be worked out as a percentage and expressed as a coefficient of variation (CV). Care should be exercised in using relative values as the units are lost and there is a chance that results with different units may be compared and from this the wrong assumptions made. The standard deviation of the mean is obtained by dividing the standard deviation by $\sqrt{n}$, where n is the number of replicates. If a further set of replicates were to be made then this is the variation you would expect to see in the mean values. The distribution of the mean values will be narrower than the distribution of single results by a factor of $\sqrt{n}$.

In Fig. 1.1, the effect of varying the number of conditions during a precision measurement is illustrated. It can be seen that with a greater variation of conditions, the precision (σ) becomes increasingly worse. When the precision estimate is obtained by the same analyst making repeat measurements one after another, in the same laboratory, and on the same instrument, this is known as Repeatability. Alternatively, when repeat measurements are made on a sample

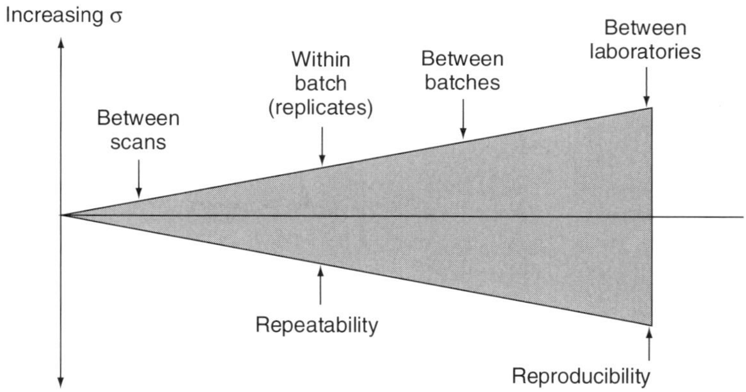

Fig. 1.1 Effect of varying conditions.

by different analysts, on different days and in different laboratories, the precision estimate reflects the variation from all possible sources and represents the widest extreme of precision measurement. This is known by the term Reproducibility.

To get a representative estimate of the precision of an analytical method, the replicate determinations carried out must be sufficiently independent. This is best illustrated by the process shown in Fig. 1.2. In this example, the determination of the acid soluble Mg content of a sample is considered. A portion of the sample under test was weighed, treated by boiling in dilute acid, filtered to remove insoluble matter, then the filtrate made up to a known volume. An aliquot of this solution was taken and diluted to fall into the calibration range of the method. The prepared solution was then measured by inductively coupled plasma atomic emission spectrometry (ICP-AES). Presenting the same solution six times to the instrument (as in step A, Fig. 1.2) provides only the run-to-run precision of the measurement step and is not sufficiently independent to give a representative estimate of the method's precision. To make the replicate tests independent, then six individual portions of sample should have been taken, weighed, boiled in acid, filtered, made to volume, diluted and then each solution measured by ICP-AES. In this way all steps in the process are included and accounted for and each replicate is now independent (as in step B, Fig. 1.2).

Repeatability, as mentioned earlier, is the tightest extreme of precision. It has value in that with knowledge of the Repeatability Limit (r) (limit tests are described later), it provides a basis for determining whether two samples measured one after the other are truly different or if any difference is due to the normal variation to be expected within the method. A precision estimate obtained under Repeatability conditions often does not adequately reflect the

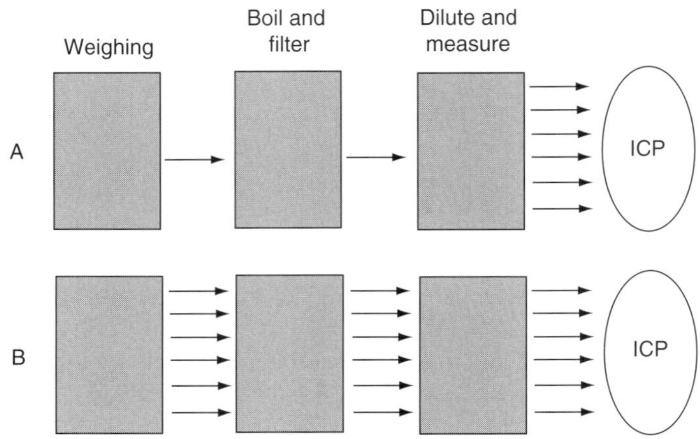

Fig. 1.2 Independence of replicates.

way the method will be used in the laboratory. For example, when different analysts carry out an analytical test, possibly using different instruments, but in the same laboratory. This is the situation often found in a QC laboratory on a chemical production plant. Their requirement is to be able to use the precision estimate to set realistic control limits for a routine test. For the precision estimate to be realistic, the replicate measurements should be made by each analyst over a number of days and, if appropriate, on any duplicate instrumentation. This estimate of precision covers the way the method will be typically used and will inevitably be less precise than the Repeatability precision. This level of precision estimate is known as *Intermediate Precision* or *within-laboratory Reproducibility*, although these terms are not as widely adopted as Repeatability or Reproducibility.

In planning the precision experiments, a decision has to be made as to how many replicates are necessary. This is a problem faced by every analyst carrying out method validation. The aim is to do the minimum number of repeat measurements that will achieve reliable precision values, but without spending too much time and effort generating more data than is necessary. Often it is decided that the analyst should carry out ten replicates in order to generate the precision data. But why should ten be enough, why not twenty or more? It is the case that the more replicates made then the closer the sample standard deviation gets to the population standard deviation. It can be seen from the graph shown in Fig. 1.3 that below six replicates the precision estimate is very poor and above about fifteen replicates there is little to gain by doing more. Therefore, whenever possible it is recommended to do a minimum of six replicates but it is probably not worth doing more than fifteen. It is important not to try to carry out too many replicate measurements in too short a time

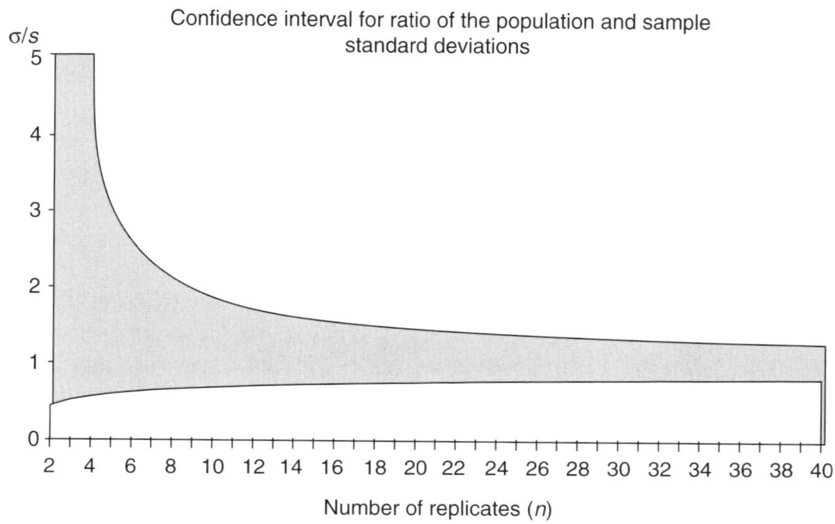

Fig. 1.3 How many replicates?

interval. In doing this, the precision obtained would be unrealistic and may lead to an expectation of a tighter precision than will be actually obtained in the normal day-to-day use of the method. In other words, a Repeatability study is rarely sufficient for evaluating the likely precision of a method in routine use. A study of intermediate precision or Reproducibility is generally also required as part of method validation.

Having decided how many replicates to make, a decision on the conditions under which the precision experiments will be carried out is required. Repeatability is evaluated by carrying out experiments under Repeatability conditions (described previously). When planning a study of intermediate precision, careful consideration needs to be given to the conditions which are to be varied. Ideally, the study of intermediate precision should be a true reflection of the way the method will be used once it has been validated and placed into operation within the laboratory. Repeatability does not often reflect the operation of the method in routine use, as discussed previously. There is therefore little leeway in choosing the conditions for a Repeatability study. There are, however, lots of options for planning an intermediate precision study (different analysts, instruments, etc.). Unlike Reproducibility precision estimates, which need collaboration with other laboratories, Repeatability and Intermediate precision can be determined solely within the same laboratory. In deciding on the conditions for the tests, consider if one or more analysts will be used and if there is duplicate equipment such as automatic pipettes, balances, etc. To get a good estimate of the intermediate precision, it is important to include sufficient variables in the replicate experiments to cover the typical operation of the method. Where different analysts have carried out the replicate testing, this will inevitably lead to a number of data sets being generated. The task then is how to consolidate the data to form a precision estimate appropriate to the method. In this situation, where a number of conditions are being investigated, or if it is impractical to collect large sets of data under Repeatability conditions, then some form of pooling data (i.e. combining smaller sets) may be possible. An invaluable statistical tool for this is to use analysis of variance (ANOVA). ANOVA helps to show if data sets are from the same population or if they are from different populations. If ANOVA indicates the data sets are from the same population then the sets may be combined to form one large data set and the standard deviation of the combined data set calculated to provide an estimate of the precision. ANOVA can also help to pool data from different populations. This technique is an efficient method for obtaining estimates of Repeatability, Intermediate precision and Reproducibility with sufficient degrees of freedom. The procedure for doing this is explained in Chapter 9.

Having obtained the precision estimates, these values may then be used in a number of practical applications. For example, two methods may be compared to determine which gives the lowest spread of results. Alternatively, the precision estimates may be used to compare the performance of individual analysts or laboratories using the same analytical method. Another useful application of precision data is that it can be used in significance testing as a means of detecting significant differences between data. This approach uses a concept known as *precision limit*. By correct application of the precision limit it is possible to say if data are actually in agreement or significantly different. An application of precision limits is shown in Fig. 1.4. In this example, a QC analyst has analysed two samples by ICP-AES, under Repeatability conditions. Are the results (A and B), different due to the normal Repeatability of the method or are they different analyte concentrations? If the precision of the method has a standard deviation $s = s_1$, then the results could be from the same population and the difference is accounted for by the normal variation in the method. However, if the method's standard deviation is $s = s_2$, this method with much tighter precision would indicate that there is a definite concentration variation between the two samples. Between these two precision values s_1 and s_2, there is a value at which there is an equal probability of being able to conclude that the observed difference is significant or not. Likewise, for any precision value it is possible to state a critical difference such that larger differences are significant and smaller differences are not. This critical difference is the precision limit and the equation used to calculate this is based on the Student's *t*-test.

A general form of the precision limit is

$$r = t_{(\nu,\, \alpha)} \sqrt{2} s_{\mathrm{r}} \tag{1.1}$$

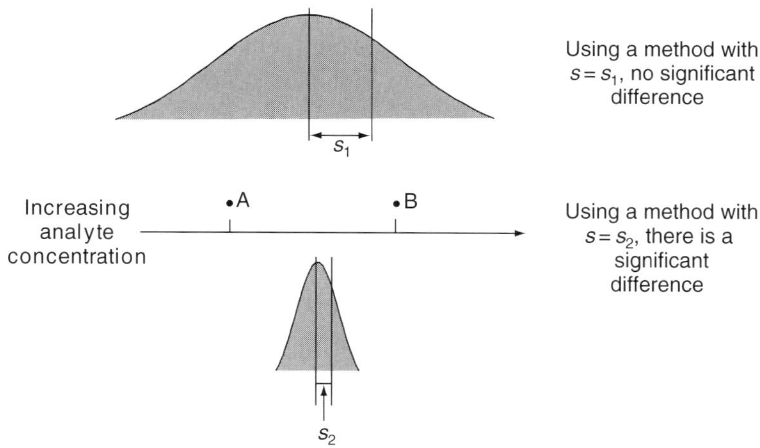

Fig. 1.4 Precision limits.

where r is the Repeatability limit, s_r is the standard deviation under Repeatability conditions, $t_{(\nu,\,\alpha)}$ is the student's t for ν degrees of freedom and α is 0.05 for 95% confidence. The degrees of freedom relate to the estimate of s_r.

In the validation plan for the method, the requirement specification may state that it is important for the method to be able to detect a minimum difference between results. If so, then the limit calculation can be rearranged to obtain the minimum acceptable standard deviation (s_r) required to be able to detect the difference (r). This is shown in Equation 1.2.

$$S_r = \frac{r}{t_{(\nu,\,\alpha)}\sqrt{2}} \tag{1.2}$$

Reproducibility precision provides an estimate of the kind of variation that different laboratories might expect to see when they analyse the same sample using the same validated method. Unfortunately, when the two laboratories use different methods, the Reproducibility estimate is not sufficient in itself to decide if the results have any significant difference or not. Problems can arise when one laboratory may claim a sample is within normal specification, whereas the other believes that the material is *out of spec* as measured by their own method. There is a way to compare the results generated by laboratories using two different validated methods, but this requires more than precision estimates. To compare different methods, the combined standard uncertainty estimate for each method needs to be considered. Uncertainty assessment is briefly covered later in the chapter.

1.5 Bias

In carrying out analytical measurements there will be both random and systematic errors. The random effects are measured by precision experiments. Systematic errors will also be present in the experiments and they too need to be considered in the method validation process. The systematic effects are associated with the property of *trueness* of a measurement.

The ISO 3534 definition of trueness [5] reads:

'The closeness of agreement between the average value obtained from a large set of test results and an accepted reference value' with a note that 'The measure of trueness is normally expressed in terms of bias.'

Under current ISO definitions, accuracy is the property of a result and comprises both precision and bias. Therefore to assess the accuracy of a result the systematic errors need to be quantified. But how can the systematic errors be detected in the method validation process, and if the bias is significant how should it be handled?

The key factors to be considered in carrying out bias tests are:

- The most appropriate reference material should be used.
- The experiments should have sufficient precision to detect practically significant bias.
- The full scope of the method should be properly covered.

In measuring bias, repeat tests are carried out and the average value found is compared to a reference value. Ideally, a CRM which matches closely the composition of test samples should be used to provide the reference value. A CRM is a sample or test material that is produced and characterised to a high standard, and is carefully checked for stability and homogeneity. CRMs provide the means by which a large number of laboratories can obtain comparable results by checking their values against the certified reference value. The use of CRMs is mandatory for some regulatory work. In method validation experiments, the use of natural matrix CRMs is best, as long as they resemble closely the matrix and analyte levels of real samples. Pure reference materials can be used as standards for method calibration and in spiking studies. When buying CRM, check that the uncertainty given on the certificate is small enough for the intended purpose of the method. This is important, as no measurement can be proved to be more accurate than its reference value.

Sometimes appropriate CRMs are unavailable for the scope of the method. In this case, *in-house* reference materials are required. These are prepared by mixing standard materials or typical products to make a sample of known composition. Although it is possible to prepare materials in this way to match a typical sample, the values produced will not guarantee comparability with other laboratories.

Sometimes, a laboratory competent in the operation of a validated standard method chooses to develop a simpler method for routine use. In this, their requirement is to prove they can match the results of the more time-consuming *standard method* with those produced by the new procedure. Studies of bias would involve operation of both methods and then a statistical comparison of the results to see if there is any practically significant bias in the new method. Obviously for this to be good practice, the original standard method should have been validated for bias with an appropriate CRM.

Apart from using a fully characterised reference material as a bias check, the addition of a known amount of analyte to a sample is an important part of the method validation process. This is known as *spiking*. In spiking experiments the sample is tested through the procedure, then an addition is made to increase the level of analyte in the sample and the measurement repeated. It is not good practice to add the spike to the prepared sample just before measurement. The best way to carry out a spike test is to add the spike to the sample before any processing is carried out. In this way the spike is subject to the same effects as the analyte during the sample preparation stages. The spike provides a useful check on losses during extraction, fusion, ashing, wet digestions, etc. Another valuable use of spiking is to test for signal suppression such as that caused by matrix effects in atomic spectroscopy. It will not, however, highlight problems due to overlapping signals, as only the effect of the additional response is indicated. In using spike experiments to estimate bias, it should be realised that the added standard might not behave exactly like the analyte originally present in the sample. Added material often extracts more easily leading to an underestimate of the bias. Therefore, it is important to allow the spike to equilibrate sufficiently before processing. Also, if the chemistry of the sample is unknown to the analyst then the spike might be added in a different form to that in the original sample. If this occurs then the interaction of the spike with the matrix may not match the incurred analyte in the sample. It has been shown that a difference between analyte oxidation state, where one form is more volatile than another, can lead to variation in nebuliser efficiency in ICP-AES or ICP-MS [6, 7]. If an effect such as this is not anticipated, it could result in the spike or calibration standards behaving differently to the analyte in the sample. This would obviously lead to inaccurate results if not picked up.

At the outset of the method validation process, as part of the requirement specification, a decision is made as to the level of bias that would be acceptable for the purpose of the method. In assessing bias, the use of statistical tests will show if the bias at a chosen confidence level is statistically significant or not. Having established the significance at a statistical level, the bias then needs to be assessed for significance at the practical level. This is an important concept as the method may still be shown to be fit for purpose if the estimated bias is deemed to be acceptable for the scope of the method. If however the level of bias is found to be practically significant, then the method must be further refined in order to meet the requirements specification. It is also important to remember that the ability of the statistical test to detect a bias will be influenced by the number of measurements made and the magnitude of the method precision compared to the bias.

Bias checks need to be sufficient to cover the full scope of the method. It may be that the method will only be used for one sample type with a narrow variation in analyte concentration. In this case, the bias check may be made on a typical sample assuming that it represents the normal operation of the method. On the other hand, the method may have a very wide scope, covering many materials and a large dynamic concentration range. To check for bias over a wide scope, the bias tests could be carried out by spiking a range of typical samples taken at random. Alternatively, it may be preferable to do a smaller number of test experiments carefully chosen to cover the extremes of sample type and at a range of concentration appropriate to the application. Failure to assess the bias across the full scope of the method will probably result in an overestimate of the performance of the method and this may lead to problems in the future.

1.6 Ruggedness

In the application on a routine basis of an analytical method, there are likely to be operating conditions that differ from those which occurred during the method development stage. For example, the analysis of liquid samples by X-ray fluorescence (XRF) would probably be carried out in a sample cup fitted with a polymer film window. For each new measurement the sample cup would have a fresh piece of film fitted. Eventually the roll of film will be replaced by a new batch and this may have an effect on the blank due to variation in the impurities between the batches of film. If the sample as presented to the XRF is at high concentration levels, then the effect of the small amount of impurities normally present in the film may not be noticeable. However, it is important to check if a new batch of film does have any significant effect or not. Otherwise, in normal use the method may fail to match the expected performance. Therefore, during method development certain key experimental parameters need to be identified and then assessed to see what effect small changes in those parameters might have on the performance of the method. This is known as robustness or ruggedness testing.

A definition of ruggedness testing has been made by the AOAC [8]: ‘Intralaboratory study to examine the behaviour of an analytical process when small changes in the environmental and/or operating conditions are made, akin to those likely to arise in different test environments.’

For a ruggedness test, experiments are defined and carried out to see which parameters have any significant effect on method performance. Then any parameter or condition showing a significant effect must be controlled if the method is to meet the requirements specification during routine use. If a parameter has a great effect on performance, but cannot be adequately controlled during routine application of the method, then the method is not rugged and must be further developed.

There are many parameters that can affect method performance e.g. pH, temperature, ratio of reagent to sample, pump rate, extraction time, etc. Those that are appropriate to study in the ruggedness test will depend on the individual method. Any parameter that is not being tested must be tightly controlled while another is being subjected to the ruggedness test.

Testing ruggedness requires that some criterion be chosen to see how the variation in method conditions affects it. A common approach is to monitor any change in the measured concentration of the analyte during the test. Other parameters are useful in this way too, such as changes in peak retention time or peak resolution. This approach would apply to methods using chromatography linked to ICP for speciation studies, for example.

One way to carry out ruggedness testing is to vary one parameter while holding all others constant. However, this would be both time consuming and inefficient, especially if a large number of variables were tested. It is far better to use a form of experimental design to enable a set of parameters to be varied in a specific and logical way, and this will greatly speed up the evaluation process. The AOAC [9] describe how to set up ruggedness experiments and there are other publications available to provide guidance on experimental design [10].

On completion of the ruggedness tests, any parameters that show a significant effect on the method's performance must be tightly controlled such that any variation (within these control limits) is no longer a problem. If ruggedness experiments demonstrate that it is not possible to control the method parameters within these limits then the method requires further development.

1.7 Limit of detection, limit of quantification and selectivity

The method's performance at the lower reaches of the calibration range is generally covered by two concepts, the LoD and the LoQ. In this region the ability of the method to quantify reliably, or even to detect the presence of the analyte is brought into doubt. It is part of the method validation process to define this region and to state these limits. The LoD relates to the level of analyte at which the probability of not detecting it, when it is actually present in the sample, is sufficiently small. The LoQ is a limit below which the uncertainty of the method becomes unacceptable. The LoQ is, therefore, only of use when associated with an estimate of measurement uncertainty and is usually equal to the lowest standard in the calibration set.

There are two types of error associated with the detection limit. These are the occurrence of false positives and false negatives. In the case of a false positive, the analyst declares that the sample contains the analyte when in fact it does not. This type of error might lead to a costly clean up of a chemical process or loss of a production batch wrongly thought to be contaminated. In the case of a false negative error, the analyst declares that the sample does not contain the analyte when in fact it does. The consequences of these errors depend very much on how the result will be used. The LoD is usually specified so that both the false positive and false negative rates are adequately controlled.

Evaluation of the detection performance of a method is normally carried out for two purposes. The first, in the case of general method validation, is to indicate the likely lower operating limit for the method. Often this does not need to be very accurately determined, particularly if it is known that the concentration of the analyte in test samples will be significantly greater than the lower operating limit of the method. The other reason why evaluation of detection performance is required, is for use in applications where a critical value or decision limit needs to be set at which some action is taken. This kind of analytical method is used when it is important to know if the sample contains the analyte or not and the analyte may be present at levels approaching the lower operating limit of the method. In this case, a statistical assessment of the lower operating limit of the method must be carried out and from this an appropriate limit value assigned. After the validation process is over and the method placed into routine use, the LoD should be based on the current performance of the method, as this may have changed since the method validation stage.

To measure the LoD and LoQ, six to ten replicates of the blank should be tested. The measurements must be independent, i.e. through the whole procedure, not just one blank solution measured ten times by the instrument. Then, as a general guide, three times the standard deviation of the blank will be a useful value for the LoD. This will give a reasonable estimate of the LoD for applications where the detection limit is not critical. For the LoQ, the acceptable value is a matter of judgement. Again, as a guide, ten times the standard deviation of the blank will provide a reasonable value for the LoQ for normal use. However, a standard should be included at this level in the calibration set as a check during routine calibration. For some applications where detection limits are critical, a full and more rigorous statistical study will be required to ensure a fit for purpose method.

Whenever blank corrections are made it should be realised that the effect of random errors in the blank value will increase the variation in that of the corrected sample values. If a single blank value is applied to a set of samples measured as a batch, the variation (standard deviation) of the sample results will be the same as if no blank correction had been made. However, any new batches of samples would be corrected by a different blank reading. Therefore random errors in the blank will affect sample readings long term. Consequently, when estimating the LoD it must be known whether any blank correction was applied during the validation study, and whether the study applied the procedure for blank correction that will be used routinely. If corrections were made during the method development stage, then this factor must be taken into account during the application of the method for routine use.

Selectivity is evaluated in the presence of interfering components. It is related to the accuracy of the method as the presence of interferants will contribute to the bias. A requirement of selectivity testing is that the chemistry of the samples and the analytical method is known and understood. Selectivity experiments are carried out to study the effect of potential interferences on the expected response of the analyte or the test. It is important to identify any possible source of interference, and then to carefully plan experiments to determine if there is any significant effect on the method. One way to do this is to design the experiments to take the form of a ruggedness test, using the potential interferents as factors, and to cover a wide range of typical materials. An example of a selectivity interference is a peak overlap due to another component in the sample, resulting in signal enhancement of the measurand.

1.8 The working range and linearity

A definition of the working range of the method is 'the set of values of measurands for which the error of a measuring instrument is intended to lie within specified limits' [11]. That is, the working range defines the range of values over which the method can be used. It is part of the method validation process to evaluate and confirm that the working range meets the requirement specification. The lower concentration limits of the working range are defined by the LoD and the LoQ. The upper limit is defined by concentrations where anomalies in sensitivity occur. An example of a sensitivity anomaly is the plateauing effect seen at high absorbance due to detector saturation. Between these boundaries lies the working range of the method, and in certain sectors (e.g. pharmaceutical) it is normal to develop methods such that the working range extends by at least $\pm 20\%$ beyond that of the measured analyte level. In Fig. 1.5, the working range of a calibration is shown. It may also be seen from this figure that the linear range is that part of the working range where the response is directly proportional to the change in concentration. If the concentration range over which the method is to be applied is not very large, then the linear range may be assessed visually. However, if the method is to be used over many orders of magnitude it is better to split the assessment of the linear range over a few ranges and check each in a separate study. A quantitative

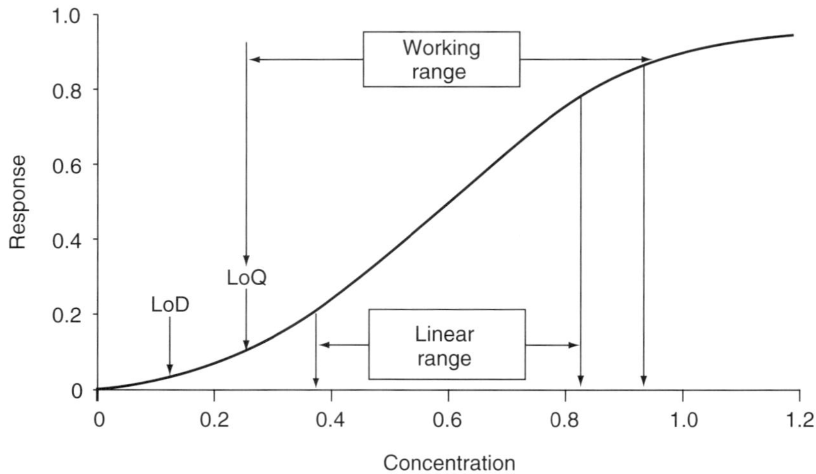

Fig. 1.5 Working and linear range.

evaluation of the linear range may be obtained by statistical methods associated with fitting linear or higher functions to the response data. Usually more standards and replicates are required to establish the working range of the method than will be used during the routine application. But, unlike the determination of precision described previously, when establishing linearity the standards do not need to be independent.

The instrument linearity and the whole method linearity should be assessed separately. Checking both instrument and method linearity is useful as the instrument may show a linear response for pure standards but when CRMs are used and taken through the method, the response can be non-linear due to inferences. As part of the method validation process it is important to study the working and linear range for each sample matrix covered by the scope of the method.

Linearity of a method is always assessed visually and backed up by statistical assessment. The common statistical assessment is based on least squares regression and includes the correlation coefficient, the residual standard deviation, and the gradient and intercept of the calibration curve. A plot of the residuals can also be useful for evaluating linearity.

To achieve a reliable study of the working range of the method there are a few basic rules to follow. It is important to use standards that not only cover the range of interest but also extend beyond it. The calibration set should be chosen so that the typical value for a real sample will lie near the mean concentration of the calibration set. This will reduce the uncertainty in concentration predictions made from the regression line. The values chosen for the calibration set should be evenly spaced and include any specific levels required by the intended application of the method. For example, if there is a specific threshold limit for a regulatory compliance, then that concentration level should be included in the set. At least six concentration levels should be used and between two and five replicates for each level in the set. Once the linearity is established, a much simpler calibration routine may be used for normal method application. Some analytical methods may produce a non-linear calibration, but still follow a mathematical function. In these cases the linearity of the method can still be assessed by analysing a range of reference materials and plotting the predicted values against actual concentrations of the reference materials. An examination of the linearity of this plot may then be made and judged against the requirement specification.

1.9 Equipment qualification

During method development and validation work, any equipment used must be performing correctly, otherwise the quality of the work may be compromised. Equipment qualification is a formal process that provides documented evidence that the instrument is truly fit for purpose. As with the other aspects of validation, it has to be decided how much effort needs to be put into equipment qualification to ensure confidence is achieved, but without doing unnecessary work. Instruments vary greatly in their complexity and the required performance characteristics must be considered. It is the need to demonstrate the fitness for purpose that determines the level of effort to apply. Quality standards require that any equipment used in the analysis is suitable for its intended use and accreditation bodies need to see supporting evidence that this is the case.

The protocol used by the laboratory for the equipment qualification process must, in itself, be fit for purpose. Guidance, developed by the Valid Analytical Measurement (VAM) Instrumentation Working Group [12], is available and is based on four stages of qualification. The VAM working party included manufacturers, suppliers and users along with representation from accreditation bodies and regulatory authorities. The four stages of equipment qualification are:

- Design qualification (DQ)
- Installation qualification (IQ)
- Operational qualification (OQ)
- Performance qualification (PQ)

A summary of each stage is presented here but reference to the guidance publication is recommended.

DQ is the planning stage of the qualification process. At this stage, consideration is given as to what will be fit for purpose and what the instrumentation will be required to do. For a new instrument this stage may be quite complex. The specification would be defined by key performance characteristics and would involve identification and selection of suppliers. If an existing instrument is to be used then the DQ process is simpler. It may be no more than a check to see if the instrument is capable of operating over the required concentration range and if it can provide the level of selectivity, precision and accuracy needed for the method. Often it will be a compromise between the ideal and what is available. This stage, even with existing equipment, should be documented to provide evidence that it was carried out.

IQ is primarily for new instruments but it may be useful to carry this out on existing equipment following upgrades or if the instrument is relocated. IQ confirms that the instrument is received by the laboratory as originally specified. A check should be made that the correct hardware and software is supplied and is undamaged, although unpacking the new instrument without the supplier's engineer present may invalidate the warranty. The IQ stage also confirms that the environment is suitable for the correct operation of the instrument.

OQ establishes that the instrument functions according to its specification. It is a check on the key operating parameters of the instrument to see if the actual performance matches the specification. OQ is carried out after installation and at regular intervals thereafter. The frequency of this depends on events that may affect performance. Factors such as repair or replacement of key parts, the level of use (infrequently or heavily used), the environment (dusty or clean), the nature of the samples used (high dissolved solids or clean waters), will have an influence on the frequency of OQ. Also, there may be recommended intervals prescribed by the instrument manufacturer. Not all performance tolerances laid out in the DQ need to be checked during OQ unless problems are indicated during performance checks during routine use.

PQ checks that the entire instrument continues to perform correctly and is fit for purpose during routine use. The evidence for this may be obtained by everyday method-related checks such as calibration and analytical QC tests or by an independent test procedure. This procedure, could be a standard, reliable method that can be operated on the instrument to provide results to compare with those generated previously using the same test. PQ checking is important, as the instrument is likely to degrade over time from its optimum performance level, and it is necessary to show that even with a loss of performance it is still fit for purpose. It is useful to have a warning and an action threshold defined for performance, so that if the performance degrades beyond the action threshold, some action is taken to bring the instrument back to an acceptable level. For general applications, the use of in-house developed PQ checks is normally sufficient to confirm correct performance. However, where an instrument is used to determine absolute values for parameters or to determine the accuracy of a parameter, traceable standards and calibrated apparatus should be used to verify instrumental performance. In practice, whether traceable standards can be used may ultimately depend on how easy it is to obtain them.

It might be considered that, if PQ is carried out and the performance of the instrument shown to be fit for purpose, then it is not necessary to carry out the OQ. This is not correct. OQ confirms that the instrument operates as it was *designed to do* and this is particularly important after the initial installation of a new instrument and following routine service, maintenance or relocation.

1.10 Assessment of uncertainty

Before attempting to assess uncertainty it is strongly recommended that appropriate guidance is sought and followed, especially for accredited laboratories. With this in mind only a brief overview of uncertainty measurement is given here. The ISO guide to the Expression of Uncertainty in Measurement [13] and the EURACHEM/CITAC guide on Quantifying Uncertainty in Analytical Measurement [14] are useful reference publications.

The ISO definition of measurement uncertainty is 'A parameter, associated with the result of a measurement, that characterises the dispersion of the values that could reasonably be attributed to the measurand'. In other words, this is the plus or minus figure associated with the reported result.

In making an assessment of uncertainty, the first step is to identify and record all factors associated with the method that may cause significant variation in the results. These factors will be sources of uncertainty. To do this effectively the analyst must have a thorough understanding of the instrumentation and the methodology. Some components will have a larger contribution than others to the overall uncertainty, and they need to be identified, assessed and controlled.

In general, factors to consider include:

- Uncertainty of standards and their preparation
- Uncertainty associated with sampling and sample preparation
- Uncertainty of the calibration
- Long-term precision of the method
- Bias uncertainty
- Analyst effects (this is a source of information rather than a contribution to the uncertainty)
- Uncertainty associated with laboratory apparatus and equipment.

In stating measurement uncertainty it is crucial that the uncertainty value quoted combines all uncertainties, whether they are from random or systematic effects. Most of the data required to make an estimate of uncertainty will have been collected during the method validation study, but all factors that influence the result must be considered. Those factors found to have a significant effect on the result must then be combined following a set of rules. ISO recommend that uncertainty components are characterised by the estimated variance s^2 or standard deviation s, and the combined uncertainty should be characterised by the numerical value obtained by applying the combination of variances. Also, that the combined uncertainty and its component parts should be expressed in the form of standard deviations. Uncertainties expressed as a standard deviation are known as standard uncertainties. Each of the uncertainty components considered significant needs to be quantified, either by carrying out a set of experiments to produce standard deviations, or by other methods. These other methods include information from calibration certificates, CRM certificates, historical data, instrument specifications and allowances based on the analyst's experience. Having quantified the effect of each factor, the general approach is to express each component as a relative standard deviation and combine them by taking the square root of the sum of their squares. This will give a combined standard uncertainty in the form of a relative standard deviation. The combined standard uncertainty is usually expanded to provide an expanded uncertainty, using a coverage factor. For approximately 95% confidence this coverage factor is normally taken as 2.

It is a requirement of laboratories working to ISO/IEC 17025 [1] that they make an assessment of uncertainty. Normally, calibration and metrology laboratories will require a formal uncertainty budget to be prepared by a thorough study of the method and the materials to be analysed. However, for industrial testing laboratories or their equivalent, often an estimate of uncertainty has to be made based on much less data. For this, a working estimate of uncertainty is made on the basis of the bias and precision studies carried out during method validation, and by taking other factors into account such as spike recovery results. Using this simplified approach is better than making no attempt at all, but it should only be used when the analysis requirement does not need a full evaluation, or if it is not possible to carry out a detailed study due to insufficient sample or lack of reference materials, etc. As with the other aspects of method validation, uncertainty estimation can be time consuming and therefore expensive. The degree of effort to be placed in uncertainty assessment depends on the requirements of the laboratory, accreditation and regulatory bodies, or the needs of the customer. Caution should be exercised in adopting a simplified procedure for uncertainty estimation as this may not be sufficient for accreditation purposes. Failure to properly consider and quantify all of the possible sources of uncertainty, may lead to an underestimate of the method's uncertainty.

1.11 Has the method met the requirements set out in the validation plan?

This is the final stage, but the analyst must be aware at all times during method development and validation, that if the precision, bias, etc. do not match the requirements, then the process needs to be halted and improvements made to the method. If the method does not meet the intended specification then it will not be possible to validate it as fit for purpose and an alternative needs to be sought.

After carrying out what may be an extensive set of experiments, amassing a large amount of experimental results and other information, it is recommended that the output (raw data, calculations, computer printouts, etc.) be collated. It is important to be able to show evidence of any validation work and having it collected in an accessible form and kept in a secure place will enable the laboratory to show auditors or customers that the method is fit for purpose and the results are reliable. It is easy to carry out work and be happy with the outcome at the time, only to find at some point in the future that a vital piece of evidence is missing and cannot be recovered. This may lead to non-compliance with regulatory authorities or cause unnecessary problems during audits.

It is also important to fully document the validated method in the form of an SOP, so that it will be applied correctly in the future. The SOP should include:

- Scope of method
- Safety issues related to chemicals equipment and apparatus
- Principle of method
- Reagents
- Apparatus
- Method procedure (including any calibration procedures)
- Quality control and quality assurance
- Expression of results
- Performance data
- Measurement uncertainty

A properly developed, validated and documented method is good science and will provide the laboratory with confidence that the results they produce are reliable and truly fit for purpose.

Acknowledgements

This chapter was developed from material that appears in the LGC analytical quality training course on Method Validation. The authors are also grateful to LGC for permission to reproduce figures from the course material.

References

[1] BS EN ISO/IEC 17025:2000 General requirements for the competence of testing and calibration laboratories.

[2] 'The Fitness for Purpose of Analytical Methods – A Laboratory Guide to Method Validation and Related Topics', EURACHEM, ISBN: 0–948926–12–0.

[3] 'Quality Assurance for Research and Development and Non-routine Analysis', EURACHEM/CITAC CG2, 1998, ISBN: 0–948926–11–2.

[4] 'Statistics for Analytical Chemists', Caulcutt, R. & Boddy, R., Chapman and Hall, London, ISBN: 0–412–23730–X.

[5] ISO 3534–1: 1993 Statistics – Vocabulary and symbols – Part 1: Probability and general statistical terms.

[6] Summerhays, K.D., Lamothe, P.J. & Fries, T.L., *Applied Spectroscopy*, 1983, **37**, 25.

[7] Cox, J.R., Pickford, C.J. & Thompson, M., *Journal of Analytical Atomic Spectrometry*, 1992, **7**, 635.

[8] AOAC Peer-Verified Methods Program Manual on Polices and Procedures, 1998, AOAC International.

[9] 'Statistical Manual of the AOAC', Youden, W.J. & Steiner, E.H. 1975, AOAC, Arlington, VA, USA, ISBN: 0–935584–15–3.

[10] 'Efficiency in Research, Development and Production: The Statistical Design and Analysis of Chemical Experiments', Davies, L., RSC publication, 1993, ISBN: 0–85186–137–7.

[11] 'VIM* – International Vocabulary of Basic and General terms in Metrology', 2nd edn, 1993, ISO Geneva, ISBN:92–67–01075–1.

[12] Bedson, P. & Sargent, M., 'The Development and Application of Guidance on Equipment Qualification of Analytical Instruments', *Journal of Accreditation and Quality Assurance*, 1996, **1**, 265–274.

[13] 'The Expression of Uncertainty in Measurement' (ISO 1993, 2nd edn, 1995).

[14] 'Quantifying Uncertainty in Analytical Measurement', EURACHEM / CITAC CG4, 2000, ISBN: 0–948926–15–5.

2 Inductively coupled plasma mass spectrometry

Ed McCurdy and Don Potter

2.1 Introduction

Inductively coupled plasma mass spectrometry (ICP-MS) was developed as a commercial analytical technique in the early 1980s and has since been applied to the determination of trace, minor and major elements in almost every analytical field. Virtually all elements can be measured, including alkali and alkaline earth elements, transition and other metals, metalloids, rare earth elements, most of the halogens and some of the non-metals. High sensitivity and low background signals combine to give very low detection limits (sub-ng/L in most cases) and the rapid scanning quadrupole analyser means that three replicate measurements of a full suite of elements take only about four minutes.

ICP-MS can be considered as a viable alternative to ICP-atomic emission spectrometry (ICP-AES) for fast measurement of higher concentration elements (μg/L to mg/L or ppb to ppm concentrations), while it also rivals or, in many cases, exceeds the detection capability of graphite furnace atomic absorption spectrometry (GFAAS) for the determination of trace and ultra-trace elements (ng/L or ppt concentrations). The fact that ICP-MS can measure all these elements in a single multi-element acquisition, provides simple mass spectra with low interference levels, accepts almost any sample type and also provides isotopic information, helps to explain the widespread acceptance of ICP-MS, with many well-equipped trace metals laboratories regarding ICP-MS as indispensable.

ICP-MS samples, typically as solutions, are introduced into a nebuliser, where a stream of argon gas breaks up the liquid into a fine aerosol and sweeps the smaller droplets into the central channel of an argon ICP. The very high temperature of the ICP (>8000°C) causes the sample droplets to be dried, decomposed and dissociated into individual atoms, which are then ionised before being extracted, through an interface, into the vacuum system. The vacuum system consists of a series of differentially pumped chambers, the final stages containing the mass spectrometer and ion detection systems. In the vacuum system, electrostatic lenses keep the ions focused and separate the positively charged ions from photons, electrons and neutral particles. Analyte ions are separated by the mass spectrometer (MS), which scans rapidly across the mass range, allowing each mass of interest to pass sequentially to the electron multiplier (EM) detector. The counts measured at the detector are passed to a data processor, where they are converted to elemental concentration or isotope ratio data.

ICP-MS instruments can generate fully quantitative data (external calibration or method of standard additions), semi-quantitative data (concentrations derived by comparison with a known internal reference element), qualitative scans (for screening) and isotopic information (isotope ratio measurement and isotope dilution analysis).

2.2 History and development

In the 1970s, Dr Alan Gray of Applied Research Laboratories in Luton, UK, working with a capillary DC arc plasma coupled to a quadrupole mass spectrometer, published early results and the first mass spectra acquired from an analytical plasma [1]. This work stimulated research into the use of inductively coupled RF plasmas (ICPs), and a landmark 1980 publication by Houk *et al.* [2] demonstrated the possibilities offered by the ICP-MS technique. The first commercial systems followed, from VG in the UK (based on a development of Gray's original prototype) and Sciex in Canada (after initial work using a microwave-induced plasma (MIP) [3]) in the early 1980s.

These systems were the successful marriage of two existing technologies – the argon ICP, already in use in ICP-AES, and the quadrupole mass spectrometer, then being applied in fields such as gas chromatography mass spectrometry (GC-MS). Although some changes were necessary to allow the ICP to operate in physical contact with the grounded spectrometer interface, the characteristics of these existing technologies were well matched and the performance of the first systems was impressive.

Since the first commercial ICP-MS systems were launched, major developments have occurred in sample introduction, plasma efficiency, ion transmission, interference removal and dynamic range. Even so, the major components of a modern ICP-MS instrument can be traced back to the early systems, proving the original concept to be an inspired combination.

2.3 Instrumentation

The main components of a typical commercial ICP-MS instrument are shown in Fig. 2.1 and are outlined in the following sections.

2.3.1 Sample introduction

Many different forms of samples can be introduced into the ICP-MS, including liquids, solids or gases. Only the more commonly used are discussed here in any detail, but an excellent review of the many methods that are used to deliver samples to the ICP is contained in reference [4]. The analysis of liquid samples

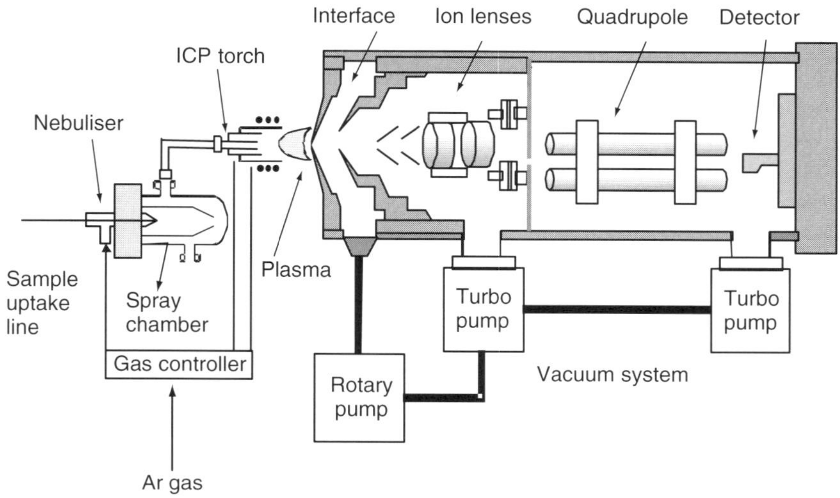

Fig. 2.1 Schematic showing the main components of an ICP-MS instrument.

is normally by continuous pneumatic nebulisation of the sample solution, followed by aerosol droplet separation using a spray chamber. Alternatively, the sample may be delivered to the plasma using flow injection (FI), aerosol desolvation or direct injection. The 1988 reviews of nebuliser and spray chamber design and performance by Sharp [5, 6], although some years old, provide an excellent overview of the principles involved.

Solid sampling methods include laser ablation (LA), spark ablation and slurry nebulisation, with LA being the most used in analytical laboratories. Gas sampling has been utilised as a method of separating volatile elements (in the gaseous state) from complex sample matrices using hydride or vapour generation. Currently there is great interest in the direct coupling of GC and ICP-MS for speciation measurement. The linking of a graphite furnace to an ICP-MS has also received attention, although the limitations of the graphite furnace inhibit the fast multi-element capabilities of ICP-MS, so that this combination is rarely used in routine applications.

In common with ICP-AES instruments, the key parts of the sample introduction system for liquid sample analysis are the peristaltic pump, nebuliser, spray chamber and plasma torch. When used in conjunction with a sample introduction device that does not deliver a liquid sample (such as GC or LA), the pump, nebuliser and spray chamber are bypassed and the sample material is introduced directly at the base of the plasma torch. In early ICP-MS instruments, liquid sample introduction systems readily available from existing ICP-AES instruments were used, with little modification. However, the guiding principle for designing a sample introduction system for ICP-MS should be the maintenance of a high and stable plasma temperature, to ensure efficient and consistent conversion of atoms into ions [7]. The main factors that affect the performance of the sample introduction system are those that determine the amount of sample material that reaches the plasma and the opportunity that the plasma has to decompose and ionise the sample effectively. These factors are the sample transport efficiency, the water vapour content of the aerosol, the plasma energy available in the central channel and the effective residence time of the sample in the plasma.

Sample flow rate is controlled either by a peristaltic pump or by self-aspiration, depending on the design of the nebuliser. Modern ICP-MS systems typically use a sample flow rate of between 0.1 and 0.5 mL/min, which is significantly lower than that used for ICP-AES. Nebuliser types used with ICP-MS include concentric, cross-flow and v-groove designs, which differ in their efficiency and tolerance of dissolved and suspended solids. An ultrasonic nebuliser (USN), with or without a desolvation device, produces a very high proportion of transportable aerosol (very small mean droplet size) and the desolvator can reduce the cooling effect of the USNs dense aerosol. The inherent high sensitivity of ICP-MS means that the further sensitivity enhancement available from a USN is rarely needed. Several references from the early work on ICP-MS instrument development describe the relationship between the spray chamber temperature and the level of interfering polyatomic ions in the mass spectrum [8]. The role of the spray chamber in removing excess water vapour is well understood and most manufacturers now use a cooled spray chamber, either utilising a re-circulating water chiller or electronic temperature control via a Peltier device.

In the ICP, the plasma is formed in a *torch*, which consists of three concentric tubes, typically of quartz, each carrying a regulated flow of high purity argon. The torch design used in most ICP-MS instruments is based on the Fassel design previously used in ICP-AES systems [9]. Electrons in the argon gas stream are excited by a radio frequency (RF) electrical field, by inductive coupling from the load coil, which carries a high power rf voltage and is positioned around the outside of the open end of the plasma torch. The plasma torches used in commercial ICP-MS instruments vary in the internal diameter of the central (carrier or injector) gas tube, which affects both the aerosol density in the plasma central channel and the speed of the carrier gas flow. For a given gas flow rate and aerosol loading, a larger injector diameter will give a more diffuse and slower moving aerosol, both of which lead to improved matrix decomposition in the plasma.

2.3.2 Plasma

One definition of a plasma is ‘a substantially ionised gas, the properties of which depend significantly on the ionisation; the gas remains electrically neutral overall'. In atmospheric pressure plasmas, the gas density is high and so the mean free path of electrons and ions is short. The number of collisions per atom per second is also high, with the result that the kinetic energy of the charged particles is distributed uniformly throughout the gas cloud. The majority of the electrons, atoms and ions in the gas cloud have high energy, so a high pressure plasma is an intense source of light and heat [10].

The plasma is formed in the argon plasma or coolant gas, flowing at approximately 15 L/min, with a secondary or auxiliary gas flow (approximately 1 L/min) being used to push the plasma away from the torch injector tip. The third gas flow is the carrier flow, which transports the sample through the centre of the plasma, where its proximity to the high temperature plasma discharge causes the sample to be dried, decomposed, dissociated, atomised and ionised, before entering the vacuum system through the interface cones. The torch and load coil arrangement used in commercial ICP-MS instruments and the processes which the sample undergoes as it passes through the plasma are illustrated in Fig. 2.2.

Two frequencies of RF generator are used in commercial ICPs, operating at a nominal frequency of 27.12 or 40.68 MHz, with the frequency controlled by a piezoelectric crystal and the power amplified to output levels at which an analytically useful plasma can be sustained [10]. There is little analytical difference between the two frequencies, but 40.68 MHz generators are more susceptible to a secondary or *pinch* discharge between the plasma and the spectrometer interface [11], and are always used in conjunction with a means of reducing the plasma potential. Commercial systems use a balanced or

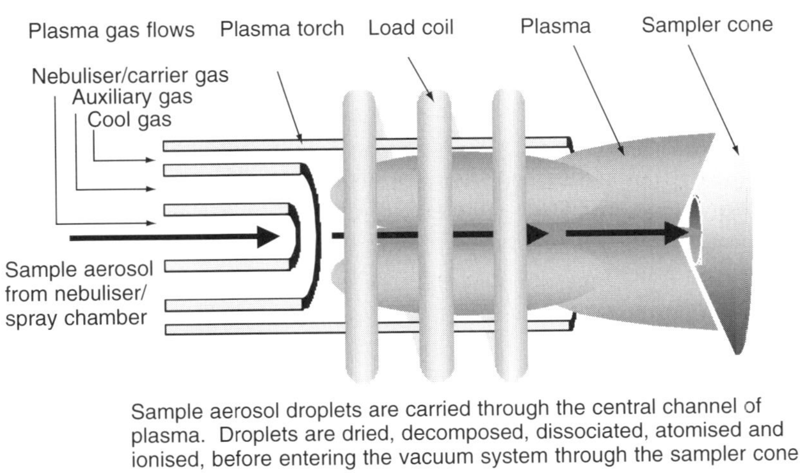

Fig. 2.2 Schematic showing the components of the load coil, plasma torch and plasma and processes of sample decomposition in the plasma.

compensated load coil design [12], which reduces the plasma potential to below 5 electron volts (eV) [13] and hence controls the secondary discharge. If this discharge is not controlled, it erodes the interface cone, leads to high and variable background signals and induces high ion energy. 27.12 MHz generators are commonly used in commercial ICP-MS instruments, and the secondary discharge is controlled (at about 5 eV), even when a simple end-grounded load coil is used [14]. A further benefit of the 27.12 MHz frequency is that the hottest part of the plasma extends closer to the plasma central channel [15], which means that the aerosol travelling through the central channel is typically exposed to a higher temperature, leading to a lower CeO^+/Ce^+ ratio – a measure of matrix decomposition and plasma robustness.

Shielded load coil designs, which remove the capacitive coupling of the load coil to the plasma, were discussed in early papers on ICP-MS [16] and Yokogawa developed the technology commercially in the 1990s under the trade name *ShieldTorch* [17]. Shielded torch designs offer two benefits compared to incompletely grounded plasma configurations. Firstly, the reduction of the plasma offset potential to close to zero volts ensures that the secondary or pinch discharge is eliminated, even at low forward power settings. This simplifies operation under *cool plasma* conditions, as first reported by Houk *et al.* in an application to measure potassium isotope ratios [18]. Cool plasma operation has been widely adopted in the semiconductor industry, where the removal of the Ar-based polyatomic interferences on the major isotopes of potassium, calcium and iron allows ICP-MS to be applied to the accurate quantification of these elements, as well as many others, at ng/L levels in high purity process chemicals [19]. In addition to removing Ar-based polyatomic ions from the mass spectrum, cool plasma conditions also effectively eliminate the background signal for easily ionised elements, such as Na and Li, improving detection limits for these elements. Cool plasma conditions do introduce some disadvantages to general analysis, most particularly poorer matrix tolerance and less effective decomposition of polyatomic oxide interferences.

The second benefit of very low plasma potential is that the ion energy is reduced, which improves the efficiency of ion focusing, since the ions follow a more consistent trajectory, and energy-dependent deflection is minimised. The improvements can be dramatic, with a significant increase in sensitivity achieved with a shielded torch configuration, compared to the same instrument without the shield fitted [20].

2.3.3 *Interface*

The effect of good plasma/vacuum interface design on the overall performance of the ICP-MS instrument cannot be overstated. Indeed, since both the ICP and the quadrupole MS were already in use at the time ICP-MS was developed, the development of the interface between these two existing technologies could be viewed as the key event that led to ICP-MS as a viable analytical technique. The ICP-MS interface consists of a step-down vacuum stage and a pair of cooled metal *cones*, each containing a small orifice. The first and second cones are referred to as the sampler cone and the skimmer cone, respectively. A schematic of the principal components and function of the interface is shown in Fig. 2.3.

The role of the interface is to extract a representative sample of the plasma ion population and transfer it efficiently to the higher vacuum regions. This cannot be achieved efficiently with a single step from atmospheric pressure to high vacuum, so several vacuum chambers are used, typically three stages in all. The step-down vacuum in the interface vacuum chamber (the first of the three stages) is from atmospheric pressure of approximately 1 bar to 2 mbar. This sudden pressure decrease leads to the supersonic expansion of the extracted ion beam, with the result that the composition of the ion beam is effectively *frozen* and a representative portion of the extracted plasma is transferred through the interface [3].

The ICP-MS interface must be robust, resistant to matrix deposition and constructed of a material that is a good electrical and thermal conductor. Matrix deposition on the interface cones was a serious problem of early ICP-MS instruments [21], but improvements in sample introduction and ICP design on most modern instruments have led to more complete sample decomposition and therefore reduced matrix deposition, while the design of the operating

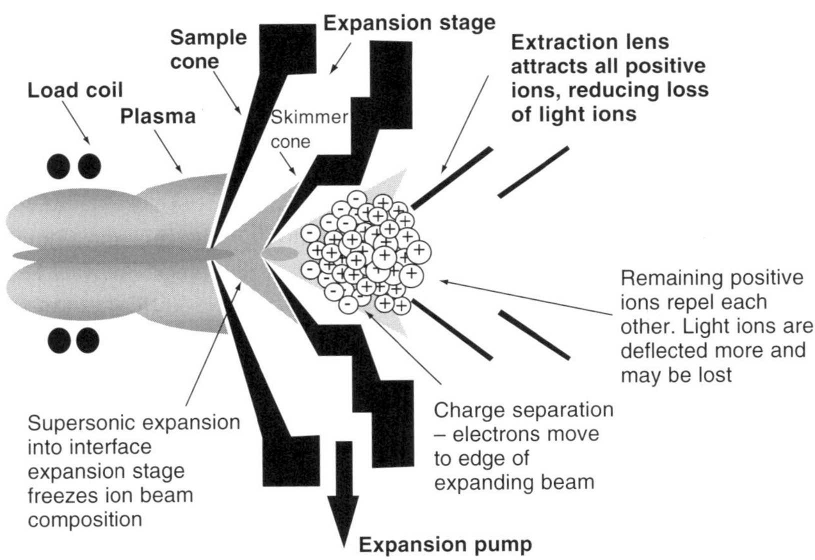

Fig. 2.3 Schematic of the ICP-MS expansion stage and the processes of ion extraction and charge separation.

temperature of the interface cones also plays a key role in minimising matrix deposition. In addition, experience of the operation of ICP-MS instruments has led to a general acceptance of the maximum level of total dissolved solids (TDS) that can be aspirated in routine analysis, typically quoted as 0.2% or 2000 ppm. The combination of interface vacuum pressure, sampler to skimmer cone spacing, and skimmer cone temperature ensures that matrix deposition does not occur at the skimmer tip, where it could cause signal drift [22].

The sample cone orifice diameter is between 1.0 and 1.1 mm in almost all current commercial ICP-MS instruments, while the skimmer cone orifice varies from 0.4 to 0.9 mm [13]. A larger skimmer orifice size transmits a greater number of analyte ions into the high vacuum regions and so should provide correspondingly higher sensitivity (assuming equivalent efficiency in other parts of the instrument). This higher sensitivity is, however, accompanied by an increase in the transport of sample matrix into the high vacuum regions and the requirement to increase the pumping capacity of the high vacuum pumps, to maintain the analyser vacuum.

2.3.4 *Vacuum system*

The vacuum system of an ICP-MS instrument has the dual role of maintaining a high vacuum (low pressure) in the analyser/detector region, while also allowing the transfer of ions from the plasma to the detector. A low pressure in the analyser region is essential in order to reduce the background and scattering effects that a high level of residual gas molecules would cause. The preferred configuration in both early and current commercial instruments is a 3-stage, step-down vacuum system comprising of the interface stage (discussed above), followed by two further vacuum regions of progressively higher vacuum, containing the ion lenses and mass spectrometer, respectively [23].

The typical vacuum pump configuration for commercial ICP-MS instruments is for the expansion stage to be pumped using a rotary vane pump, which is switched off when the ICP-MS is in *standby* mode, to allow access to the interface cones and ion lenses for maintenance. The intermediate and analyser vacuum stages are pumped by two separate turbo-molecular pumps or by a single, dual-stage pump. A *backing* rotary pump, which may be separate or the same unit as the expansion rotary, pumps away the exhaust from the turbo-molecular pumps. The second and third vacuum stages are isolated from the interface region by a gate valve, which seals the high vacuum region when the interface pump is switched off, to allow routine maintenance of the interface cones to be carried out.

2.3.5 *Ion focusing*

Typically, ICP-MS sensitivity increases across the mass range (7–238 amu) on a weight/weight basis, as a result of the preferential transmission of higher mass ions through the ion focusing system, due to *space-charge* effects. Once the ions formed in the plasma have been extracted into the high vacuum region beyond the interface, a process known as charge separation occurs. This refers to the separation of the positively charged ions from the electrons, which make up the corresponding negatively charged component of the extracted plasma. Since the ions have much greater mass and therefore kinetic energy than the electrons, the electrons tend to diffuse rapidly to the margins of the expanding gas cloud, with the result that the remaining material develops an overall positive charge. The ion beam therefore tends to repel itself, due to space-charge, leading to de-focusing and loss of the lighter ions, which have lower kinetic energy and diffuse to the margins of the beam [24].

In order to reduce the loss of ions from the beam, electrostatic ion *lenses* are used to focus the ions and transfer them efficiently to the mass spectrometer entrance aperture. While several different ion lens designs have been used in ICP-MS, the typical design is one or more cylindrical tubes, to which a voltage can be applied, the voltage setting being adjusted to allow the electrostatic field to be optimised or *tuned*. The ions pass through the electrostatic field generated around the ion lens and are deflected, so the ion lens settings can be adjusted to steer the ions in the required trajectory. In addition to guiding and focusing the ions, the ion lens system is responsible for separating the ions (which must be transferred efficiently to the mass spectrometer for analysis) from the neutral species and photons (which must be prevented from reaching the detector if the random background signal is to remain low). This separation occurs as the ion lens fields deflect the ions (due to their charge), but not the photons and neutral species (which are uncharged), so the ions can be bent onto a different trajectory from the other species. The ion lens system may consist of a simple, single cylindrical electrostatic lens, which has the virtue of low cost and simple operation, but has limited flexibility. Alternatively, a multi-component ion lens design may be used, which increases cost and complexity but allows greater flexibility of optimisation. Early designs of lens systems utilised a grounded metal disc, known as a *photon* or *shadow* stop, on the axis of the instrument, to block the direct line from plasma to detector. An alternative design uses an off-axis ion lens arrangement, as this gives higher ion transmission, particularly at low mass, compared to the photon stop configurations [25].

Space-charge effects and the loss of lighter ions should ideally be minimised, as it is beneficial for ions to be transmitted with close to equal efficiency through the ICP-MS system, making comparison of sensitivity and detection limits more consistent across the mass range. A compound ion lens system, or a mass dependent single lens voltage [26], reduces the *bias* between isotopes of the same element and so reduces the corrections that must be used in applications involving isotope ratio measurement [27]. A *flat* mass response is also beneficial for semi-quantitative determinations, where calibration is based on the relative sensitivity between elements, which may be several mass units apart.

In recent years, many commercial ICP-MS instruments have utilised multipole ion guides in place of or in addition to some of the traditional ion lens components. The multipole ion guide has the advantage of high ion transfer efficiency, simple operation and, most importantly, the ion guide may be housed in an enclosed chamber, which can be pressurised with an inert or reactive gas. The gas type and pressure can be adjusted to control the environment through which the ion beam passes – a configuration known as a collision cell.

2.3.6 Collision cells

Collision cells (also called collision/reaction cells – CRC) in ICP-MS were derived from technology used in chemical physics and organic mass spectrometry, where organic molecules are either fragmented (using an inert gas, such as Ar or Xe), to give information about structure, or modified by reaction (chemical ionisation using a reagent gas such as methane, isobutane or ammonia), to provide information on ion molecule reaction processes. In both cases, the collision cell is used to derive information about the parent organic molecule, in addition to what could be deduced from the simple mass spectral information derived from, for example, a single quadrupole analyser.

In ICP-MS, the primary function of the cell is to modify the composition of the ion beam in such a way that interfering polyatomic species are moved or neutralised, or the analyte itself is moved to an alternative, interference-free mass. Douglas investigated a collision-only cell for ICP-MS in 1989 [28], but the usefulness of this device was limited due to the fact that the rate of removal of polyatomic species by collisional dissociation was comparable to the rate of loss of analyte ions due to scattering. Rowan and Houk also investigated collision cells in reaction mode early in the history of ICP-MS [29], and Koppenaal *et al.* demonstrated successful removal of target polyatomic interferences, initially as a result of reaction with water vapour and later using an ion trap or multipole cell pressurised with H_2 [30, 31]. Application of a collision cell in a commercial ICP-MS instrument was first reported from Micromass in 1997 [32], and other commercial devices followed from leading manufacturers of conventional (non-cell) systems.

There are three main processes of interference removal in the cells used in commercial ICP-MS instruments. They are collisional fragmentation or dissociation (sometimes called collision-induced dissociation or CID), energy discrimination (ED, otherwise called kinetic energy discrimination – KED) and reaction. CID processes account for a relatively small proportion of the polyatomic ion interference removal in ICP-MS, because these processes require the collision energy for the polyatomic ion/cell gas pair to be significantly above the dissociation energy for the polyatomic ion. This is the case for relatively few common polyatomic ions, but does occur to a certain degree for species such as ArC^+, ArO^+, $ArNa^+$, $ArMg^+$ and $ArCa^+$ [33].

ED has two key roles to play in the ICP-MS collision cell. First, if the initial ion energy for both analyte and polyatomic species entering the cell can be controlled (using shielded torch technology, for example), then the analyst can benefit from the fact that polyatomic ions will lose more energy than analyte ions, as a result of their larger apparent cross-section, which means they suffer more collisions with the cell gas. This loss of energy through collision can then be used to discriminate between lower energy (polyatomic) ions and higher energy (analyte) ions, such that the lower energy ions are excluded from the quadrupole by a bias voltage [33]. The principle of ED is illustrated in Fig. 2.4.

ED is of enormous benefit for the analysis of complex unknown samples, because the ED process will affect any polyatomic ion, regardless of its level, composition or reactivity. This means that ED is the most useful process for the determination of multiple elements in unknown, multi-component matrices, where the level of matrix elements can vary unpredictably, as is often the case for environmental, food, agricultural and many clinical samples. ED also rejects ions formed in the cell itself, the so-called *product* ions, which result from reactions between sample components and the cell gas, as they always have relatively low energy, defined by the cell bias voltage. For KED to operate successfully, the energy of the analyte ions must not be reduced (through collisional

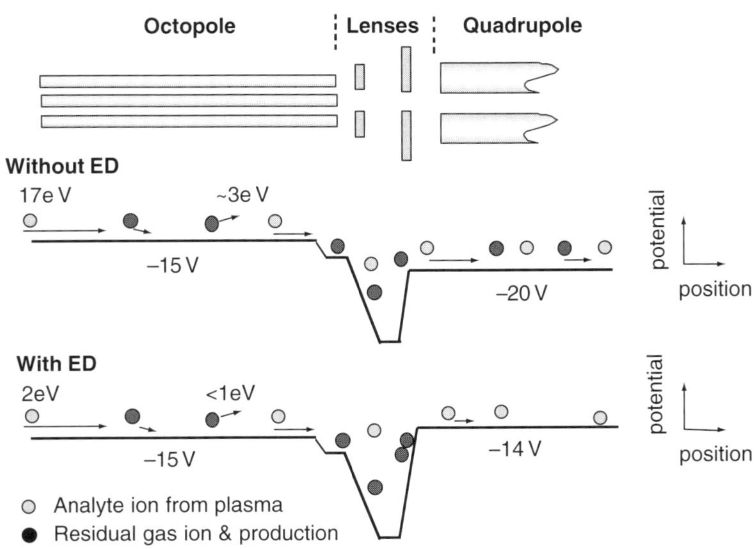

Fig. 2.4 Principle of energy discrimination (ED) in the collision reaction cell. Low energy product ions (formed in the cell) and reduced energy (after collision) polyatomic ions from the plasma can be stopped at the quadrupole entrance by a positive bias step from the octopole to the quadrupole.

damping) to such an extent that the analyte ions cannot be separated from the polyatomic and cell product ions, so collision cells that operate with ED tend to utilise relatively light gases, where the energy loss per collision is relatively low [34]. Dexter and co-workers [35] have identified that two distinct processes contribute to the removal of interfering species, namely the ion energy derived from the plasma (ion kinetic energy effect or IKEE) and the discrimination used to separate low energy interfering species ions from higher energy analyte ions at the cell exit. Yamada *et al.* [36] showed that control of the IKEE (using a ShieldTorch) allows the operator to discriminate effectively against both low energy, cell-formed species and plasma-derived polyatomic species, using the cell conditions to selectively reduce the ion energies for interfering polyatomic species, allowing their separation from higher energy, nonatomic analyte ions.

Reaction processes occur in any collision cell, when a reactive gas, such as H_2, O_2, CH_4, NH_3, etc., is used to pressurise the cell. Very reactive gases, such as NH_3 and CH_4, can remove polyatomic interferences with a high degree of efficiency, and are principally of use for determinations in high purity samples, such as semiconductor process chemicals, as they are relatively non-specific and will react efficiently with many polyatomic ions, analyte ions and matrix ions, based on the thermodynamics of the reaction processes which might occur between any given ion/molecule pair [37]. As a result, the use of very reactive (and non-specific) cell gases can lead to the formation of a large number of new polyatomic ions, when complex samples are measured. To overcome this limitation and permit the use of these non-specific reaction gases, one commercial collision cell configuration, known as a dynamic reaction cell (DRC) from PerkinElmer, uses a quadrupole as the cell ion guide, which allows the stability regions within the cell to be controlled so as to reject ions a set mass below and above the target analyte mass. This is known as a *band-pass* filter and its function is to stop the newly formed cluster ions from appearing in the mass spectrum and causing new interference overlaps [38], although it does not stop the formation of such ions, which means that the reported high reaction rate of analytes (such as As^+ and Ni^+) with highly reactive gases (such as NH_3) can lead to analyte loss [37].

Hydrogen, which is highly reactive, but only with a limited number of species, is an ideal reaction gas for ICP-MS. The two main reaction processes in a H_2 cell are H atom transfer and charge transfer [33]. Both processes can be exothermic for important Ar-based interfering ions, such as Ar^+, ArH^+ and ArO^+, but are strongly endothermic for most metal ions, so H_2 can be used to selectively attenuate target polyatomic interferences, without the associated problem of reaction losses of analyte ions. This is an important reason why more specific reaction gases (such as H_2) tend to be preferred in instrument configurations where multiple elements are analysed in complex matrices [39]. The reactivity of H_2 with several important polyatomic species is equivalent to that of NH_3 [37], and H_2 is a lighter gas, so higher pressures of H_2 may be used without severe analyte loss due to scattering.

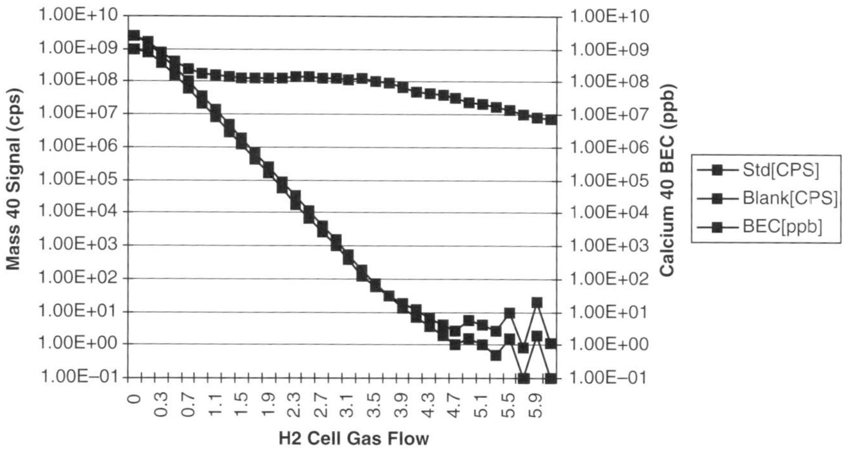

Fig. 2.5 H_2 cell gas optimisation giving 9 orders reduction of $^{40}Ar^+$ background.

The reaction efficiency of H_2 is illustrated in Fig. 2.5, which shows the attenuation of the $^{40}Ar^+$ signal to allow the measurement of Ca at its main isotope, also at mass 40. This plot illustrates that the Ar^+ background can be reduced by 9 orders of magnitude, which is equivalent to the best reported interference removal efficiency achieved with other reaction gases [38]. The plot in Fig. 2.5 is based on the laser analysis of a glass standard reference material (NIST 612), in which Ca is a major element at 8.4% (11.8% CaO). Calibrated against this Ca concentration, the background equivalent concentration (BEC) for Ca at the optimum H_2 flow rate was 1–2 ppb in the solid sample. As with the use of cool plasma conditions to allow ppt level analysis of K, Ca and Fe in high purity samples, the use of collision cells in ICP-MS has led to the application of the ICP-MS technique to several new analytical challenges, which previously would have required alternative sample preparation or analysis methods. Examples include the measurement of trace levels of first row transition metals in open ocean seawater [39] and the determination of difficult trace elements, such as Se at low levels in clinical samples [40].

After passing through the conventional ion lens system or collision cell, the ions enter the final vacuum stage in which the mass analyser and EM are situated.

2.3.7 *Mass analyser*

The mass analyser used in the vast majority of commercial ICP-MS instruments is the quadrupole mass spectrometer (ICP-QMS). The quadrupole mass spectrometer is ideally suited to inorganic mass spectrometry. The high scan speed is appropriate for the measurement of both steady state and transient signals; the atomic mass unit (amu) range covers all the natural and radiogenic isotopes of interest (typically a mass range from 4 to 260 amu); the resolution is sufficient to separate adjacent isotopes 1 amu apart and the manufacturing and operating conditions for the analyser make it suitable for routine, mass produced instrumentation.

The spectrometer rapidly scans over the mass range and separates ions on the basis of their mass to charge ratio (m/z). In operation, the voltages applied to the two opposite pairs of quadrupole rods give a fluctuating electromagnetic field in the space between the rods. This fluctuating field induces a vibration in the ions travelling axially through the quadrupole, such that ions lighter or heavier than the *set mass* enter unstable trajectories and are rejected from the quadrupole. Only ions at the set mass travel in stable trajectories and pass through the quadrupole to reach the detector. The *set mass* of the quadrupole is determined electronically, by the settings of the RF and direct current (DC) voltages applied to the two opposite pairs of rods. The layout of a quadrupole and the voltages applied to the rods are shown in Fig. 2.6. Rods with a hyperbolic profile generate the ideal electric field distribution for ion separation [41], but round rods are typically used for the quadrupoles in commercial ICP-MS instruments, since they are simpler and cheaper to manufacture. Since the quadrupole voltages can be adjusted very rapidly, the whole mass range from 5 to 240 amu can be scanned very quickly (>10 times per

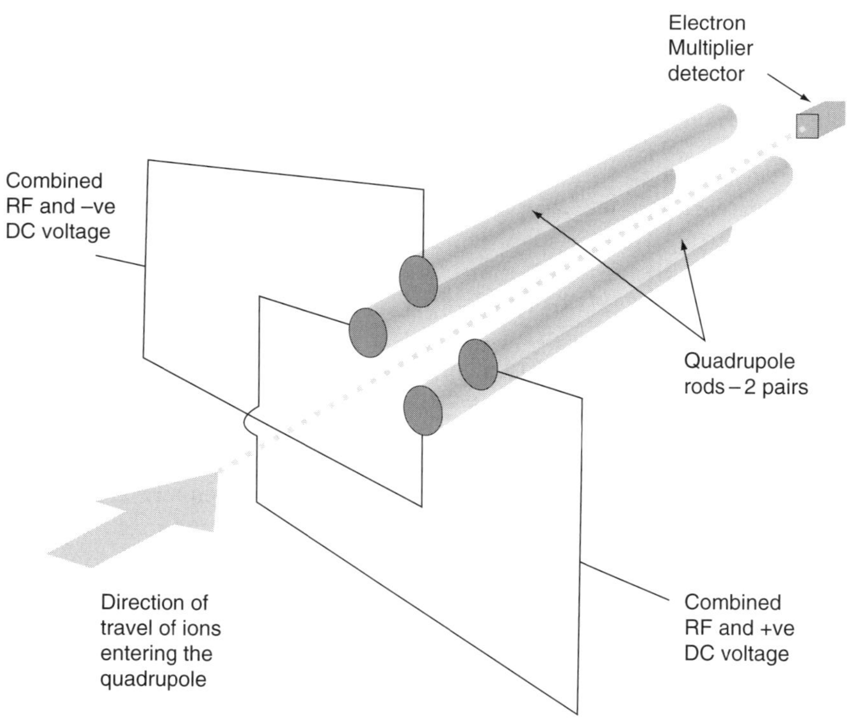

Fig. 2.6 Schematic showing alignment of and voltages on the quadrupole rods.

second), giving a mass spectrum that is built up from multiple sweeps across the mass range of interest.

Ion transmission through the quadrupole is affected by the conditions (energy and trajectory) under which the ions enter the quadrupole, the residual gas pressure, which controls analyte ion scattering, and the mass filtering effect, which determines the peak separation. The performance of the quadrupole is dependent on the length and shape of the quadrupole rods and the frequency of the quadrupole RF voltage applied, as a higher frequency and a longer rod length expose the ions to more RF cycles, thereby improving the filtering efficiency. Similarly, a hyperbolic rod profile gives an improved field distribution, improving transmission at a given resolution.

While a quadrupole mass spectrometer is used in the majority of commercial ICP-MS instruments, some instruments utilise a magnetic sector analyser, typically employed where mass resolution significantly higher than unit resolution is required. High resolution (HR, also known as sector field (SF)) ICP mass spectrometers were first introduced by VG Elemental in 1989 [42] and offered the analyst the opportunity to separate or resolve analyte peaks from polyatomic overlaps, providing improvements in detection limits and measurement certainty for some difficult applications. A schematic diagram of a commercial high resolution ICP mass spectrometer is shown in Fig. 2.7. A resolution setting of up to 10 000 ($M/\Delta M$) allows the separation of analyte/interferent pairs such as $As^+/ArCl^+$ at mass 75 and Cr^+/ArC^+ at mass 52, which are not separable in quadrupole-based ICP-MS instruments.

A further benefit of the high-resolution approach is that the design of the curved ion flight path gives a combination of high ion transmission and very low random background (from photons and neutral species) of around 0.2 counts per second. As a result, when operated in normal unit resolution mode (*low* resolution), the detection limits achievable on a high resolution ICP-MS can be a factor of 5 to 10 times lower than on quadrupole-based instruments. Sensitivity (ion transmission) does decrease significantly with increasing resolution, such that a resolution setting (R) of 4000 results in ion transmission of about 10% of that at R of 300 (unit mass), but the low random background and high inherent sensitivity, combined with the unequivocal separation of analytes from polyatomic interferences mean that high resolution is an extremely valuable tool for research and demanding applications. Some applications, which might previously have required high resolution ICP-MS, are now being addressed using quadrupole ICP-MS instruments equipped with collision/reaction cell technology, which allows quadrupole-based instruments to address the removal of polyatomic interferences in complex matrices.

Time of flight (TOF) mass spectrometers have also been used in ICP-MS and have demonstrated some advantages in applications where many masses are measured in short lived transient signals [43]. This is because the TOF mass

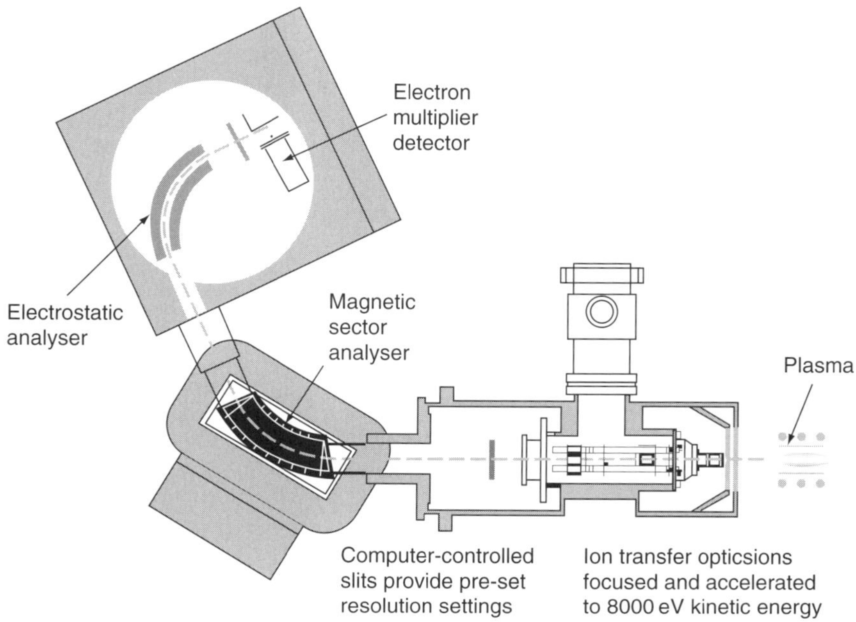

Fig. 2.7 Schematic of a commercial high resolution ICP-MS instrument (Courtesy of Thermo Finnigan).

spectrometer separates the ions and delivers all masses to the detector with a very short time delay, allowing many thousands of full mass scans to be acquired per second, giving a pseudo-simultaneous measurement. ICP-TOF-MS has not yet proved a viable alternative to ICP-QMS for routine applications, due to its limited sensitivity, inability to *skip* very high intensity background peaks and the fact that the mass calibration is dependent on the analyte ion energy, which can vary with sample introduction conditions.

2.3.8 Detector

Detectors used in ICP-MS have developed over the years, to offer a powerful combination of high sensitivity and wide dynamic range. The high sensitivity is derived from the fact that the detector used in almost all commercial ICP-MS instruments is an EM detector. The EM detector is so called because the ion striking the first detector dynode causes the release of electrons from the dynode surface; these electrons are attracted to the second dynode, where each incident electron causes the release of further electrons and so on down the multiple detector dynodes. By the end of the EM detector, the cascade of electrons has multiplied to the point where a signal *pulse* can be detected above the electronic noise, using a threshold value determined by a *discriminator* cut-off setting. In this way, the EM detector measures individual ions as they arrive (known as ion- or pulse-counting), which combined with the low random background gives very low limits of detection.

The simple EM detector, operating in pulse-counting mode only, has a dynamic range limited to about 6 orders of magnitude. Above ~2 million incident ions per second (6 orders of magnitude higher than a nominal background of 2 counts per second (cps)), the detector is unable to register any signal increase correlating to the increased number of ions arriving at the detector. The usable measurement range of modern EMs has been extended by the development of technology that switches the detector automatically into *analogue* mode at higher count rates, where the current generated by the ions passing down the detector is measured, rather than the individual ions [44]. In this way, the same *dual mode* detector can measure high concentrations at the same time as very low concentrations – a dynamic range of up to 9 orders of magnitude, using a cross-calibration to correct the measured signal to cps equivalent. In practice, this means that single ppt levels of trace elements can be measured at the same time as 100s ppm of major elements, without requiring any change in instrument operating conditions. The detector electronics allow the mode switch to be carried out automatically, so that the detector selects the appropriate mode for each peak as the quadrupole sweeps across the mass range, resulting in a single data file with simultaneous measurement of peaks in both detector modes.

2.4 Operation

In practical analyses, there are several strategies that the analyst can use to improve the stability, integrity and accuracy of the measurements made on an ICP-MS instrument. Some of the most critical parameters are outlined in the following sections, which emphasise the practical steps that can be taken to improve the analysis.

2.4.1 Sample preparation and introduction

It has been suggested that all liquid samples prepared for ICP-MS analysis should be in a matrix of ultra-pure water, stabilised only with a trace of nitric acid [45]. In this way, the potential sources of polyatomic interferences can be minimised, the only species present being Ar (from the plasma gas), O and H (from water), N (from HNO_3) and traces of C (from CO_2 and organic residues). While it is certainly true that the addition of other sample solvent components, such as H_2SO_4 and HCl, will create the potential for formation of new polyatomic interferences, in most cases the interferences formed are at relatively low level and overlap elements which are not typically measured at ultra-trace levels. Obvious exceptions to this rule do exist in, for example, the determination of trace V, Cr and As in samples containing high levels of chloride. $^{51}V^+$, $^{52}Cr^+$ and $^{75}As^+$ suffer from polyatomic interferences from $^{35}Cl^{16}O^+$, $^{35}Cl^{16}O^1H^+$ and $^{40}Ar^{35}Cl^+$, respectively. However, in the determination of As in environmental samples, for example, the formation of $ArCl^+$ can be controlled at a level which allows As to be measured at the levels typically required in this application [20]. The main parameter to affect the level of several of these matrix-based polyatomic ions is the optimisation of the plasma conditions, which controls the degree of decomposition of the polyatomic species, while collision cells can also be used to attenuate background polyatomic species.

2.4.2 *Plasma optimisation*

In its simplest terms, the purpose of the ICP is to form positively charged ions from all of the analyte atoms present in the sample aerosol carried from the spray chamber. There are many stages to this process and the efficiency of the plasma can be influenced significantly by system design and appropriate optimisation.

The formation of ions from the sample analyte atoms is achieved by the removal of one electron from each atom. This occurs with varying ease and efficiency for different elements, depending on how tightly held the most loosely bound electron is. This variation is known as the *ionisation efficiency* for each element, based on the first ionisation potential of the element (the energy required to remove one electron from a neutral atom), together with estimated values for plasma electron temperature and density. Even a small change in plasma temperature may have a dramatic effect on the degree of ionisation of poorly ionised elements, which are often those elements required to be measured at the lowest levels, such as As, Se and Hg. The ionisation potential is specific for each element, but the plasma temperature is highly dependent on plasma conditions, as discussed in Sections 2.3.1 and 2.3.2. In addition to the ionisation of poorly ionised elements, the plasma also has a major impact on the occurrence of polyatomic ions in the mass spectrum. Generally, if a high effective plasma temperature is maintained, many matrix oxide and other potential polyatomic interferences will be reduced, often to levels where, in practice, they become negligible [20].

Although the vast majority of routine applications of ICP-MS use only argon as the plasma and carrier gas, research has investigated the effects of gas additions and *mixed-gas* plasmas. As well as the addition of oxygen, to decompose the carbon matrix in organic sample analysis, helium has been added to the carrier gas to increase sensitivity in LA-ICP-MS [46], and various gases, particularly organic gases, have been added to the plasma to control interferences in solution analysis [47].

2.4.3 Control of interferences

While ICP-MS is clearly enormously powerful, any analytical technique (ICP-MS included) has some limitations. In common with other atomic spectrometry sources, the ICP-MS plasma can suffer from ionisation suppression effects, where the energy required to decompose or ionise a high matrix (especially a high level of an easily ionised element) can reduce the energy available for ionisation of the poorly ionised analyte atoms, resulting in the loss of sensitivity for these analytes [48]. Sample introduction systems may also suffer from the effects of physical differences between samples, such that more concentrated or viscous samples may form transportable aerosol droplets less efficiently than less viscous samples, which may be aspirated more efficiently by the nebuliser [49]. The first of these problems is largely overcome by the use of a high temperature plasma as the ion source (the plasma is much hotter than the chemical flames used in flame atomic emission spectrometry, for example), matrix matching or calibration using the method of standard additions, while viscosity effects can be mitigated by an understanding of the role of internal standards to correct for sample transport variations.

The main sources of spectral interferences in ICP-MS are:

(i) Direct overlap from a different element with an isotope at the same nominal mass, known as an isobaric interference, e.g. $^{114}Sn^+$ overlap on $^{114}Cd^+$.
(ii) Overlap from a polyatomic ion formed from the combination of species derived from the plasma gas and/or liquid samples, H_2O, e.g. $^{40}Ar_2^+$ overlap on $^{80}Se^+$.
(iii) Overlap from a polyatomic ion formed from the combination of species derived from the sample solvent and/or sample matrix, e.g. $^{40}Ar^{23}Na^+$ overlap on $^{63}Cu^+$.

Direct isobaric overlaps cannot be separated easily using spectrometric methods; even high resolution ICP-MS instruments do not have sufficient resolving power to differentiate between them, but some direct overlaps can be removed efficiently using alternative methods. The $^{40}Ar^+$ overlap on $^{40}Ca^+$ represents one of the most serious isobaric interferences for ICP-MS, which, until the early 1990s, limited the performance of the technique in the semiconductor industry. The determination of Ca, K and Fe at low levels in semiconductor process chemicals was compromised by, respectively, the $^{40}Ar^+$ overlap and the $^{38}Ar^1H^+$ and $^{40}Ar^{16}O^+$ polyatomic ions, all of which are major background species in conventional ICP-MS spectra. These problems were largely eliminated in 1994, when technology was developed that allowed cool plasma operation on a routine basis, as discussed in Section 2.3.2 [17].

A further benefit of cool plasma conditions for high purity sample analysis is that the ionisation of argon is virtually eliminated. This means that the ion beam passing through the interface and intermediate vacuum chambers is free from the bulk of the positively charged ions (mainly Ar^+) normally present, which in turn means that the space charge effects normally associated with ion transmission [24] are reduced, resulting in extremely efficient transmission and very high sensitivity for the light elements that would normally be lost from the ion beam.

Direct isobaric overlaps can also be addressed with collision cell devices, one example being the use of specific cell gas conditions to promote reaction of one isobar, while leaving the other isobar unaffected. Although there are a few examples where isobaric overlaps from other elements can cause serious problems, there are relatively few cases where the selection of an alternative isotope does not give an acceptable detection limit. Indeed, the only naturally occurring element that does not have at least one isotope free from direct isobaric overlap is Indium (^{113}In is isobaric with ^{113}Cd, while ^{115}In is isobaric with ^{115}Sn).

In most analytical situations, it is the formation of polyatomic species from the plasma and sample components which poses a far greater problem for practical sample analysis. Polyatomic species, such as the ArH^+ and ArO^+ species mentioned above, are a part of the normal background mass spectrum, when a liquid aerosol is passed into the plasma. Additional polyatomic species may form from other components of the solution introduced, leading in some cases to severe overlaps on preferred analyte masses.

Some of the potentially problematic interferences derived from the sample matrix components of natural seawater are shown in Table 2.1. This table illustrates that interfering ions may contain multiple atoms, although diatomic ions are the most common. It also illustrates the point that an analyte isotope mass may suffer from combinations of interferences from several polyatomic ions derived from different components of the matrix. The degree of formation of polyatomic ions, particularly those derived from the sample matrix, is highly dependent on the efficiency of the sample introduction and plasma systems, so can vary significantly between instruments and under different operating conditions.

Generally in ICP-MS laboratories, some attempt is made to monitor and minimise the formation of polyatomic interferences and the parameter most often monitored for this purpose is the refractory oxide ion of Cerium (Ce),

Table 2.1 Some potential matrix-based interferences resulting from a seawater matrix

Isotope	Interfering species
^{51}V	$^{35}Cl^{16}O$, $^{37}Cl^{14}N$, $^{40}Ar^{11}B$
^{52}Cr	$^{36}Ar^{16}O$, $^{40}Ar^{12}C$, $^{35}Cl^{16}OH$, $^{37}Cl^{14}NH$
^{55}Mn	$^{40}Ar^{14}NH$, $^{40}Ar^{15}N$, $^{39}K^{16}O$, $^{23}Na^{32}S$, $^{37}Cl^{18}O$
^{54}Fe	$^{40}Ar^{14}N$, $^{40}Ar^{16}O$, $^{37}Cl^{16}OH$, $^{40}Ca^{14}N$
^{56}Fe	$^{40}Ar^{16}O$, $^{40}Ca^{16}O$
^{59}Co	$^{36}Ar^{23}Na$, $^{24}Mg^{35}Cl$, $^{42}Ca^{16}OH$, $^{23}Na^{35}ClH$
^{58}Ni	$^{40}Ar^{18}O$, $^{23}Na^{35}Cl$, $^{42}Ca^{16}O$
^{60}Ni	$^{23}Na^{37}Cl$, $^{25}Mg^{35}Cl$
^{63}Cu	$^{40}Ar^{23}Na$, $^{40}Ca^{23}Na$
^{65}Cu	$^{40}Ar^{25}Mg$, $^{40}Ar^{24}MgH$
^{64}Zn	$^{40}Ar^{24}Mg$, $^{40}Ar^{23}NaH$, $^{32}S^{16}O_2$, $^{32}S_2$
^{66}Zn	$^{40}Ar^{26}Mg$, $^{34}S^{16}O_2$, $^{32}S^{34}S$, $^{33}S_2$
^{68}Zn	$^{40}Ar^{14}N_2$, $^{34}S_2$, $^{40}Ar^{28}Si$
^{75}As	$^{40}Ar^{35}Cl$, $^{40}Ca^{35}Cl$
^{78}Se	$^{40}Ar^{38}Ar$, $^{40}Ar^{37}ClH$, $^{38}Ar^{40}Ca$
^{80}Se	$^{40}Ar_2$, $^{40}Ar^{40}Ca$, $^{32}S^{16}O_3$, ^{79}BrH
^{98}Mo	$^{40}Ar^{23}Na^{35}Cl$
^{111}Cd	$^{79}Br^{32}S$
^{112}Cd	$^{96}Mo^{16}O$
^{114}Cd	$^{98}Mo^{16}O$
^{123}Sb	$^{88}Sr^{35}Cl$, $^{36}Ar^{87}Sr$

Reproduced from Ref [37] by permission of The Royal Society of Chemistry.

expressed as the signal ratio CeO^+/Ce^+, in percentage. Ce is used as it forms one of the most stable oxide bonds of any element and can, therefore, be used as a *worst-case* indicator. The CeO^+/Ce^+ ratio has been demonstrated to act as a good indicator of levels of several reported matrix-derived polyatomic interferences [20]. The reduction in the level of matrix interference resulting from an optimised (low) CeO/Ce ratio is illustrated in Fig. 2.8 for the sulphur-based interferences on Zn at mass 66, which shows the apparent additional Zn concentration which results from the variation in decomposition of the S_2^+ and SO_2^+ interferences at mass 66, under CeO/Ce tuning conditions between 0.3 and 3.0%.

The introduction of collision cell technology has been immensely valuable in expanding the application of ICP-MS to analyses that could previously not be addressed directly. The most exciting developments are in the capability to reduce or remove interferences arising from argon and matrix-based polyatomic interferences, first reported by Rowan and Houk [29]. See discussion in Section 2.3.6. The collision cell devices used in commercial ICP-MS instruments are primarily used to attenuate the signals for polyatomic species, which would otherwise overlap the measurement mass of a target analyte. Some cell designs use very specific, single-element conditions to selectively and efficiently remove a single interfering species, using highly reactive cell

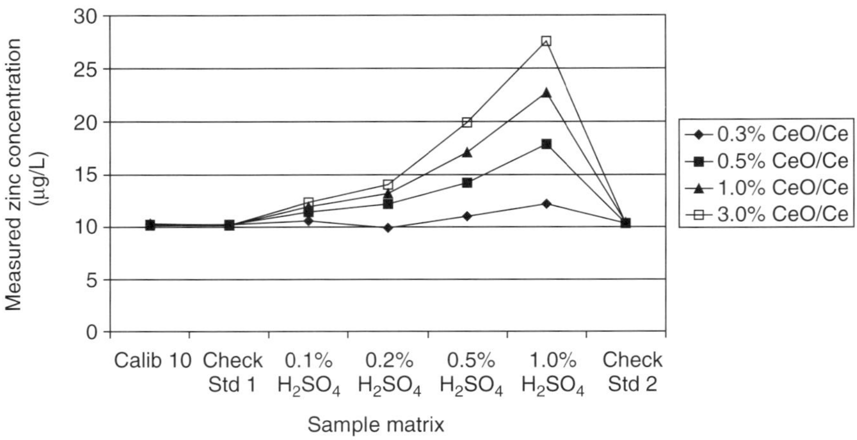

Fig. 2.8 Sulphur-based polyatomic interference on Zn at mass 66, reduced by optimisation of the CeO/Ce ratio.

gases and complex but theoretically predictable reaction pathways. These instruments are intended as an alternative to the cool plasma approach in the removal of Ar-based interferences on ultra-trace contaminants in the semi-conductor industry, where the sample matrix is low, predictable and constant and the analyte concentrations are typically also very low. Other approaches use less reactive gases but more generic interference removal methods, such as ED, which has proved suitable for the measurement of multiple analyte elements in complex and variable sample matrices, where the source and level of potential interferences cannot easily be predicted in advance [39].

2.4.4 Data acquisition

The quadrupole ICP-MS is suited to scanning rapidly across the mass range, building up a mass spectrum based on multiple *sweeps* over the masses of interest. Alternative measurement protocols include single ion monitoring (where the quadrupole mass is static and data are collected against a time scale), time resolved analysis (where several masses are measured and the mass spectrum from each sweep across the mass range is stored in a separate *time slice*) and isotope ratio measurements (where the quadrupole measures two or more isotopes of the same element, to provide a precise ratio of the isotopic abundances).

Commercial ICP-MS instruments typically provide the operator with a range of measurement strategies, such as *scanning*, where a mass range is defined and the quadrupole steps sequentially across all masses in the range, and peak jumping (or hopping), where the operator selects the actual isotopes to be measured and the intervening masses are *skipped* (not acquired). In general terms, scanning may be the most suitable acquisition method for screening of unknown samples, where the composition of the sample matrix is unknown and the presence of potential analytes may need to be confirmed through the comparison of the measured isotopic pattern against a reference database of the theoretical isotope pattern for the analyte. Peak jumping is typically used when the target analytes are known and additional information is not required (the presence of other analytes, for example).

Time resolved analysis is used when the signal of interest for one or more elements changes with time. Typical applications include the separation and analysis of elemental species using a chromatographic step prior to the ICP-MS measurement and the monitoring of the signal for specific elements as they vary with time and sampling position in LA-ICP-MS analysis. Isotope ratio measurements are used both for isotopic analysis itself and for isotope dilution quantification, which is used to quantify elements based on the change in ratio that results from the spiking of an unknown sample with a spike solution enriched in one isotope of the target analyte.

In each case, the ions that are passed through the mass analyser reach the detector and generate a signal, which is integrated, stored and calibrated against a reference signal (usually from a standard material). The instrument software provides a range of options for processing the raw count-rate data measured at the detector, and a range of calibration options for converting the measured counts into analytically useful concentration or ratio results.

2.4.5 Calibration and quantification

The most typical quantification method used with ICP-MS measurements in liquid samples is external calibration. In this method a calibration plot is constructed, based on the measured signal for the elements of interest against their concentration in a known solution. The known solution is typically prepared from single or multi-element stock solutions, but reference materials can also be used, provided the reference values are sufficiently well defined to act as a calibration standard. In theory, single point (plus a blank) calibrations could be used, since the response of ICP-MS is linear with concentration but, in practice, several standards (typically 3–5 standards, plus a calibration blank solution) are normally used to define the calibration plot in the concentration range of interest. Internal standards are commonly used in ICP-MS, particularly where samples vary in composition from the calibration standards, since changes in sample transport, nebulisation and long-term drift would all lead to errors, which may be corrected if an element with similar behaviour is used as a reference.

ICP-MS is ideally suited to semi-quantitative analysis, in which a reference element is used to *calibrate* the measured signal from another element, providing an accurate result for many elements, using standards that contain only a few. Semi-quantitative calibration routines must make allowances for differences in elemental sensitivity, by applying corrections for the instrument mass bias, the abundance of the isotope measured and any variation in the ionisation efficiency of the element.

The other measurements typically carried out by ICP-MS are those that determine the relative abundances of two or more isotopes of the same element, known as isotope ratio measurements. Simple isotope ratio analysis is a key benefit in mass spectrometric methods, as compared to optical measurements, and was identified as one of the important potential benefits of ICP-MS early in the history of the technique [50]. As well as applications where the natural variation of an element's isotopic composition *per se* is of interest [51], the measurement of isotope ratios is also used as the calibration method for isotope dilution analysis [52], where the high precision of the ratio measurement is utilised to provide accurate quantification for elements where an isotopic spike can be purified and added to the unknown sample. Since the alteration of the original isotope ratio is measured, rather than the response for the element, this method provides excellent precision and accuracy.

Quadrupole ICP-MS instruments have a single detector, which means that isotope ratio measurements are made sequentially, but commercial ICP-MS instruments have been developed with a detector array, whereby each isotope of interest is measured using a dedicated detector, known as multi-collector (MC) ICP-MS [53]. These instruments measure isotope ratios with a precision equivalent to thermal ionisation mass spectrometry (TIMS) and about 10 times better precision than achieved routinely using a single collector instrument.

2.5 Applications

While space in this volume does not permit an exhaustive discussion of the many and varied applications of ICP-MS, a brief overview of a selection of the analytical challenges that have been addressed by the technique follows.

2.5.1 Environment, agriculture and food

The range of applications for ICP-MS in the analysis of environmental, agricultural and food samples reflects the diversity of sample types encountered in these fields. Analytical challenges range from the determination of low-level trace elements in clean samples (such as rain water, ice cores and atmospheric aerosols) or high matrix samples (such as open ocean, seawater [39]) through to the measurement of much higher concentration elements in waste water, soil and sediment digests. The wide elemental coverage and detector dynamic range of ICP-MS make it highly suitable for the simultaneous measurement of all the required inorganic sample constituents in applications such as the routine monitoring of trace and mineral elements in drinking water [54]. The advent of extended dynamic range EM detectors has allowed ICP-MS instruments to measure high level analytes at the same time as the trace elements. ICP-MS detectors are now able to measure concentrations up to 500 mg/L or ppm for a single element, without requiring the use of element-specific attenuation or measurement of a low abundance isotope. The requirement to monitor the levels of elements in the environment and related samples is increasing, with additional elements being added to the required list of analytes and the levels at which these elements must be controlled being reduced, as our understanding of potential toxicity improves.

Trace elements can have a major role in human nutrition, and the elemental content of materials is monitored at almost every stage of food production. Furthermore, industrial food production practices and the high proportion of processed food in many diets mean that the natural levels of many essential dietary components must be regulated artificially, through the use of food additives and supplements. ICP-MS has been applied to the monitoring of trace elements in food materials [55, 56]. with applications including monitoring of seasonal and geographical trends in elements with dietary significance [57], determining isotope ratios in elements provided as trace element supplements [58] and measuring metals in proteins to monitor elemental absorption from the diet [59]. In some cases, the trace element content of a food material may provide information about its geographical origin, which may be of interest in cases where the value of the food product is related to its region of production. Trace element *fingerprinting* of food materials, such as wine, is a useful method in this field, as trace element levels may be less subject to alteration during food processing or storage than the organic constituents.

In these fields, it is becoming much more important to be able to determine not just the total amount of potentially toxic elements, but also its chemical form, since this can have a dramatic impact on the element's availability, mobility and toxicity [60]. Tributyl and triphenyl tin compounds have been used extensively as biocides (TBT in marine anti-fouling paint, for example) and the contamination of the environment with these long-lived and highly toxic species has been widely reported [61]. Recently, interest has centred on the potentially harmful effects of other organotin products used as plasticisers and UV stabilisers in the production of plastics such as PVC and as fungicides in hygiene products such as antibacterial insoles. While many organotin species of interest can be measured using LC coupled to ICP-MS [62], a wider range of species can be measured using GC-ICP-MS, a technique now receiving widespread interest [63, 64]. The key design criteria for a GC-ICP-MS interface are the maintenance of a high and constant temperature throughout the transfer line and injector, to ensure that species with a high boiling point do not condense in the transfer line, leading to species loss or disruption of the chromatographic separation.

An area of current research interest is the use of GC-ICP-MS for the monitoring of pesticides. In this application, the advantages of high sensitivity and specificity of the ICP-MS are combined with the efficiency of the GC separation to determine elements such as P, S and the halogens. The uniform response of the ICP-MS and the fact that the ICP-MS response for each element is essentially unaffected by co-eluting compounds, unlike some other GC detectors, means that a simple elemental calibration can be generated from a range of compounds with known composition, using the relative concentration of each element in each compound. This technique, known as compound independent calibration (CIC), allows a calibration to be derived from a single analysis of a mixed standard, despite the fact that the different calibration levels are measured in different compounds and at different retention times.

In addition to LC-ICP-MS and GC-ICP-MS, other chromatographic techniques are also compatible with ICP-MS, including ion chromatography (IC) and capillary electrophoresis (CE), or capillary zone electrophoresis (CZE).

2.5.2 Semiconductor

The elements typically monitored in semiconductor materials include the alkali and alkaline earth metals, the transition and heavy metal contaminants and elements that are added deliberately as dopants, to control the conductivity of specific parts of the chip [65]. Until the development of cool plasma ICP-MS, the measurement of the elements K, Ca and Fe could not be achieved by ICP-QMS at the levels required by the industry. High-resolution ICP-MS could resolve the $^{40}Ar^{16}O^{+}$ interference on Fe^{+} at mass 56 [66], but does not have sufficient resolving power to separate $^{40}Ca^{+}$ from $^{40}Ar^{+}$. Since cool plasma ICP-MS became a routine analytical method in 1994, the use of ICP-MS instruments in semiconductor laboratories has increased dramatically for measurement of these elements, together with the wider suite of metals monitored routinely at ng/L and sub-ng/L levels in ultra-pure water (UPW) [19] and the other semiconductor process chemicals [67].

An area of increasing interest is the control of trace metal contamination in the organic process chemicals such as isopropyl alcohol (IPA) and methanol (MeOH), and the photoresist and stripper materials associated with the wafer masking and etching processes [68]. Demand for multi-element certification of process chemicals, including organic chemicals, at lower levels of metal contamination has led to the adoption of ICP-MS for the analysis, to provide both multi-element determination and lower limits of detection. The trace elements measured in organic samples are essentially the same as the elements monitored in inorganic process chemicals and the levels measured in high purity solvents are also of the same order as the single and sub-ng/L or ppt levels measured in acids and UPW [68]. ICP-MS is also suitable for the determination of trace elements in the silicon wafer itself, using a method called vapour phase decomposition (VPD) or surface metal extraction (SME).

2.5.3 Clinical and pharmaceutical

Elemental concentrations may be measured in body fluids and tissues in order to study occupational exposure, infection, poisoning and the treatment of disease. ICP-MS offers rapid multi-element analysis at very low concentrations with isotopic capability and is favoured for many routine and research applications in this field. As well as monitoring of trace elements in body fluids [69, 70] and other tissues [71], isotope ratio measurement is of great interest, where stable isotopes can be used to mark or label compounds of interest, which are then traced as they undergo metabolism and biological transformation [58, 72].

Trace metal analysis of clinical samples may be required for routine determination of exposure levels for toxic heavy metals (typically Cd, Hg and Pb), elements associated with occupational (Cr, Ni, Pt), or dietary exposure (Sn, Hg) and essential elements (such as I and Se), where insufficient dietary intake may lead to clinical deficiency. Other elements may be used as indicators of renal failure or other disease while levels of trace elements may be important in certain clinical treatments, such as Al in the pure water used for kidney dialysis. The development of high-resolution ICP-MS and collision/reaction cell devices for quadrupole ICP-MS has improved the determination of previously difficult elements such as Cr, As and Se in biological materials, allowing accurate determination even at the low levels at which these elements may be present [40]. Speciation of trace elements in clinical samples is also of growing interest and includes the analysis of Se species in natural diets and in dietary supplements, using LC or CE coupled to ICP-MS.

In the pharmaceutical industry, analysis of potentially harmful metals – the US Pharmacopoeia (USP) heavy metals – is required in drug product bulk materials. There is also a diverse range of applications for trace metal analysis in drug development and testing, again utilising LC to separate species prior to ICP-MS analysis.

2.5.4 Geological and nuclear

Measurement of trace elements in geological materials was one of the first applications to make use of ICP-MS when the technique was in its infancy [73]. The low detection limits, multi-element capability and simple ICP-MS spectra, particularly for elements with complex emission spectra, such as the rare earth elements (REE) [74], led to the widespread use of the new technique for geochemical analysis. In addition to the determination of trace element levels in bulk rock and mineral samples [75], ICP-MS has been applied very successfully to the analysis of the distribution of elements in geological materials, using LA of sample material from thin polished slices (known as sections) of the natural rock sample [76]. The development of new lasers operating in the far UV region has allowed LA-ICP-MS to analyse even transparent and easily cleaved minerals, such as quartz, mica and calcite [77].

Nuclear applications may require the determination of the isotopic composition and trace element levels in the uranium fuel, monitoring of nuclear power station cooling water and measurement of levels of radioactive materials released to the environment.

2.5.5 *Forensic and archaeological*

The application of trace metal analysis in forensic science has received less attention and investment than the parallel science of biological sample analysis (DNA fingerprinting). However, there are many instances where the level, ratio or isotopic composition of the elemental content of a sample can provide conclusive evidence of its source. Trace element composition can be used in applications including the fingerprinting of gold samples to identify the source of natural or refined gold [78], identification of gunshot residues and differentiation between glass fragments with identical refractive indices. Many other types of *scene of crime debris* may be suitable for analysis by LA-ICP-MS, where trace element pattern matching may be more important than any quantification of the sample components. Liquid sample analysis may be appropriate in some cases, such as in suspected poisoning, but LA-ICP-MS has enormous potential in forensic and archaeological applications, where the sample size may be small, other physical and chemical tests may need to be carried out on the same sample and bulk sample destruction may not be an option, as is typically the case with valuable archaeological artefacts. LA-ICP-MS is also suitable for applications where element distribution is studied, such as in the growth rings of long-lived bivalve shellfish and coral, or the distribution of trace elements in human archaeological remains, such as teeth.

2.5.6 *Chemical, petrochemical and material testing*

The industrial applications of ICP-MS are wide ranging and include analysis of metal concentrations in oils, plastics, ceramics, pure metals and alloys, and the acid, solvent and nuclear materials discussed in previous sections. In the petrochemical industry, applications of ICP-MS include direct analysis of metals in oils, oil fractions and fuel products [68, 79], analysis of solid polymer materials by LA-ICP-MS [80] and, more recently, investigations into the synthetic materials used as catalysts and substrates for research into hydrogen fuel cells [81]. The form or species of elements may also be important in the petrochemical industry, both for the environmental and process impact of the species. For example, the elemental form of Hg can corrode aluminium heat exchangers and pipe work, so it is important to know the form of Hg present in naphtha, fuels and gas condensates. Gas chromatography coupled to ICP-MS has proved extremely sensitive for this application [82] and the specificity of the MS measurement permits both unambiguous elemental identification and the application of isotope dilution for quantification.

Ceramics may be difficult to digest using conventional acid dissolution, so LA-ICP-MS can be beneficial in this application. Trace metals in pure metal samples or alloys are analysed routinely using spark source optical emission spectrometry (OES), X-ray fluorescence (XRF) and glow discharge (GD) OES. In this application, the analysis of dissolved metal solutions or direct ablation of solid metals using LA-ICP-MS can prove beneficial, yielding detection limits that are typically significantly lower than the traditional analytical techniques.

The range of applications of ICP-MS seems likely to continue to grow, with more elements being measured at lower levels of detection. In addition, the potential for routine monitoring of trace element species, using external calibration or isotope dilution, is an application that ICP-MS, coupled to LC, GC or CE, is already addressing successfully and this seems destined to continue.

Acknowledgements

The authors gratefully acknowledge the comments and suggestions made by Professor R.S. Houk.

References

[1] Gray, A.L., *Analyst*, 1975, **100**, 289–299.

[2] Houk, R.S., Fassel, V.A., Flesch, G.D., Svec, H.J., Gray, A.L. & Taylor, C.E., *Anal. Chem.*, 1980, **52**, 2283–2289.

[3] Douglas, D.J. & French, J.B., *Anal. Chem.*, 1981, **53**, 37–41.

[4] Montaser, A., Minnich, M.G., McLean, J.A., Liu, H, Caruson, J.A. & McLeod, C.W., Chapter 3: Sample introduction in ICPMS, in *Inductively Coupled Plasma Mass Spectrometry*, A. Montaser (ed.), Wiley-VCH, 1998.

[5] Sharp, B.L., *J. Anal. Atom. Spectrom.*, 1988, **3**, 613–652.

[6] Sharp, B.L., *J. Anal. Atom. Spectrom.*, 1988, **3**, 939–963.

[7] Gray, A.L. & Fresenius, Z., *Anal. Chem.*, 1986, **324**, 561–570.

[8] Hutton, R.C. & Eaton, A.N., *J. Anal. At. Spectrom.*, 1987, **2**, 595–598.
[9] Dickinson, G.W. & Fassel, V.A., *Anal. Chem.*, 1969, **41**, 1021–1024.
[10] Turner, I.L. & Montaser, A., Chapter 4: Plasma generation in ICPMS, in *Inductively Coupled Plasma Mass Spectrometry*, A. Montaser (ed.), Wiley-VCH, 1998.
[11] Park, C.J. & Noh, S.M., *J. Anal. At. Spectrom.*, 1998, **13**, 715–720.
[12] Douglas, D.J. & French, J.B., *Spectrochim. Acta*, 1986, **41B**, 197–204.
[13] Montaser, A., McLean, J.A., Liu, H. & Mermet, J.-M., Chapter 1: An introduction to ICP spectrometries for elemental analysis, in *Inductively Coupled Plasma Mass Spectrometry*, A. Montaser (ed.), Wiley-VCH, 1998.
[14] Jarvis, K.E., Gray, A.L. & Houk, R.S. (eds), Chapter 1, Origins and development, in *Handbook of Inductively Coupled Plasma Mass Spectrometry*, Blackie, Glasgow, 1992.
[15] Greenfield, S. & Montaser, A., Chapter 1: Inductively coupled plasma emission spectroscopy, A. Montaser (ed.), Wiley-VCH, 1989.
[16] Gray, A.L., *J. Anal. At. Spectrom.*, 1986, **1**, 247–249.
[17] Sakata, K. & Kawabata, K., *Spectrochim. Acta*, 1994, **49B**, 1027–1038.
[18] Jiang, S.J., Houk, R.S. & Stevens, M.A., *Anal. Chem.*, 1988, **60**, 1217–1221.
[19] Hoelzl, R., Fabry, L., Kotz, L. & Pahlke, S., *Fresenius J. Anal. Chem.*, 2000, **366**, 64–69.
[20] McCurdy, E.J., Wilbur, S., Woods, G.D. & Potter, D., in *Plasma Source Mass Spectrometry: The New Millennium*, G. Holland & S.D. Tanner (eds), The Royal Society of Chemistry, Cambridge, 2001, 45–60.
[21] Williams, J.G. & Gray, A.L., *Anal. Proc.*, 1988, **25**, 385–388.
[22] Williams, J.G., Chapter 3, Instrument options, in *Handbook of Inductively Coupled Plasma Mass Spectrometry*, K.E. Jarvis, A.L. Gray & R.S. Houk (eds), Blackie, Glasgow, 1992.
[23] Gray, A.L., *Spectrochim. Acta*, 1985, **40B**, 1525–1537.
[24] Douglas, D.J. & Tanner, S.D., Chapter 8: Fundamental considerations in ICP-MS, in *Inductively Coupled Plasma Mass Spectrometry*, A. Montaser (ed.), Wiley-VCH, 1998.
[25] Jarvis, K.E., Gray, A.L. & Houk, R.S. (eds), Chapter 2, Instrumentation for ICP-MS, in *Handbook of Inductively Coupled Plasma Mass Spectrometry*, Blackie, Glasgow, 1992.
[26] Denoyer, E.R., Jacques, M.D., Debrah, E. & Tanner, S.D., *At. Spectrosc.*, 1995, **16**(1), 1–6.
[27] Jarvis, K.E., Gray, A.L. & Houk, R.S. (eds), Chapter 11, Isotope ratio measurement, in *Handbook of Inductively Coupled Plasma Mass Spectrometry*, Blackie, Glasgow, 1992.
[28] Douglas, D.J., *Can. J. Spectrosc*, 1989, **34**, 38–49.
[29] Rowan, J.T. & Houk, R.S., *Appl. Spectrosc.*, 1989, **46**, 976–980.
[30] Koppenaal, D.W., Barinaga, C.J. & Smith, M.R., *J. Anal. Atom. Spectrom.*, 1994, **9**, 1053–1058.
[31] Eiden, G.C., Barinaga, C.J. & Koppenaal, D.W., *Rapid Commun. Mass Spectrom.*, 1997, **11**, 37–42.
[32] Turner, P., Merren, T., Speakman, J. & Haines, C., in *Plasma Source Mass Spectrometry, Developments and Applications*, G. Holland & S.D. Tanner (eds), The Royal Society of Chemistry, Cambridge, 1997, 28–34.
[33] Yamada, N. & Takahashi, J. Effects of Impurities in the Cell Gases on the Octopole Reaction Cell ICP-MS, poster FP-07 presented at the 2002 Winter Conference on Plasma Spectrochemistry, Scottsdale, Arizona, 6–12 January, 2002.
[34] Tanner, S.D., Baranov, V.I. & Bandura, D.R., in *Plasma Source Mass Spectrometry: The New Millennium*, G. Holland & S.D. Tanner (eds), The Royal Society of Chemistry, Cambridge, 2001, 99–116.
[35] Dexter, M.A., Reid, H. & Sharp, B.L., *J. Anal. Atom. Spectrom.*, 2002, **17**(7), 676–681.
[36] Yamada, N., Takahashi, J. & Sakata, K., *J. Anal. Atom. Spectrom.*, 2002, **17**(10), 1213–1222.
[37] Hattendorf, B. & Gunther, D. *J. Anal. Atom. Spectrom.*, 2000, **15**, 1125–1131.
[38] Baranov, V.I. & Tanner, S.D., in *Plasma Source Mass Spectrometry: The New Millennium*, G. Holland & S.D. Tanner (eds), The Royal Society of Chemistry, Cambridge, 2001, 34–45.
[39] Leonhard, P., Pepelnik, R., Prange, A., Yamada, N. & Yamada, T., *J. Anal. Atom. Spectrom.*, 2002, **17**, 189–196.
[40] Marchante-Gayon, J.M., Feldmann, I., Thomas, C. & Jakubowski, N., *J. Anal. Atom. Spectrom.*, 2001, **16**(5), 457–463.

[41] Turner, P.J., Mills, D.J., Schroder, E., Lapitajs, G., Jung, G., Iacone, L., Haydar, D.A. & Montaser, A., Chapter 6: Instrumentation for low- and high-resolution ICP-MS, in *Inductively Coupled Plasma Mass Spectrometry*, A. Montaser (ed.), Wiley-VCH, 1998.
[42] Bradshaw, N., Hall, E.F.H. & Sanderson, N.E., *J. Anal. Atom. Spectrom.*, 1989, **4**, 801–803.
[43] Myers, D.P., Li, G., Mahoney, P.P. & Hieftje, G.M., *J. Am. Soc. Mass Spectrom.*, 1995, **6**, 411–420.
[44] Hutton, R.C., Eaton, A.E. & Gosland, R.M., *Appl. Spectrosc.*, 1990, **44**(2), 238–242.
[45] Jarvis, I., Chapter 7, Sample preparation for ICP-MS, in *Handbook of Inductively Coupled Plasma Mass Spectrometry*, K.E. Jarvis, A.L. Gray & R.S. Houk (eds), Blackie, Glasgow, 1992.
[46] Gunther, D. & Heinrich, C.A., *J. Anal. Atom. Spectrom.*, 1999, **14**, 1363–1368.
[47] Ebdon, L., Ford, M.J., Hutton, R.C. & Hill, S.J., *Appl. Spectrosc.*, 1994, **48**, 507–516.
[48] Olivares, J.A. & Houk, R.S., *Anal. Chem.*, 1986, **58**, 20–25.
[49] Olesik, J.W., Stewart, I.I., Hartshome, J.A. & Hensman, C.E., in *Plasma Source Mass Spectrometry, New Developments and Applications*, G. Holland & S.D. Tanner (eds), The Royal Society of Chemistry, Cambridge, 1999, 3–19.
[50] Date, A.R. & Gray, A.L., *Analyst*, 1983, **108**, 159–165.
[51] Date, A.R. & Gray, A.L., *Int. J. Mass. Spectrom. Ion. Phys.*, 1983, **48**, 357–360.
[52] Heumann, K.G., in *Inorganic Mass Spectrometry*, F. Adams, R. Gijbels & R. Van Grieken (eds), Chemical Analysis Series, 95, Wiley, New York, 1988, 301–376.
[53] Walder, A.J., Koller, D., Reed, N.M., Hutton, R.C. & Freedman, P.A., *J. Anal. Atom. Spectrom.*, 1993, **8**, 1037–1041.
[54] Woods, G.D. & McCurdy, E., in *Plasma Source Mass Spectrometry, New Developments and Applications*, G. Holland & S.D. Tanner (eds), The Royal Society of Chemistry, Cambridge, 1999, 108–119.
[55] Munro, S., Ebdon, L. & McWeeny, D.J., *J. Anal. Atom. Spectrom.*, 1986, **1**, 211–219.
[56] Dean, J.R., Crews, H.M. & Ebdon, L., Applications in food science, in *Applications of Inductively Coupled Plasma Mass Spectrometry*, A.R. Date & A.L. Gray (eds), Blackie, London, 1989, 141–168.
[57] Larsen, E.H., Knuthsen, P. & Hansen, M., *J. Anal. Atom. Spectrom.*, **14**, 41–44.
[58] Whittaker, P.G., Barrett, J.F.R. & Williams, J.G., *J. Anal. Atom. Spectrom.*, 1992, **7**, 109–115.
[59] Owen, L.M.W., Crews, H.M., Hutton, R.C. & Walsh, A., *Analyst*, 1992, **117**, 649–655.
[60] Sutton, K.L. & Caruso, J.A., *J. Chromatogr.*, 1999, **856**, 243–258.
[61] Donard, O.F.X., Lespes, G., Amouroux, D. & Morabito, R., Chapter 8: Organotin compounds in the environment: still a critical issue, in *Trace Element Speciation for Environment, Food and Health*, L. Ebdon, L. Pitts, R. Cornelis, H. Crews, O.F.X. Donard & Ph. Quevauviller (eds), The Royal Society of Chemistry, Cambridge, 2001, 142–175.
[62] Wahlen, R. & Fairman, B., *Spectroscopy Europe*, 2001, **13**(5), 16–22.
[63] Tao, H., Rajendran, R.B., Quetel, C.R., Nakazatu, T., Tominaga, M. & Miyazaki, A., *Anal. Chem.*, 1999, **71**, 4208–4215.
[64] Rajendran, R.B., Tao, H., Nakazato, T. & Miyazaki, A., *Analyst*, 2000, **125**, 1757–1763.
[65] Taylor, H.E., Huff, R.A. & Montaser, A., Chapter 9: Novel applications of ICP-MS, in *Inductively Coupled Plasma Mass Spectrometry*, A. Montaser (ed.), Wiley-VCH, 1998.
[66] Walsh, A., Potter, D., McCurdy, E. & Hutton, R.C., in *Applications of Plasma Source Mass Spectrometry*, G. Holland & A.N. Eaton (eds), The Royal Society of Chemistry, Cambridge, 1991, 12–24.
[67] Shive, L.W., Ruth, K. & Schmidt, P., *Micro.*, 1999, **17**(2), 27–31.
[68] McCurdy, E., Woods, G.D. & Mizobuchi, K., Practical Considerations in the Routine Analysis of Organic Solvents by ICP-MS, Paper O-9 presented at the European Winter Conference on Plasma Spectrochemistry, Hafjell, Norway, 4–8 February 2001.
[69] Vandecasteele, C., Vanhoe, H. & Dams, R., *J. Anal. Atom. Spectrom.*, 1993, **8**, 781–786.
[70] Sieniawska, C.E., Mensikov, R. & Delves, H.T., *J. Anal. Atom. Spectrom.*, 1999, **14**, 109–112.

[71] Roberts, N.B., Walsh, H.P.J., Klenerman, L., Kelly, S.A. & Helliwell, T.R., *J. Anal. Atom. Spectrom.*, 1996, **11**, 133–138.
[72] Janghorbani, M. & Ting, B.T.G., *J. Nutr. Biochem.*, 1990, **1**, 4–19.
[73] Date, A.R. & Gray, A.L., *Spectrochim. Acta*, 1985, **40B**, 115–122.
[74] Jarvis, K.E., *Chem. Geol.*, 1988, **68**, 31–39.
[75] Garbe-Schoenberg, C.D., *Geostand. Newsl.*, 1993, **17**, 81–97.
[76] Pearce, N.J.G., Perkins, W.T., Abell, I., Duller, G.A.T. & Fuge, R., *J. Anal. Atom. Spectrom.*, 1992, **7**, 53–57.
[77] Jeffries, T.E., Jackson, S.E. & Longerich, H.P., *J. Anal. Atom. Spectrom.*, 1998, **13**, 935–940.
[78] Watling, R.J., Herbert, H.K., Delev, D. & Abell, I.D., *Spectrochim. Acta*, 1994, **49B**(2), 205–219.
[79] Van Heuzen, A., ICP-MS in the petroleum industry, in *Applications of Inductively Coupled Plasma Mass Spectrometry*, A.R. Date & A.L. Gray (eds), Blackie, London, 169–188.
[80] Marshall, J., Franks, J., Abell, I. & Tye, C., *J. Anal. Atom. Spectrom.*, 1991, **6**, 145–150.
[81] Westheide, J.T., Becker, J.S., Jaeger, R., Dietze, H.-J. & Broeckaert, J.A.C., *J. Anal. Atom. Spectrom.*, 1996, **11**, 661–666.
[82] Bouyssiere, B., Bako, F., Savary, L. & Lobinski, R., *Oil Gas Sci. Tech. Rev. IFP*, 2000, **55**(6), 639–648.

3 Inductively coupled plasma atomic emission spectrometry

Stephen J. Hill and Andrew Fisher

3.1 Introduction

Since its inception in the 1960s and the introduction of the first commercial instruments in 1974 [1], inductively coupled plasma – atomic emission spectrometry (ICP-AES) has become one of the most popular analytical techniques for the quantitative determination of trace elements. There are a number of reasons for this. Most modern instruments are capable of determining approximately 70 elements, with a linear range which covers 5 orders of magnitude, i.e. 2 or 3 orders more than rival techniques such as flame atomic absorption spectrometry. The technique also offers rapid analysis, because modern instruments operate in either simultaneous mode (i.e. interrogate several different analytical lines simultaneously) or rapid sequential mode (i.e. automatically moving from one analytical line to the next in succession). In addition, ICP-AES is also relatively free from interferences and the instrument is less expensive when compared with techniques such as inductively coupled plasma – mass spectrometry (ICP-MS). The advent of programmable autosamplers and more powerful computers which readily facilitates the programming of quality control procedures, re-calibration and re-analysis where necessary, also enables the instrument to be used without the constant attention of an operator.

3.2 Theory

3.2.1 How a plasma is formed?

A schematic diagram of a modern ICP-AES instrument is shown in Fig. 3.1, the plasma being formed at the top of a plasma torch. Argon gas is passed through the torch, the top of which is surrounded by a water-cooled copper coil (the load coil) which is connected to the RF generator. When RF power is applied to the load coil, an alternating current is set up which oscillates at a frequency governed by the generator. This oscillating current then sets up a magnetic field in the region at the top of the torch. If a spark is then applied to the gas by a Tesla coil, electrons are stripped from some of the argon atoms. These electrons are then accelerated by the magnetic field and collide with other argon atoms

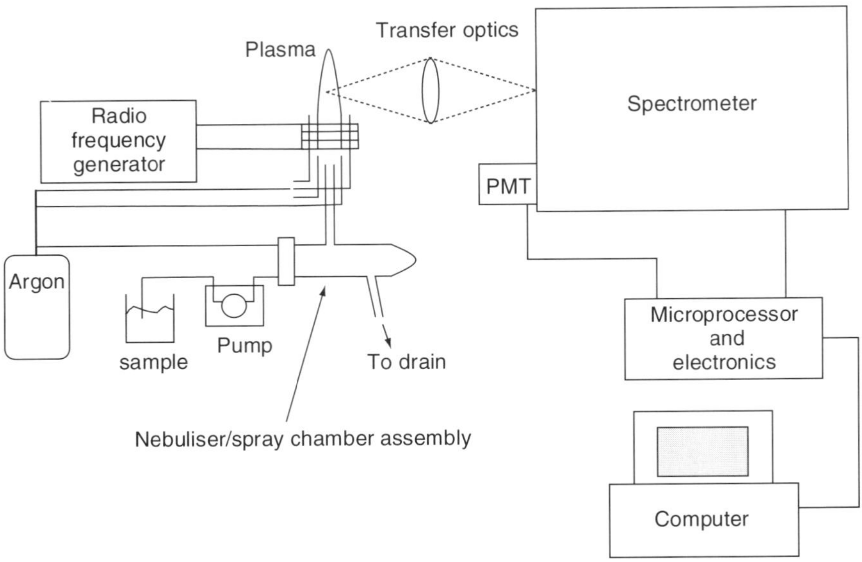

Fig. 3.1 Schematic diagram of a generalised ICP-AES instrument.

causing further electrons to be ejected and an avalanche effect. This process is almost instantaneous and self-sustaining, the plasma formed consisting of neutral argon atoms, ionised argon and electrons. The magnetic field causes the ions and electrons to flow in closed circular horizontal paths, such that the eddy currents so produced heat the neutral argon atoms by collisional energy exchange to produce a white hot fireball. Overall, the plasma is neutral in terms of charge, with the number of positive ions being the same as the number of free electrons.

The nebuliser (injector) gas flow is placed at the centre of the torch assembly so that the gas punches a hole through the centre of the plasma. Thus the plasma is toroidal in shape. It is through this hole (the annulus) that sample is introduced. A schematic diagram of a plasma is shown in Fig. 3.2 which includes an indication of the approximate temperature found in each region. It is these high temperatures (5000–6000 in the tail flame) that cause the analytes introduced to the plasma to become excited and emit radiation. These high temperatures are also the reason that so few interferences occur, since they are sufficient to dissociate the vast majority of analyte compounds.

3.2.2 Atomic emission theory

A detailed description of the basic theory of atomic emission is not really required here, but a brief overview may be helpful to aid the understanding of some of the terminology used in later sections. A far more detailed and theoretical description of the processes within the plasma may be found elsewhere [2].

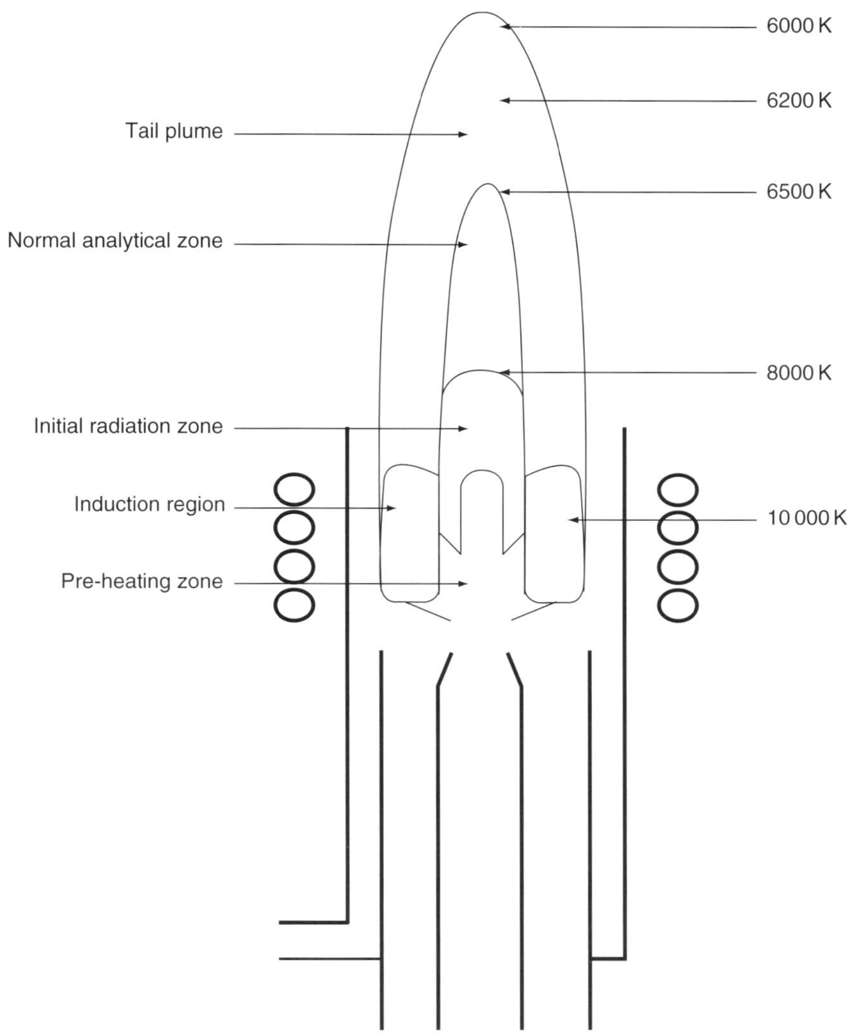

Fig. 3.2 Schematic diagram of a plasma.

An atom consists of a nucleus, containing protons and neutrons, that is surrounded by clouds of electrons that occupy defined energy levels. When the atom is at its least energetic state, i.e. all the electrons are resting in the lowest energy states (closest to the nucleus), the atom is described as being in its *ground state*. If the ground state atoms are then exposed to heat or electromagnetic radiation, or are involved in a collision with an excited atom or ion, several phenomena may result. The most likely effect is that either the increase in kinetic energy of the atom will make it move faster, or the atom will absorb the energy and this will excite an electron to a higher energy state. The atom is now regarded as being in an *excited state* which is less stable and hence the electron will attempt to return to the more stable ground state or a lesser excited state. This may result from either collision with another particle (e.g. an atom or ion) or by the release of a discrete batch of electromagnetic radiation, known as a photon. The intensity I_{em} of a spontaneous emission line is given by the equation:

$$I_{em} = A_{ji} h v_{ji} N_j \tag{3.1}$$

where A_{ji} is the Einstein coefficient for spontaneous emission, h is Planck's constant, v is the frequency of the radiation for the transition ji, and N_j is the number of atoms in the excited state from the Boltzmann Distribution Law.

If the energy absorbed by the atom is sufficiently high, an electron may even be ejected from the atom, forming a positive ion. The extent to which this occurs is dependent on the ionisation energy of the analyte, but ranges from close to 100% for caesium to 0% for fluorine. Once formed, the ion also has a ground state and several excited states and is therefore also capable of emission. The possibility of ions as well as atoms producing emission is the reason that in older literature ICP-AES is often referred to by the alternative name of inductively coupled plasma – optical emission spectrometry (ICP-OES).

The relaxation of electrons from different excited energy levels back to the ground state will yield photons of light of different energy, i.e. the wavelength of the light emitted will be different for each transition. A characteristic spectrum for each element will therefore be formed. The relationship between energy and wavelength is described by the Planck's equation which relates the energy difference between two levels:

$$\mathrm{E} = hv \tag{3.2}$$

If v is substituted with c/λ where c is the speed of light and λ is the wavelength, Planck's equation may be expressed as:

$$\mathrm{E} = h\, c/\lambda \tag{3.3}$$

It can be seen from Equation 3.3 that at higher energy, a lower wavelength is produced and that high wavelengths indicate that a less energetic transition has occurred. The process is depicted graphically in Fig. 3.3.

From the discussion above it is clear that the energy level to which an electron is excited is governed by the transition probability, i.e. it is known that under certain conditions a known proportion of electrons will be excited to any given energy level. It is the transition probabilities that govern the sensitivity of a wavelength (an analytical line). If the probability of a transition is low, then

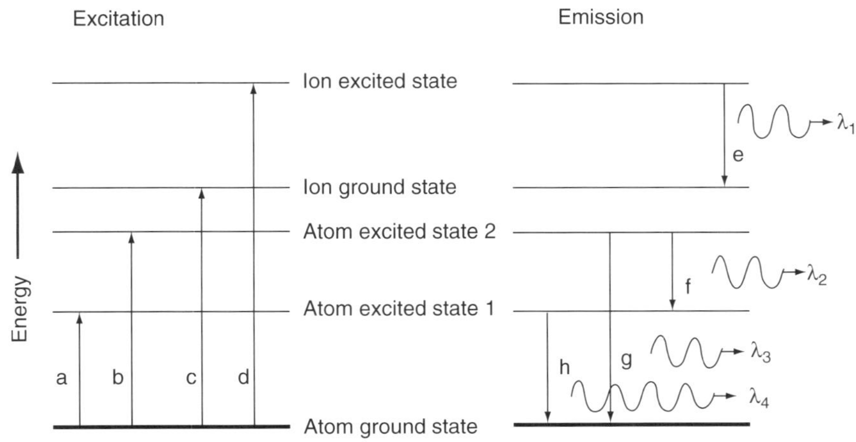

Fig. 3.3 Graphical depiction of excitation and emission. a and b represent excitation, c represents ionisation, d represents ionisation and excitation, e represents ion emission, f represents atomic emission from an excited state to a lesser excited state and g and h represent atomic emission back to the ground state.

the number of analyte atoms (or ions) in which the electrons are excited to that energy level will also be low. This means that the number of atoms/ions emitting light at the corresponding wavelength and hence the overall sensitivity will be low. Therefore, to obtain a detectable signal, the concentration of that analyte in the sample will have to be higher. Each line of an analyte therefore has a different sensitivity. In practical terms, this can be useful analytically because each line will have its own linear range. If the expected range of concentration of an analyte in a sample is known, then the most appropriate analytical line may be chosen, thereby negating the need for large sample dilutions and hence keeping sample manipulation to a minimum and decreasing the likelihood of dilution errors or contamination.

Table 3.1 gives a list of elements that may be determined using ICP-AES together with an estimate of the limit of detection (LoD) obtained using a conventional nebuliser/spray chamber assembly and the analyte's most sensitive line. The values given in Table 3.1 compare well with flame atomic absorption, often being better by a factor of 10 or more, but compare less favourably with LoDs obtained using ICP-MS.

3.2.3 *Interferences*

There are a number of different types of interference that may affect the accuracy and precision of an analysis. Instrumental drift, although not really an interference effect, can lead to exceptionally poor precision and inaccuracy. Most modern instruments operate optimally in a temperature controlled

Table 3.1 Limits of detection (LoDs) obtainable using the analyte's most sensitive line and a conventional nebuliser/spray chamber assembly to introduce the sample

Analyte	LoD / $\mu g\ L^{-1}$	Analyte	LoD / $\mu g\ L^{-1}$
Al	5	Hg	20
Sb	40	Mo	1
As	30	Ni	7
Ba	0.5	Pd	70
Be	0.1	Pt	30
Bi	10	K	80
B	3	P	20
Cd	2	Se	50
Ca	0.1	Si	5
Ce	100	Ag	5
Cr	5	Na	6
Co	3	Sr	0.2
Cu	0.3	S	50
Ge	100	Te	20
Au	25	Tl	50
In	100	Sn	7
Fe	1	Ti	3
La	6	W	20
Pb	7	U	400
Li	3	V	2
Mg	0.5	Zn	1
Mn	0.4	Zr	5

laboratory. An internal wavelength calibration, e.g. from a mercury line, will help maintain the spectrometer accuracy. The absence of control or correction of spectrometer drift will cause poor on-peak stability and poor precision.

Matrix effects may also lead to problems. It should be noted that an acid digest of a sample will possibly contain a higher concentration of acid than a calibration standard (often prepared in 2% acid), and a higher concentration of dissolved solids will have an effect on the viscosity. If the transport efficiency is different between samples and standards, for example if standards and samples are nebulised with a different efficiency, then inaccurate data will result. Matrix matching of samples and standards, as closely as possible, will diminish these effects. The application of an internal standard to all samples, blanks and standards may also help. The internal standard is usually an element that is not present in the sample at any significant concentration. If there is a difference in transport efficiency between samples and standards, it is reasonable to assume that the internal standard would suffer the same difference and hence a normalisation may be made.

As stated above, the temperature of the plasma is sufficient to dissociate nearly all molecules and hence many of the interference effects observed in techniques such as flame atomic absorption are overcome. The plasma is however spatially inhomogeneous. This means that for optimal performance the viewing height and operating parameters should be optimised carefully. In general, atomic emission is seen more readily lower in the plasma and ionic emission is more likely to be seen slightly higher up. There is also a possibility that atomic emission may be seen again very high in the plasma, although molecular re-combination may also occur in this region. The power and nebuliser gas flow rate also have an effect on optimal viewing height. If good sensitivity is required for the determination of an analyte, the measurement of atomic emission fairly high in the plasma should be avoided because the signal to noise ratio and sensitivity will be poor.

Major problems from spectral interferences are relatively rare in ICP-AES. When interferences do exist, there is normally a range of alternative analyte lines that may be used to circumvent any interference effects. It is worth remembering that some molecular interferences e.g. those arising from OH, SO and PO, may lead to structured interference at some regions of the electromagnetic spectrum. Standard additions and matrix matching may also assist in overcoming some types of interference. An overview of potential interference observed in ICP-AES has been presented by Zander [3].

3.3 Instrumentation

3.3.1 Generators

The generator is the device that provides the RF power to generate and sustain the plasma discharge. The typical power requirements for such a generator range between 900 and 1500 W although low power plasmas may also be used. Occasionally, it may be necessary to increase the power of the plasma to 1600 W or higher. This is especially so when organic solvents are being aspirated or alternative gases to argon are introduced. Many commercial RF generators provide up to 2000 W. The power is transferred to the argon gas via a three turn water-cooled copper coil which surrounds the top of the plasma torch.

The frequency of operation of the generator in most commercial instruments is either 27.12 MHz or 40.68 MHz. The 40 MHz generators offer improvements in coupling efficiency and a reduction in the background emission intensity, and therefore many manufacturers prefer to use this higher frequency rather than the more traditional 27 MHz generators.

There are two basic types of generator: free running and crystal controlled. Crystal controlled generators use a piezoelectric quartz crystal to produce an oscillating RF signal. This then passes through a frequency multiplier and an amplifier before it is applied to the load coil. These generators are rather large and require a matching network to maintain the electrical parameters such as the output impedance. Free running generators operate at an oscillating frequency that is dependent on both the electronic components of the generator and on the conditions within the plasma. This type of generator is therefore capable of adjusting its output automatically to compensate for any changes within the plasma.

Further details of how generators work is beyond the scope of this text, but a readable account of the different types of generator available has been published by Greenfield [4].

3.3.2 Line isolation devices

The role of the line isolation device is to isolate specific spectral lines emitted from the analyte(s) from all of the other emissions produced by other atoms, ions and molecules within the plasma. There are several devices available to perform this function, including monochromators which, as their name indicates, only isolate one line at any given instant, and polychromators and echelle-based devices that may interrogate several different lines simultaneously. The large majority of lines that are useful analytically occur in the region 160–450 nm, although there are also some lines in the region 450–900 nm, and therefore the majority of commercial instruments are capable of detecting wavelengths in this region. At very low wavelength, e.g. <190 nm, the spectrometer will have to be purged with an inert gas (argon or nitrogen) or placed under vacuum, due to the many lines in that region from the oxygen in air. If the spectrometer is not purged, severe drops in sensitivity and interferences will occur. There are numerous types of monochromator available, but the two most common commercial designs are the Ebert and the Czerny Turner configurations. Schematic representations of these are given in Fig. 3.4. In both cases, light enters through the entrance slit and is reflected off a collimating mirror onto the diffraction grating. This is a device that separates the light into its different spectral components, and passes the desired wavelength to a second collimating mirror in the case of the Czerny Turner monochromator (the Ebert design uses just one larger mirror) which focuses the monochromatic light onto an exit slit positioned in front of the detector.

The diffraction grating is a mirrored surface that has closely spaced lines ruled or etched onto its surface. The density of these lines varies between spectrometers, but is typically in the range of 600–4200 per mm. The larger the number of grooves, the higher the resolving power of the spectrometer. As the continuum light hits the grating it will be diffracted at an angle that is dependent upon the wavelength. In general, the longer the wavelength and the higher the groove density, the more the light will be diffracted. This is shown pictorially in Fig. 3.5. Since the grating is moveable, i.e. it may be rotated at different angles, the light diffracted onto the collimating mirror may be changed. This is the basic theory of a sequential scanning monochromator. Light of one wavelength is focussed onto the detector, and then the diffraction grating is adjusted so that a second wavelength may be detected. Modern spectrometers are computer controlled and so the grating position may be changed rapidly and reproducibly. However, the process may still take a few seconds, and hence if transient signals are being obtained, e.g. from an alternative sample introduction

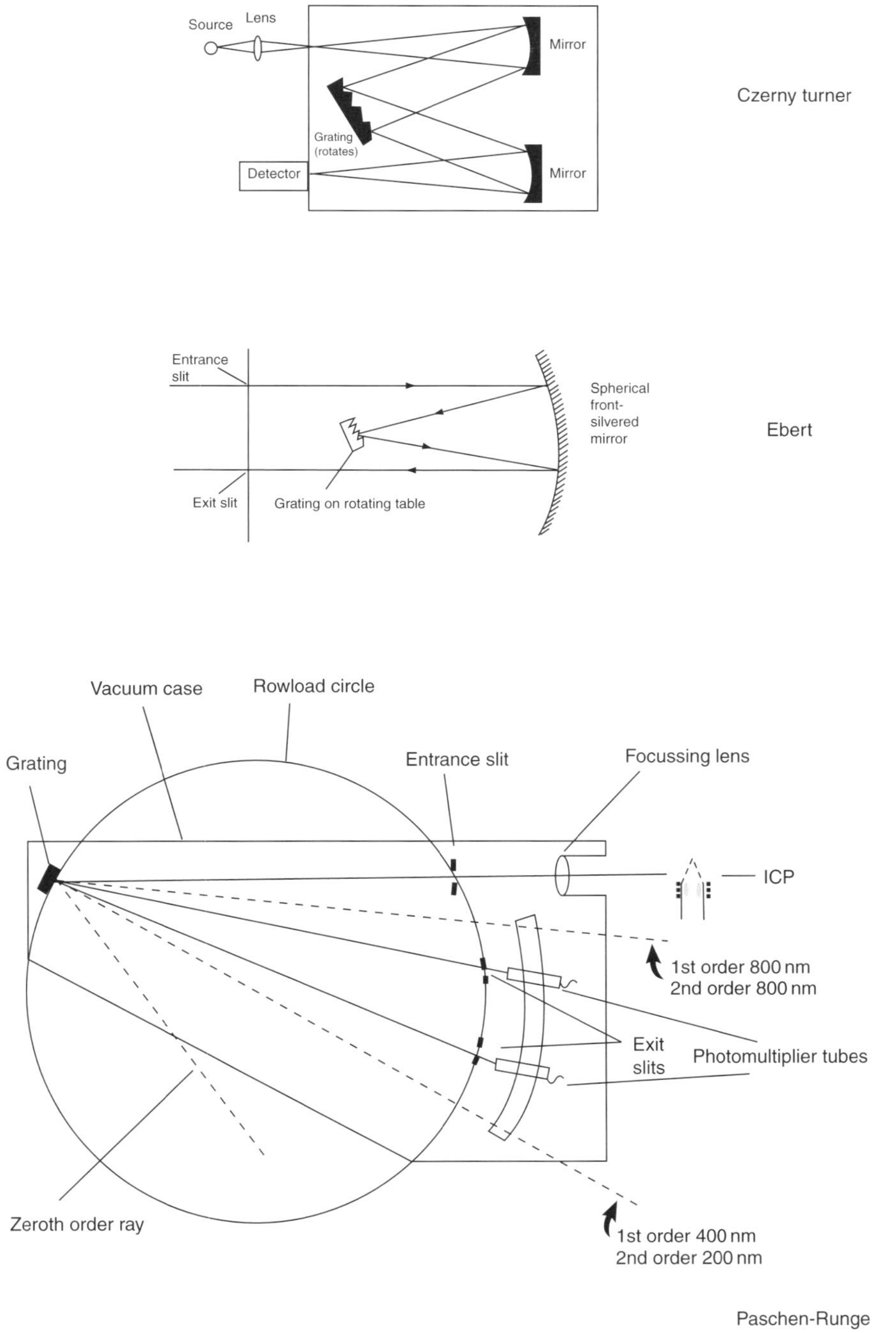

Fig. 3.4 Some commonly occurring line dispersion devices.

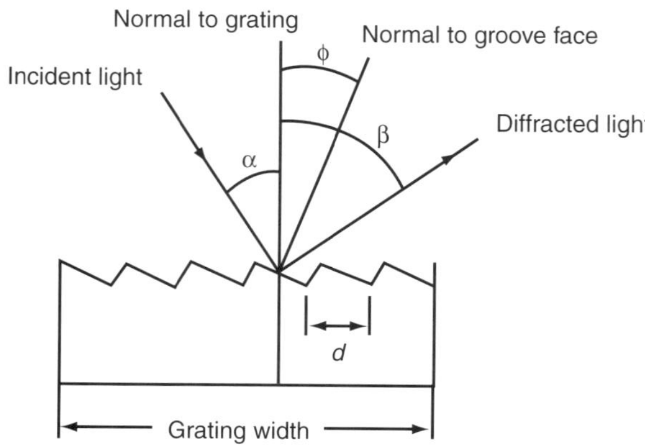

Fig. 3.5 A diffraction grating.

method such as hydride generation (HG) or electrothermal vapourisation (ETV) (see Section 3.3.5.3), then there is unlikely to be sufficient time to determine more than one element. Replicate analyses would therefore have to be performed of the same sample if more than one analyte is to be determined.

A polychromator has an advantage over a monochromator in that it may detect several wavelengths simultaneously. This means that such devices are more rapid than monochromators for multi-element analysis and it is also possible to detect several analytes in a transient signal. The possibility of using real-time background correction and internal standardisation is also an advantage. The layout of a typical polychromator is shown in Fig. 3.4. The photomultiplier (PMT) tubes (detectors) are fixed on the edge of a Rowland circle. There is obviously a limit to the number of detectors that may be fitted around the circumference of the Rowland circle, and since the grating does not rotate in this type of instrument, there is a limit to the number of lines that may be interrogated simultaneously. Although, in theory up to about 60 detectors may be used, many instruments are not capable of interrogating more than 30 lines. A further potential disadvantage of these instruments is that they are expensive and inflexible. For example, if an analyte not in the original range (30 lines) is to be determined, a large expense in both time and cost will be necessary to add this in at a later date. There are a number of slightly different designs, but many commercial instruments of this type are based on a Paschen-Runge style spectrometer.

Many of the most modern instruments now utilise an echelle-based spectrometer. A typical echelle-based spectrometer is shown in Fig. 3.6. These devices utilise a diffraction grating and a prism (although occasionally two gratings are used), set perpendicular to each other. The first grating has a lower density of grooves etched onto its surface than the gratings described previously, but this separates the polychromatic light by wavelength and produces multiple, overlapping spectral orders. It should be noted that if an analyte gives an intense

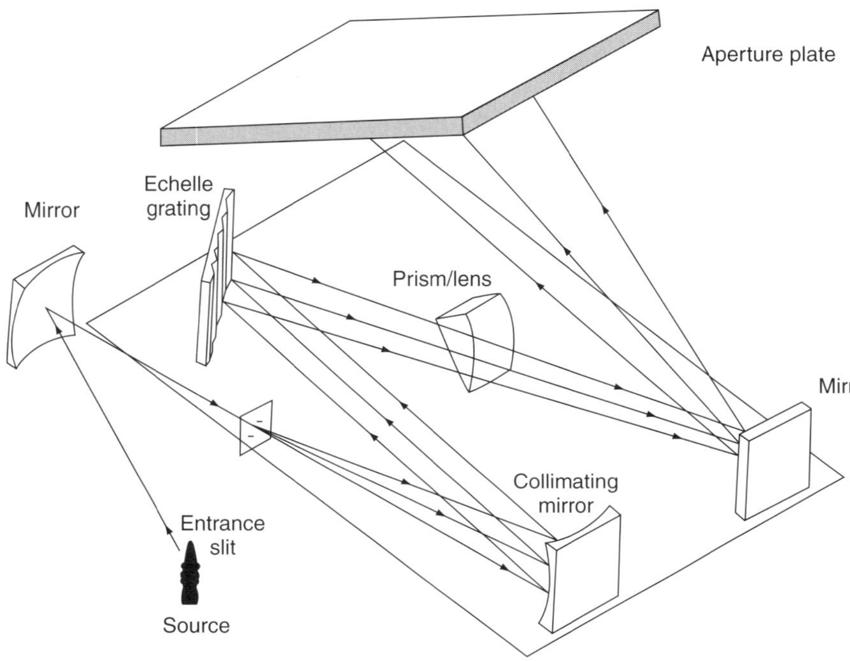

Fig. 3.6 An echelle spectrometer.

emission at a wavelength of 200 nm, there is likely to be another emission at 400 nm and possibly even at 600 nm. These other wavelengths are known as spectral orders also known as harmonics of two and three greater than the original wavelength. If another line is too close to the primary line of interest, an interference will occur because the spectrometer may not have sufficient resolving power. However, if the analyte is measured in the second order, the interfering line will be twice the distance away, hence improving the likelihood of resolution. The second dispersing device (either a prism or a second grating) separates these overlapping orders into a two-dimensional pattern, with wavelength in one direction and order in the other. This pattern, called an echellogram, may be projected onto a range of detectors. If the pattern is expanded then a PMT tube detection system may be used. However, in modern instruments the vast majority project the pattern onto solid state detectors that are capable of monitoring numerous analytical lines simultaneously.

3.3.3 Detectors

There are several different detector types used for ICP-AES instrumentation. The type used in any given instrument will depend to some extent on the type of line isolation device used. Until fairly recently, the most common detector was the PMT. These vacuum tube devices are very sensitive and cover a large wavelength range (between 160 and 900 nm). A schematic diagram of a PMT and the mechanism of action is shown in Fig. 3.7. Two basic types of PMT exist, i.e. the side-viewed (as shown in Fig. 3.7) and the end-viewed, although both operate in a similar way. Light separated by the line isolation device is allowed to pass through the transparent quartz window and impinge on the cathode of the PMT which then ejects electrons. These electrons are then accelerated down a dynode chain (each dynode being successively more positive in potential) and as each electron hits successive dynodes, further electrons are emitted. A typical PMT contains between 9 and 16 dynodes and hence a very effective avalanche effect is achieved. An estimated 10^6 electrons may be generated from a single photon strike on a PMT containing a 9 dynode series. The electrons are then collected from the final dynode by the anode. The current measured at the anode, i.e. the number of electrons striking it per unit time, is proportional to the number of photons

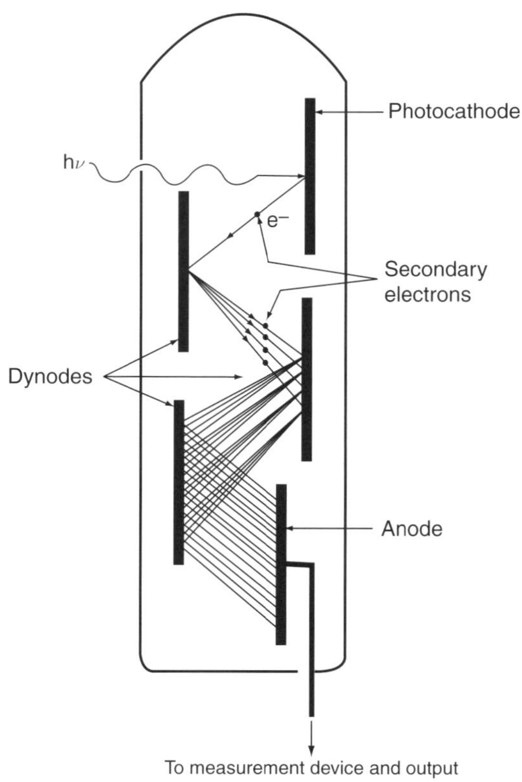

Fig. 3.7 A simplified schematic diagram of a PMT (showing only three dynodes).

striking the cathode which, in turn, is proportional to the concentration of the analyte within the plasma.

Although there are only two basic types of PMT, a number of different materials may be used in the construction of the photocathode. The material used will determine the useful wavelength range of the PMT. Gallium arsenide is one of the more common materials because it provides a reasonably uniform response over a wide wavelength range. Other materials that have been used include caesium-antimony or sodium-potassium-caesium-antimony. A number of alternative materials have been developed in an attempt to improve the response time, dynamic range and sensitivity of PMTs. The signal to noise ratio of PMTs has also been improved. A portion of the noise is produced by dark current, i.e. residual current that is produced when no photons are striking the cathode as a result of thermal excitation or by cosmic or radio-active rays. The dark current increases as the PMT is used under more extreme conditions. For example, if a less sensitive emission line is used, then the voltage applied to the PMT will have to be increased to aid detection, and this may lead to a noisier signal. Different PMTs are designed to cover different regions of the electromagnetic spectrum. The majority will be damaged if exposed to sunlight, but others, e.g. solar blind PMTs operate within the UV region and do not respond to daylight.

Solid state detectors have become increasingly popular in modern instrumentation. These are detectors such as charge coupled devices (CCD), charge injection devices (CID) and diode arrays. The CCD and CID are termed charge transfer devices (CTD). These devices have been available for over 20 years, but it is only in the last 10 years that they have been used as detectors for ICP-AES instrumentation. The sensitivity, detection limits and linear range of these devices are similar to those of PMTs, but they have the advantage of being able to detect several analytical lines and/or background measurements simultaneously. In some ways these devices may be considered to be a sort of electronic photographic plate. Each photo-active array has three areas: (1) a photosensitive area; (2) a storage register and (3) a readout register. Photons from the analytes within the plasma strike the photosensitive area and are converted into photoelectrons. These are then pulled away from the surface and stored in the storage area as an electric charge. After a set period of time the accumulated charge is transferred to the readout register, where it is moved to the output electronics and amplified. The method by which these devices work is described more fully elsewhere [5–7]. A further variation of the CCD is the segmented CCD (SCD). An SCD is used in at least one current commercial instrument. The SCD has dimensions of 13 $\times$ 18 mm and contains several thousand pixels arranged in small sub-arrays of 20–80 pixels each. There are over 200 of these sub-arrays positioned on a silicon wafer and they are positioned so that they correspond to the two-dimensional pattern produced by an echelle spectrometer. They therefore correspond to over 200 of the most frequently determined analyte lines. The use of an SCD therefore limits the number of lines that may be interrogated and therefore does not have the flexibility of a CCD (that may have hundreds of thousands of pixels), or of a PMT. However, the SCD has the advantage over the PMT of being virtually simultaneous (i.e. more rapid) and of being more sensitive at lower wavelengths than most CCDs. One potential problem with CCDs is that they are occasionally prone to blooming, i.e. if sufficient signal is accumulated, e.g. from a very intense line, it may spill out from the storage register and contaminate adjacent pixels. Considerable research has been conducted to develop anti-blooming CTDs.

3.3.4 Radial and axial ICPs

There has been a great deal of debate as to whether radial or axial spectrometers give better performance. In axial spectrometers, the torch is turned through 90° so that it is pointed towards the collection optics. To prevent damage to the optics, a shear gas is usually employed to *cut-off* the end of the tail-flame and hence dissipate the hot gases as shown in Fig. 3.8. The axial (end-on) plasmas have a better collection efficiency of the emitted light than conventional radial (side-on) plasmas, and therefore the sensitivity should be improved when compared with the radial instrumentation. However, the radially viewed plasmas are optically thin. This means that the light emitted from any analyte atom/ion has to pass only a very short distance in the plasma through other analyte atoms or ions, leading to less chance of self-absorption. Similarly, because of this optical thinness, interference effects are also less. This means that when the LoDs obtained from radial and axial plasmas are compared for a complex

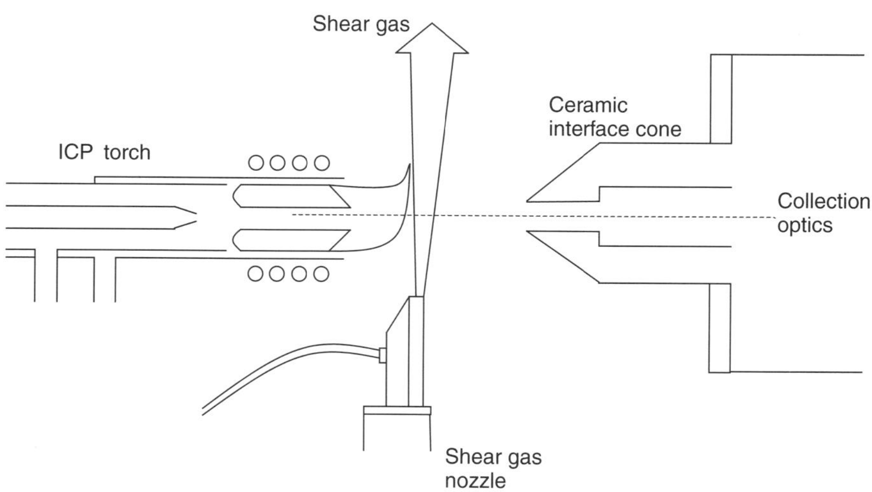

Fig. 3.8 Axial (end-on) ICP showing shear gas.

sample, they are usually very similar. In some cases radial plasmas can offer more flexibility in that if an interferent is spatially dependent (i.e. only occurs in a certain region of the plasma), then a different viewing height may be used so that the potential interference may be avoided. Although outside the scope of this text, a recent study has compared radially and axially viewed plasmas in terms of signal to background ratios and matrix effects [8].

3.3.5 Sample introduction

In the large majority of cases, samples are introduced into the ICP-AES in liquid form. Therefore solid samples must be either dissolved or acid digested in order to be transformed into a suitable form for introduction. There are, however, a number of different sample introduction techniques that may be used, some of which are capable of introducing solid samples, and these different techniques will be described below.

3.3.5.1 Nebulisers and spray chambers

Sample is normally transported to the nebuliser by a peristaltic pump. If a poor quality pump is used, i.e. a pump that has only a few rollers, then a noisy signal may be produced. Similarly, if wide bore sample tubing is used, then the pump will have to be slowed down so that a suitable uptake rate is obtained. This will also exacerbate the pump noise. In general it is better to have a pump with a large number of rollers (ten or more), narrow bore pump tubing and a very fast revolution rate of the pump. This will lead to the minimisation of pump noise.

The function of the nebuliser is to transform the stream of liquid into a nebular (an aerosol). In the majority of applications, this primary aerosol then enters the spray chamber which acts as a droplet size filter, i.e. it will discriminate against larger drops and pass these to waste, while it will allow the smallest ones to be transported to the plasma. Droplets exiting the spray chamber to the plasma are called the tertiary aerosol. The droplet size of the primary aerosol produced by the nebuliser will determine the transport efficiency of the sample to the plasma. If the droplets produced are very small, the transport efficiency will be greater and hence the signal will be larger. However, if the nebulisation process is too efficient, excessive sample transport may result and this will cause plasma perturbation and possible extinction. In general, the majority of nebuliser/spray chamber assemblies produce a transport efficiency of between 1 and 3%. The processes occurring within the nebuliser and spray chamber have been reviewed in two comprehensive publications by Sharp [9, 10]. It should be noted however that, in the majority of cases, the nebuliser and spray chamber work together to introduce the optimal amount of sample. Another paper by Luan [11] has also reviewed the mechanism of action, theory and operating characteristics of pneumatic nebulisers. A paper by McLean *et al.* [12] has described the methods used for nebuliser diagnostics, i.e. to measure droplet size etc. within the nebular, and a further publication by Browner [13] discusses aerosol generation and sample transport to the plasma.

There are a wide range of nebulisers and spray chambers that may be used to introduce liquid samples. The standard pneumatic Meinhard style (Fig. 3.9) provides an extremely stable signal because they are self-aspirating, i.e. the sample does not have to be transported to them by a peristaltic pump. However, such nebulisers are intolerant of dissolved or suspended solids. Standard pneumatic nebulisers usually work on the principle of the Venturi effect, i.e. the flowing gas passes across the tip of a tube within the nebuliser causing a drop in pressure which leads to liquid samples being drawn up and then shattered in the gas stream. Some nebulisers are made of glass (e.g. Meinhard), and are therefore intolerant of materials dissolved in hydrofluoric acid (unless this has been neutralised) or sodium hydroxide whereas others are made of inert

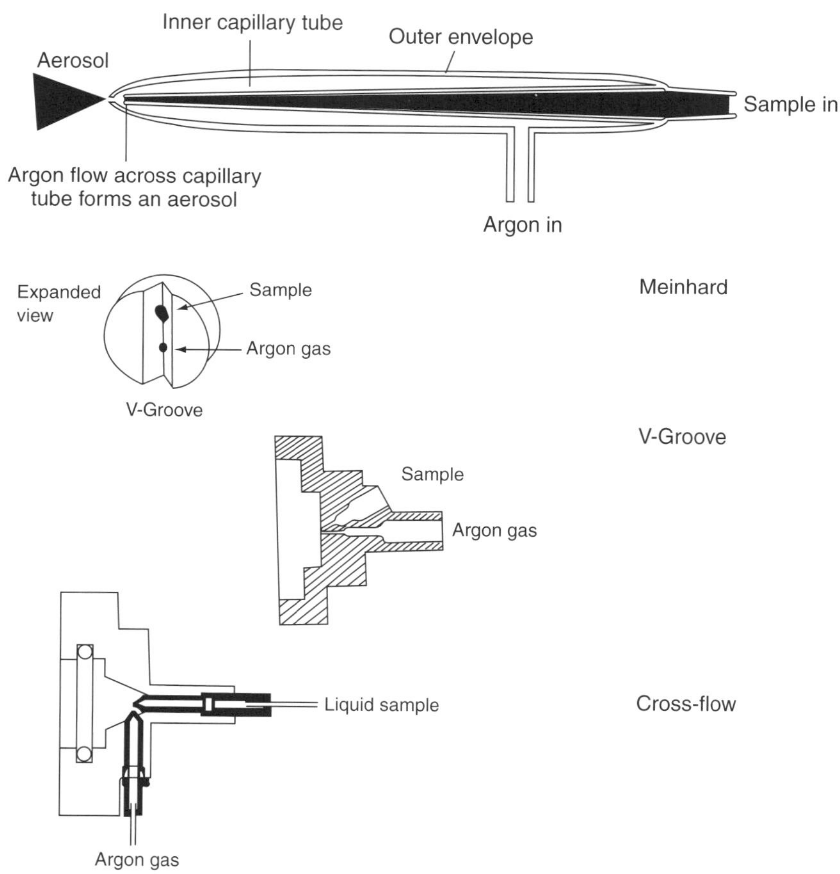

Fig. 3.9 Some commercially available nebulisers.

polymers such as Teflon, Ryton or Kel-F. The type of nebuliser used may therefore depend on the application. Examples of nebulisers that may be used to introduce samples with high concentrations of dissolved solids e.g. blood, seawater, acid digests of soils, etc., include the Burgener, the Ebdon, specialised Meinhard nebulisers, cross-flow, de Galan (V-Groove) and conespray. The capacity to aspirate dissolved solid is different for each nebuliser, but may extend to 10–20% m/v for the Ebdon, or even higher for the Burgener. The use of an inappropriate nebuliser may prove time consuming due to the problems associated with removing the blockage. Occasionally, blockages may be removed by immersing the nebuliser in a strong acid e.g. sulphuric acid, or by using an ultrasonic bath, but in general the use of very fine wires to remove blockages should be discouraged because they are more likely to do irreparable damage. It should be noted however, that the capacity to aspirate dissolved solid does not necessarily extend to an ability to aspirate suspended solid as, for example, when using the Burgener nebuliser. Many manufacturers produce nebulisers carrying their own trade name and their characteristics may be found by consulting the different manufacturers literature. A typical sample uptake rate for most types of nebuliser is between 0.7 and 1.5 mL min^{-1}, although other types of pneumatic nebuliser may operate with a much reduced flow rate (0.1–0.4 mL min^{-1}). The use of such nebulisers is obviously beneficial when there is limited availability of sample.

Attempts have been made to improve the sample transport efficiency to the plasma. Many different types of nebuliser have been developed in an attempt to achieve this. These include various types of frit style nebuliser including one developed by Layman and Lichte [14] and the Hildebrand grid nebuliser. The nebuliser developed by Layman and Lichte improved sample transport efficiency from 1–2% to between 30 and 95% due to the production of a primary aerosol containing very small droplets with a size distribution. It was however prone to blocking or salting up if solutions with high dissolved/suspended solids were introduced.

Ultrasonic nebulisers (USN) have been used with ICP-AES instrumentation for over 30 years. Sample is pumped onto the surface of a piezoelectric transducer operating at a frequency of up to 10 MHz. The ultrasonic waves shatter the film of liquid and produce a nebular of extremely fine droplets. The nebular is then transported to the plasma by a flow of argon gas. Signal intensity often improves by a factor of 10 using such a device. The improved transport efficiency means that the USN is usually used with a desolvation device so that plasma instability is avoided. The desolvation device usually consists of a heated tube to evaporate the droplets and a condenser to remove the resulting solvent vapour. The dried aerosol particles may then enter the plasma. A schematic diagram of a USN with its desolvation device is shown in Fig. 3.10. Although USNs may improve the sensitivity, they often suffer from memory effects i.e. they often require long wash out periods. In addition, they are

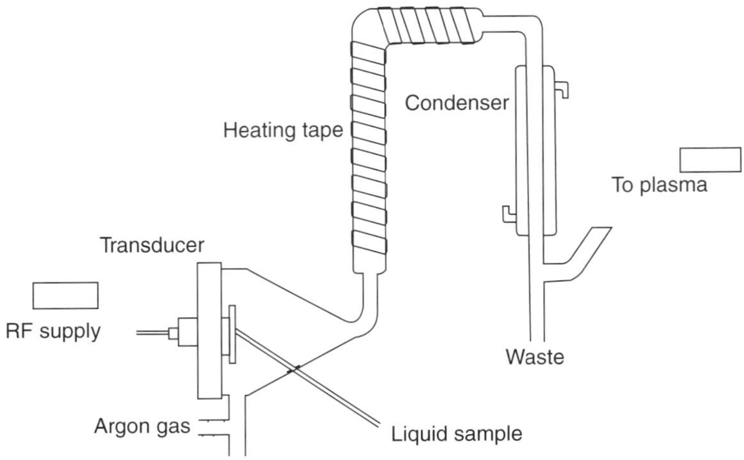

Fig. 3.10 Schematic diagram of an ultrasonic nebuliser.

expensive to purchase and traditionally regarded by many as being complex and unreliable. In recent years however, their reliability has improved. Botto and Zhu [15] and Wuilloud *et al.* [16] have recently published applications using USN as a sample introduction device for ICP-AES detection.

Direct insertion nebulisers (DIN) have a very low sample flow rate (<100 $\mu L\ min^{-1}$) and, therefore, do not require a spray chamber. Instead, as their name implies, they may be inserted directly into the injector of the torch. Sample transport efficiency is therefore close to 100%. The figures of merit from such nebulisers are normally similar to more conventional nebulisers, but the amount of sample required is very much reduced. Again, these nebulisers tend to be expensive, but they have found a niche when they are used to couple microbore liquid chromatography or capillary electrophoresis with ICP-AES. One such example has been reported by Vaughan and Claassen [17], although, there are more examples of the use of this type of nebuliser with ICP-MS detection. There is an assortment of other high efficiency nebulisers available. These include the microconcentric nebuliser (MCN) [18], a high efficiency nebuliser (HEN) [19] and a concentric capillary nebuliser (CCN) [20]. In this latter case, at a flow rate of 50 $\mu L\ min^{-1}$ the LoDs obtained were similar to conventional Meinhard and conespray nebulisers, although aspiration of samples with high dissolved solids content was reported to adversely affect the performance.

Thermospray and electrospray nebulisers may also be attached directly to the plasma torch and are finding increasing usage amongst research workers who couple chromatography with ICP detection, but their use is by no means routine at present. The mechanisms by which these nebulisers work are beyond the scope of this text, but several accounts have been reported in the literature

[21, 22]. In one recent report, a thermospray nebuliser has been used to introduce sample to ICP-AES instrumentation [23]. In this study, sample was introduced at a rate of approximately 1 mL min^{-1}. To prevent plasma extinction, a desolvation device consisting of a heated spray chamber followed by a condenser was used. Sensitivity was reported as being much greater than for a conventional nebuliser, but because of poor noise characteristics, the LoD was improved by only a factor of 2–4.

As with nebulisers, commercially available spray chambers are available in a wide range of styles. As well as acting as a droplet size discriminator, the spray chamber also acts to dampen pump noise. In general, the larger the spray chamber the better the noise characteristics. However, there is a trade-off between noise and memory effects. A large spray chamber will have a large internal surface area to which analyte may become adsorbed and it may also have regions of dead volume, where solution may become temporarily trapped. There are a number of different spray chamber designs, the more common including the Scott double pass, cyclone, single pass and single pass with an impact bead (Fig. 3.11). Spray chambers are normally made of glass, although polymer-based ones are also marketed for use with hydrofluoric acid (HF). Many spray chambers incorporate a cooling water jacket to help maintain a constant temperature (and hence constant nebulisation characteristics for any given sample type). This will lead to improved plasma stability and signal intensity. The presence of a cooling jacket around the spray chamber is even more essential when samples containing an organic solvent are introduced. These solvents tend to have lower surface tension and higher vapour pressure than water and hence they form an aerosol more readily and are transported more efficiently to the plasma. This will normally cause plasma instability and possibly extinction. To combat this, the spray chamber should be cooled to a temperature of −5 to −10°C, achieved by adding anti-freeze to the water flow through the jacket. A more detailed discussion on the introduction of organic solvents to ICPs can be found elsewhere [24, 25]. Different spray chambers may also have an impact on the magnitude of interference effects. For example, it has been shown that cyclone style spray chambers can, under certain circumstances, lead to fewer interferences than conventional Scott style spray chambers [26].

3.3.5.2 Torches

Although, originally two basic types of torch were in use, as developed by Greenfield and Fassel respectively, it is the latter design that is used in the vast majority of modern instrumentation. A comparison of Greenfield and Fassel torches is listed in Table 3.2 and an ICP torch is shown schematically in Fig. 3.12. The torch consists of three concentric quartz tubes through which the gas (normally argon) flows. The outer tube contains the plasma gas (also termed the coolant gas) and has argon flowing at typically 11–15 L min^{-1}, the

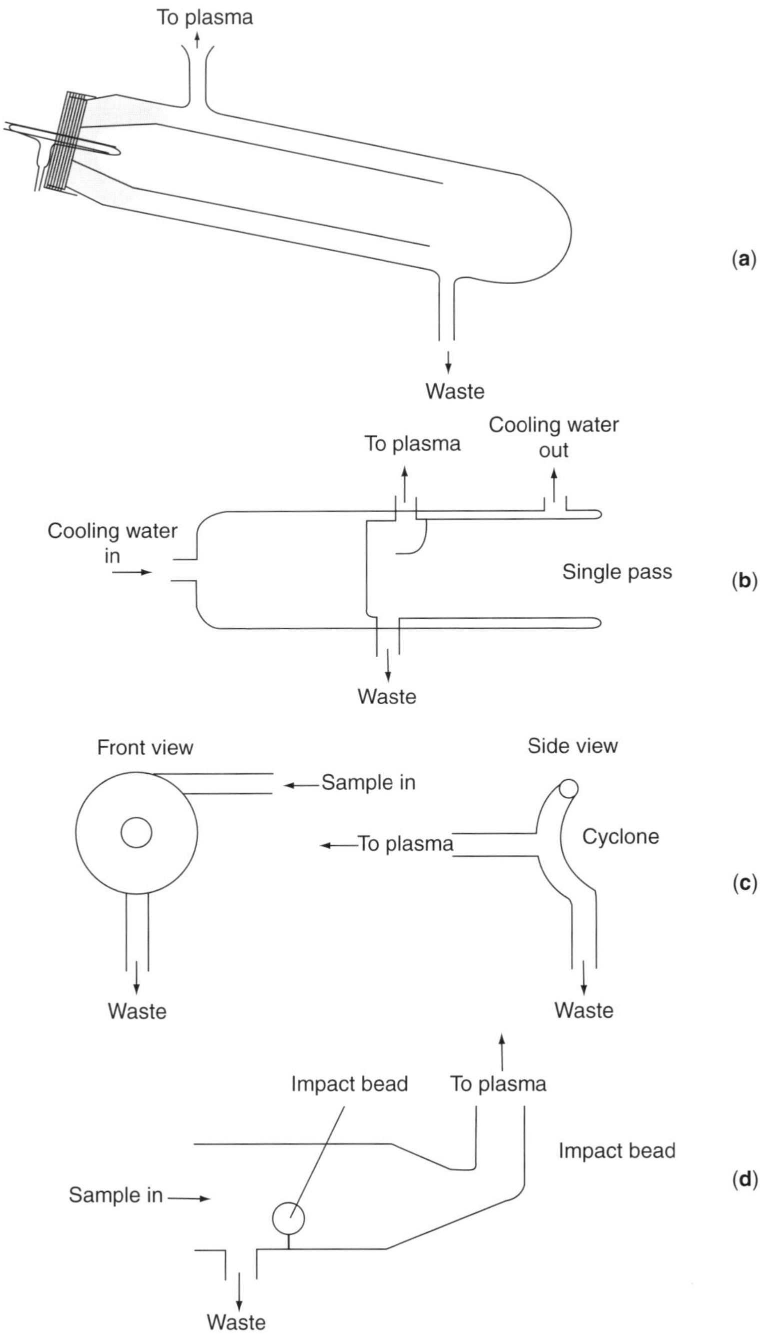

Fig. 3.11 Different types of spray chamber (a) Scott double pass (b) single pass (c) cyclone (d) single pass with impact bead.

Table 3.2 Comparison of Greenfield and Fassel torches

Fassel	Greenfield
Torch dimensions: injector = 1.8 mm bore coolant = 20 mm diameter	Torch dimensions: injector = at least 2 mm bore coolant = 29 mm
All argon cooled (usually)	Argon and nitrogen cooled
Low overall gas consumption: 13–18 L min^{-1}	Very high overall gas consumption: 12–38 L min^{-1} Ar and 20–70 L min^{-1} N_2
Low power requirements: 0.9–1.5 kW	High power requirement: several kW
Intolerant of other gases	Robust. Can cope with poor concentricity of torch tubes and other gases
Cheap running costs	Expensive running costs

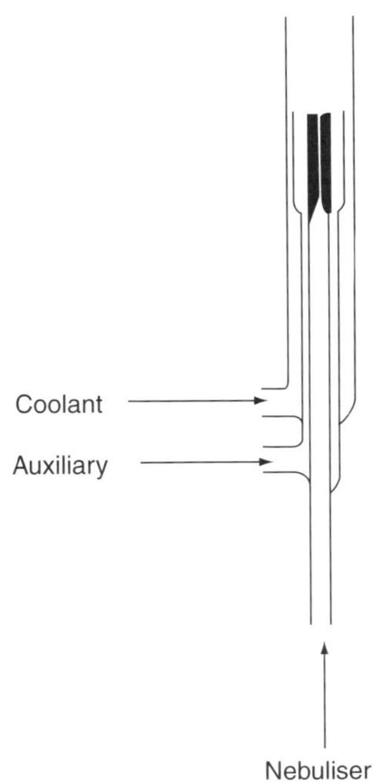

Fig. 3.12 An ICP torch.

middle tube contains the auxiliary or intermediate gas flow (typically 0.7–1.5 L min^{-1}) and the innermost tube has the nebuliser (injector) gas flow which is typically 0.6–1.3 L min^{-1}. The coolant gas flow spirals tangentially through the torch. The spacing between the outer and auxiliary tubes is narrow so that the coolant gas passes through at a high velocity. This assists in cooling the torch and hence prevents damage. The auxiliary gas flow keeps the plasma discharge away from the auxiliary and injector tubes.

Although the basic design is the same for many torches, there are many modifications in use. Some torches may be operated with low gas flows e.g. the mini- and micro-torches. The mini-torch has a diameter of 13 mm (as opposed to 18 mm for a standard torch), and an injector with an internal diameter of 0.75 mm (compared with 1.4–1.5 mm for an injector in a standard torch) [27]. This torch can operate with a power of less than 1 kW and with a coolant gas flow of 8 L min^{-1}. The micro-torches are even smaller, with an internal diameter of 9 mm [28]. For this torch, the coolant gas flow is typically 6.4 L min^{-1} and the power used is 500 W. The figures of merit for these torches are similar to those of a standard torch, but they are more prone to blocking when solutions with high dissolved salt content are aspirated.

Many modifications of the *standard* Fassel torch have been made. Semi-demountable torches where the injector may be removed and replaced with one of a different internal diameter (e.g. for when high salt solutions are to be analysed) are relatively common. These semi-demountable torches are useful because injectors of different materials may be used, e.g. some manufacturers use ceramic or alumina injector tubes, which are more robust and resilient to HF. In addition, graphite injector tubes have also been used [29]. Alternatively, the style of injector may be changed. Some of the different injectors available are shown in Fig. 3.13. Injectors that have gently tapering bore are more appropriate for solutions with high dissolved/suspended solids, because the aerosol

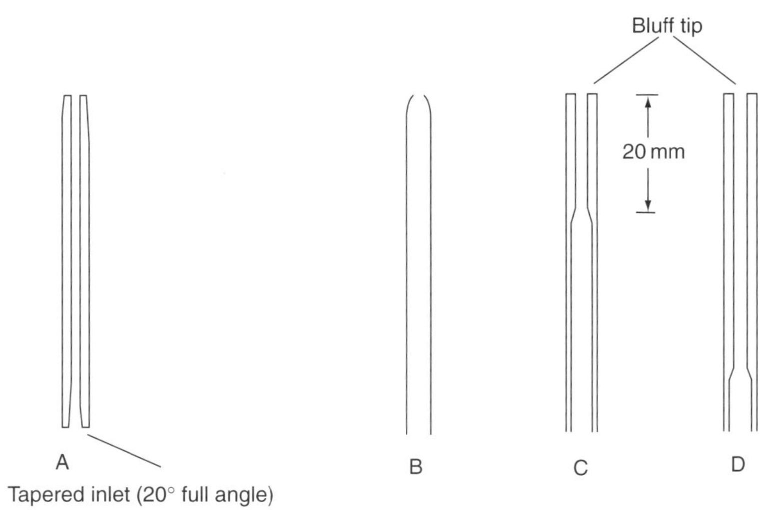

Fig. 3.13 Different types of injector: (a) stream-lined laminar flow (b) turbulent flow tapered (c) intermediate laminar/turbulent flow (d) laminar flow capillary.

particles pass more readily to the plasma. With injectors that contain more oblique angles, aerosol particles can accumulate and dissolved solids may desolvate and build up until blockage is complete. This will have the effect of giving a large signal drift until eventually, when blockage is complete, the signal will of course be lost. In addition, the nebuliser gas flow will not be able to escape and hence a build-up of pressure will occur, which may lead to the spray chamber fracturing or the nebuliser coming adrift from its mounting collar. The laminar flow injector is also prone to blocking, whilst the turbulent flow injector is more prone to de-vitrification of the tip.

Fully demountable torches have also been marketed, but these require precision engineering to ensure complete concentricity so that damage may be prevented and optimal performance obtained. Fully demountable torches are available commercially that fit together easily and in a reproducible manner. Other torches have viewing slots cut into slightly extended coolant tubes. If this slot is positioned so that light emitted from the analytes within the plasma passes through it to the collection optics, then the collection efficiency will be improved, because refraction and diffraction caused by the quartz tube do not occur.

3.3.5.3 Other sample introduction methods

There are numerous other sample introduction methods available for ICP-AES. Each may have advantages and disadvantages associated with them. The HG method improves the sensitivity and LoDs, often by a factor of 40–100. Here analytes are transformed into a gaseous compound (a hydride) by mixing the sample with a reagent such as sodium tetrahydroborate, in a reactor, and then this gaseous compound is transported to the plasma in a flow of inert gas (usually argon). The improvement in sensitivity arises through the improved transport efficiency of the gas to the plasma which is close to 100% (c.f. 1–2% for a conventional nebuliser/spray chamber assembly). In addition, because the analytes are separated from the bulk matrix and hence the majority of concomitant species, which remain in the liquid sample, fewer spectroscopic interferences are observed. The use of HG can therefore yield substantially better figures of merit when compared with conventional nebulisation. It should be noted, however, that only a limited number of analytes are capable of forming gaseous hydrides. The analytes that are most commonly determined in this manner are arsenic, selenium and occasionally tin, lead, antimony and tellurium. There are, however, also a number of disadvantages associated with the HG technique. High concentrations of transition elements e.g. copper, nickel and iron and also of the precious metals including platinum, gold and palladium in the sample matrix may interfere with the hydride forming process and lead to low recoveries when compared to pure aqueous calibration standards. These problems may be largely overcome by the addition of masking agents e.g. 1,10 phenanthroline and cysteine that complex with the potential interferants and hence preventing them from interfering. A more fundamental problem, however, relates to the oxidation state of the analyte species. Selenium in its +4 state will form a hydride, but Se^{VI} will not. Similarly, several organic selenium compounds will also not form a hydride. For arsenic, both As^{III} and As^{V} form hydrides, as do compounds such as monomethylarsonic acid and dimethylarsinic acid, but all at different rates. Therefore, a different sensitivity will be obtained for each of the species which is likely to lead to inaccurate results if compared with a calibration made from one species. Other arsenic compounds e.g. arsenobetaine are extremely inert and do not form hydrides at all. Therefore, some chemistry is often required to transform all of the analyte species into a form capable of generating a hydride. Numerous methods are available including using powerful oxidants e.g. perchloric acid and alkaline sodium persulfate, boiling with hydrochloric acid (this will reduce Se^{VI} to Se^{IV}) and using UV photolysis to fracture the organic molecules. Another potential problem may be encountered as a result of the excess hydrogen produced by the tetrahydroborate reagent. This too will be transported to the plasma and may cause plasma perturbation. It is therefore often necessary to optimise the concentrations of the reagents used, to keep this excess hydrogen to a minimum.

An analogous method to HG is cold vapour generation, most commonly for mercury. Mercury in a liquid sample is reduced to its elemental form using either sodium tetrahydroborate or, more commonly, stannous chloride. The mercury vapour is then sparged from the liquid matrix and transported to the plasma using a flow of argon in a similar manner to that described above. Other reagents used to form vapours include sodium tetraethylborate. This reagent has been used to form the volatile ethylated vapours of lead [30] and cadmium [31]. The whole subject area of vapour generation has been discussed and reviewed in detail by Dedina and Tsalev [32]. Although that book focuses on detection by atomic absorption, the fundamentals of the HG technique and many of the requirements also hold true for ICP-AES determinations. It should be noted that if batch HG is used, i.e. a discrete volume of sample is mixed with a discrete volume of reductant, a transient signal will be obtained. This means that the ICP-AES instrument will need to have software capable of detecting the transient signal. In addition, unless an echelle-based spectrometer is used, only one analyte will be detected as there may not be sufficient time for a scanning spectrometer to change wavelength to determine further analytes. This will not be the case when continuous HG is used.

ETV as a method of sample introduction can cope with both liquid and solid samples. Originally this approach was designed for use with atomic absorption spectrometry, but may also be used in conjunction with ICP-AES and ICP-MS. This approach offers another method by which the analyte is separated from the matrix and hence reduces interferences. Improved sensitivity also results, because transport efficiency of the analyte to the plasma is close to 100%. Again a transient signal will be obtained, but this time, unless a simultaneous spectrometer is used, only one analyte may be determined at a time, because the residence time of the analyte in the plasma will not be sufficient for a scanning spectrometer to change wavelength to an analytical line of a second analyte.

Laser ablation (LA) also facilitates the introduction of solid samples for ICP-AES, although it is also possible to introduce liquid samples in this way. Various lasers may be used, but the neodymium:yttrium aluminium garnet (Nd:YAG) laser is the most commonly used. Since the laser is commonly used to vapourise a small spot on the surface of a sample and the vapour is transported to the plasma in a stream of argon, this is therefore another method that produces a transient signal. LA has been renown for being difficult to standardise, since different samples will vaporise with a different efficiency and hence different amounts of sample will be introduced even when the laser conditions used to vapourise the samples are identical. To obtain the most accurate results, calibration should be achieved using standards that are homogeneous and resemble the sample matrix as closely as possible.

A more detailed overview of both ETV and LA as introduction methods for both solid and liquid samples has been presented elsewhere [33, 34].

Direct sample insertion in ICP-AES requires a specialised torch. Sample (either liquid or solid) is placed on a probe that is usually made of graphite. The probe is then inserted up the injector of the torch and positioned so that the sample vapourises into the plasma [35]. The design of some of the probes used is shown in Fig. 3.14 and the whole device is shown schematically in Fig. 3.15.

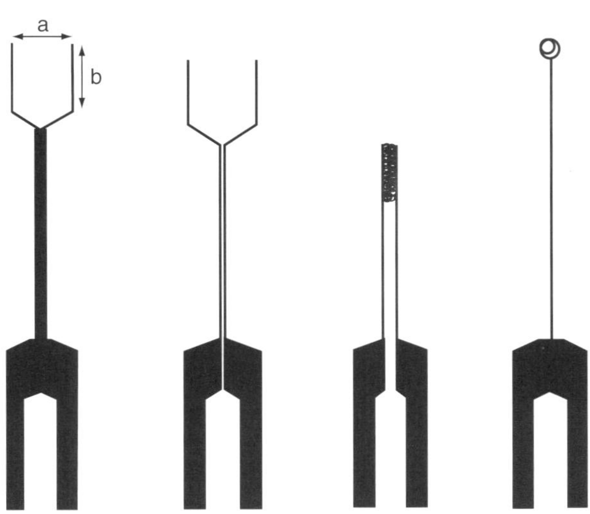

Fig. 3.14 Direct sample insertion probes. The probes on the left are made of graphite, whilst the loop is made of tungsten.

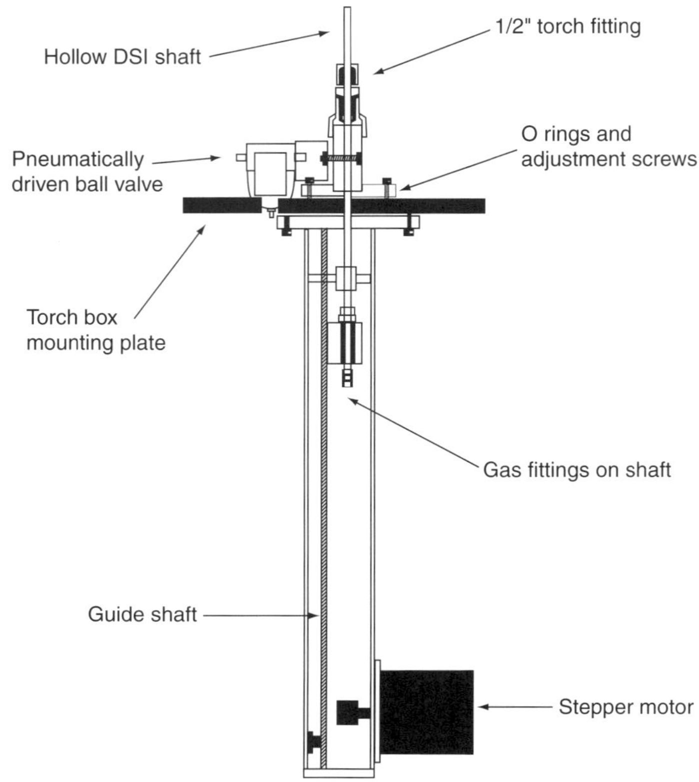

Fig. 3.15 A direct insertion device [35].

Sample transport to the plasma using probes is again virtually 100% and therefore detection limits of 5–400 pg are obtained, depending on the analyte. It should be noted, however, that severe problems may be encountered when attempting to weigh out a few mg of solid samples and therefore precision is frequently poor. In addition, some more refractory analytes, especially those that form carbides, often present troublesome memory effects.

3.4 Applications

The aim of this section of the chapter is to present an overview of potential applications of ICP-AES and should not, therefore, be considered a comprehensive review of the literature. For an update of recent advances, a good starting point is the annual reviews published in the *Journal of Analytical Atomic Spectrometry* (JAAS) [36].

3.4.1 Simple applications

Examples of simple applications of ICP-AES in the literature are many and varied. A number of authoritative reviews have also covered specific areas such as environmental applications [37], geological applications [38], and food science [39]. For many applications of ICP-AES, it is sufficient to merely nebulise a liquid sample directly into the plasma. For solid samples, an acid digestion, a fusion, a dry ashing or some other sample preparation procedure is usually required. One method that has not been described previously in this text is slurry nebulisation. This is a technique that introduces the sample as a dispersion of very fine particles in an aqueous dispersant. The advantages of the technique are that it does not require the use of hazardous acids such as HF and, because only minimal sample handling is necessary, the chances of analyte loss through volatilisation or of contamination are much reduced. Another advantage is that the slurry can be calibrated against aqueous standards. For accurate results to be obtained, it is necessary for the particle size of the slurry to be extremely small, i.e. less than 2 μm. Assuming that the particles are smaller than this, they behave in a very similar manner to the aerosols of aqueous standards. If, however, they are larger than 2 μm, the nebuliser and spray chamber assembly will discriminate against the larger particles and hence the sample reaching the plasma will not be representative. In such an instant, low recoveries will be obtained. Unfortunately, the use of an internal standard for slurry analysis is often unhelpful. Once the sample is spiked with the internal standard, it will remain in the solution of the sample rather than become incorporated into the particles, and hence its transport efficiency may be unaffected. The only sample preparation step required for slurries is the initial grinding process. One useful method is to place the sample in a clean plastic bottle with approximately 10 g of expanded zirconia beads and about 4 mL of a dispersant. After sealing the bottle, it is placed on a laboratory shaker for a number of hours, the exact time depending on the nature of the sample. More refractory samples such as soils and rocks will require a longer grinding period than softer samples such as plant or animal tissues. The dispersant used will also be dependent on the sample type. Biological materials may be dispersed with Triton X-100 or Aerosol OT, whereas refractory samples are better dispersed with sodium pyrophosphate or sodium hexametaphosphate. For instances where zirconium is one of the analytes, it is better to use a microniser mill, which uses agate rods instead of zirconia beads. Papers on a number of applications using slurry nebulisation have been published. These include a review paper by Ebdon *et al.* [40] and a paper describing the use of hydrogen as an alternative gas in an attempt to aid the volatilisation of the slurry particles [41]. This approach is especially useful for the analysis of slurries of very high refractory materials, because the presence of hydrogen increased the rotational temperatures from 2200 to 3900 K. This has been attributed to an increase in the energy transfer from the toroidal to the annular area of the plasma. For biological samples that may contain analytes at very low concentration, a pre-concentration process may be used. If the analyte is not volatile, the sample (5 g) may be placed in a muffle furnace at 400°C and ashed for several hours. After cooling, the remaining ash, which may constitute only 10–20% of the original weight, may be slurried. A 5–10-fold pre-concentration is therefore obtained.

Occasionally, a combination of sample introduction methods is used. An example is the ETV introduction of slurries of silicon carbide [42]. In this paper, an emulsion of polytetrafluoroethylene (PTFE) was used as a matrix modifier that assisted in the vapourisation of the refractory matrix.

3.4.2 Pre-concentration methods

As discussed previously, the LoDs for ICP-AES are less impressive than for ICP-MS or electrothermal atomic absorption. The technique may therefore be unsuitable for the determination of some analytes in certain natural samples unless a pre-concentration procedure is used to bring the analytes into an appropriate concentration range. For water samples this may be done in a number of ways. Freshwater may simply be evaporated. Such a technique can yield large pre-concentration factors, but it is unsuitable for seawater as the dissolved salt content would be increased to an intolerable level. Solvent extraction is another method that yields high pre-concentration factors. If a compound such as ammonium pyrrolidine dithiocarbamate (APDC) is added to 1 L of water, it complexes with many of the metal ions. These metal complexes may then be extracted into a smaller volume of solvent such as isobutyl methyl ketone (IBMK). A further pre-concentration may be achieved if a back-extraction into a small volume of acid is performed. Other reagents e.g. 8-hydroxyquinoline, may be used for the same purpose. Other reagents such as piperidine dithiocarbamate (pipDTC) may also be used. Co-precipitation has also been used as a method of pre-concentration [43]. However, it should be noted that co-precipitation is rarely selective and that a range of other ions may also be pre-concentrated. In addition, the precipitating agent e.g. lanthanum or iron hydroxide, will increase the dissolved solids content of the final sample. Another method of pre-concentration often employed is based on the use of ion exchange or chelation resins, either packed into a column or added as a small batch to the water sample. The resin is added to the water sample, stirred for a suitable period of time and then filtered off. The analytes may then be eluted from the resin by applying a small volume of acid. If 1 L of water is extracted with 1 g of resin, and the analytes eluted with 10 mL of acid, a pre-concentration factor of 100 is obtained. The obvious advantage of this technique is its speed, but there is always a chance of contamination. Alternatively, if the water is pumped through a column packed with the resin, the system is enclosed and hence the possibility of contamination is reduced. This method is, however, slower than the batch technique. The use of resins also has the advantage of isolating the analytes from concomitant ions in the sample. If seawater were to be analysed, then the analytes may be separated from the high concentration of salt and hence facilitate determination by ICP-AES. Possible problems may be encountered if an analyte is not fully retained on the column or if the acid does not elute the analyte from the resin. In either case, inaccurate results would be obtained. One example of pre-concentration by the batch and on-line techniques is the investigation into the equilibrium and kinetic properties of a fast iminodiacetate based chelating exchange resin by Hashemi *et al.* [44]. When solid samples are to be analysed, a sample dissolution or digestion may be necessary. A pre-concentration may be achieved on the solid sample itself by ashing it in a muffle furnace as described earlier. Once the ash has been formed, it may be acid digested. If a further pre-concentration is required, the acid may be reduced in volume by evaporation. If 10 g of sample is ashed, the residue may be digested in 5–10 mL of acid. If this is further evaporated and then diluted with water to a volume of 10 mL, then the overall dilution factor is 1, because 10 g of sample is now contained in 10 mL. Alternatively, the sample may be acid digested in the normal way, the acid may be neutralised, and the analytes extracted with APDC or 8-hydroxyquinoline as described previously.

3.4.3 Speciation and coupled techniques

It has been noted above that some forms of elements are more toxic than others. For environmental monitoring and toxicology purposes, it is therefore often no longer adequate to provide only the total concentration of the analyte, as this will yield no real information on the overall toxicity. Instead, it is necessary to determine the concentration of individual species of the analyte. This area of analysis, *speciation*, is now one of the most rapidly expanding areas of analytical chemistry. There are numerous methods available for speciation analysis, but ICP-AES has been used as a detector for many of the applications. If the sample is a liquid, then little sample preparation will be necessary. If, however, it is a solid, a typical speciation analysis will involve the extraction of the species from the solid sample in such a way as to ensure complete extraction whilst also ensuring that no change in the speciation occurs. It is beyond the scope of this text to describe these processes in detail, but they are covered in detail elsewhere [45, 46]. Whatever process is selected, once extracted the species must be separated from one another and this is normally achieved using chromatography. High performance liquid chromatography (HPLC) is the most frequently used technique. If the end of the column is connected to the nebuliser of the ICP instrument, then as the species elute they may be detected on-line. In principle, the coupling of HPLC with ICP-AES is very straightforward. If, however, the chromatography requires the use of organic solvents such as methanol or perhaps even acetonitrile, a desolvation system may have to be placed between the column and the nebuliser to prevent plasma perturbation or extinction. Such systems do not have to be complex and are described in the literature [47]. However, the necessity to use a desolvation device will depend on the RF generator of the instrument. Higher frequency generators may be able to cope more readily than a generator operating at 27 MHz. It should also be noted that the distance between the end of the column and the nebuliser should be kept to a minimum to prevent peak broadening arising from diffusion. The nebuliser and spray chamber used may also have an impact on the resolution of closely eluting species and consequently a reduced volume spray chamber will result in less peak broadening. Several papers have been published describing the effects of different nebuliser/spray chamber assemblies on chromatographic resolution [48]. Although the study in this paper is based on ICP-MS, the same conclusions may be drawn for ICP-AES. The coupling of HPLC with ICP-AES and ICP-MS has been reviewed in more detail by Hill *et al.* [49].

Off-line methods have also been used for speciation studies. Here the HPLC is used to separate the analyte species and then, rather than being detected on-line, fractions of up to 1 mL volume are collected and analysed individually. This type of methodology is no longer a common practice as on-line methods have been developed for many applications and clearly provide ease of use as well as better resolution. However, the off-line methods do offer the chance to perform some sample manipulations, e.g. evaporation of organic solvents. One recent example of the use of an off-line method has been reported for the speciation of Al in forest soil extracts [50].

Gas chromatography (GC) has also been coupled with ICP-AES instrumentation, but the interface arrangement is far more complex since the GC column must be extended via a heated transfer line to prevent analyte condensation. A nebuliser and spray chamber is not necessary for this coupling since the transfer line extends to the base of the torch. However, a flow of argon will be necessary to act as a make-up gas because the small flow of gas through the GC column is insufficient to *punch* the plasma. Several successful applications of this type of coupling have been reported [51, 52].

Capillary electrophoresis (CE) has also been coupled with ICP-AES, but at present this is purely in the research stages and no routine applications have been reported. The coupling of a very low flow technique such as CE with ICP-AES has led to several problems and specialised interfaces are necessary. In general a micro-concentric nebuliser is used inserted into the base of the plasma torch [53].

There is always a need to improve LoDs for speciation analysis and therefore many authors have made multiple couplings, utilising several analytical techniques. Amongst these couplings are HPLC-HG-ICP-AES [54], HPLC-UV photolysis-HG-ICP-AES [55] and HPLC-Microwave oxidation – HG- ICP-AES [56]. With the improved transport efficiency of vapour generation, the detection limits offered by these approaches may be improved by a factor of between 40 and 80.

3.4.4 Analysis of gases and high resolution applications

The approach used for the analysis of gaseous samples will depend on the analyte. The most common analyses performed in the gaseous phase is chemical vapour generation, for example, of As, Se which has been described previously (Section 3.3.5.3). However, air may contain small particulate matter that is transported as an aerosol. Depending on the origin, these particles may contain potentially toxic elements and have been the subject of several studies. One such recent publication has reviewed the collection and determination of metallic contaminants in gases [57].

Mention should also be made of high resolution spectrometers. A standard ICP-AES instrument is sufficient for the vast majority of applications, but occasionally, a high resolution instrument will be necessary. An example is when isotopic analysis is to be performed. Isotopic analysis is more often associated with ICP-MS determinations, but it is possible to use a high resolution spectrometer to separate the atomic lines of different isotopes of elements e.g. uranium and plutonium [58]. Rare earth elements and the lanthanides are commonly determined by geologists. Unfortunately, these elements have very rich line spectra and therefore the possibility of interference problems. A high resolution spectrometer may be useful in overcoming this potential interference [59, 60]. There are several different types of high resolution spectrometer including Fourier transform spectrometers, Fabry Perot interferometry (with a resolving power of $>$1 000 000) and specialist echelle- and grating-based spectrometers. Each of these are described in more detail elsewhere [61]. An alternative approach to using expensive high resolution spectrometers for overcoming the interference problems is the use of a chromatographic step to separate the analyte rare earth elements from the matrix [62]. Chromatographic procedures for the separation and determination of rare earth elements have been reviewed by Kumar [63].

3.5 Alternative plasmas

Although usually more associated with theoretical studies, the use of plasma diagnostics is occasionally necessary in order to understand the processes occurring within the plasma for practical applications. Plasma diagnostics involves such subjects as the measurement of plasma temperatures (rotational, excitation, electron and the ionisation temperatures) and the electron number density [2]. These fundamental parameters depend very heavily on the type of plasma used. The vast majority of practical applications are based on the use of an argon plasma for ICP-AES. There are, however, occasions where mixing other gases in one or more of the gas flows that form the plasma may offer advantage. In addition, the plasma may occasionally be made purely from an alternative gas. The use of mixed and alternative gas plasmas in ICP-AES has been reviewed by Montaser and Van Hoven [64] and by Ohls *et al.* [65]. The formation of such plasmas with conventional RF generators may prove problematic, but can facilitate benefit from changing the fundamental plasma parameters. The introduction of hydrogen, for example, has been reported by several workers to improve the detection limits and to increase plasma temperatures and electron density. In one application [41] the introduction of approximately 6% hydrogen was found to increase the atomisation efficiency of slurry particles of a refractory sample. The enhanced atomisation efficiency improved the accuracy of the analysis because recovery increased from approximately 60% to over 80%. The introduction of nitrogen also has a marked effect on the plasma. If it is introduced to the nebuliser gas flow, the annulus widens and the plasma shortens. This leads to less plasma–sample interaction. If, however, it is introduced to the coolant gas flow, the annulus becomes narrower and this promotes the plasma–sample interaction, hence improving the atomisation process. The analysis of slurries and solids has been demonstrated to be more successful in the presence of nitrogen in the coolant gas [66]. The background spectra of an Ar-N_2 plasma is similar to that of an all Ar plasma below 300 nm, but the spectrum is more cluttered between 300 and 500 nm. Therefore, it may prove more difficult to determine Cr at 425.4 nm and Tl at 351.9 nm in an Ar-N_2 plasma. Oxygen added to a plasma has been shown to decrease the spectral bands arising from –CN and to simplify the background spectrum between 300 and 500 nm. Again, the temperatures within the plasma increase in comparison with an all Ar plasma, but tend not to be as high as with a mixed Ar-N_2 or Ar-Air plasma. Plasmas made entirely of oxygen have also been formed and have extremely high temperatures [67]. The introduction of He or other noble gases into the plasma will increase the ionising properties of the plasma and an increase in the kinetic temperature is observed [68]. Such plasmas also have a much higher thermal conductivity and electrical resistivity when compared with argon, and therefore sustaining a helium plasma is much more difficult. Other gases that have been introduced include air, halocarbons and carbon dioxide. In general, the introduction of an alternative gas may simplify the background spectrum which, on occasion, may give some advantage as well as improve the plasma–sample interaction, thus leading to improved accuracy and LoDs.

3.6 Chemometrics and data manipulation

Another expanding area in analytical chemistry is the application of chemometric techniques. One of the many uses of chemometrics is for optimisation of performance i.e. chemometric approaches may be used to identify the optimal plasma conditions. Simplex optimisation may be used to determine the optimal viewing height, power and gas flow rates in ICP-AES [40]. An alternative is univariate optimisation, i.e. keeping all but one parameter constant and then performing the optimisation around this one parameter. The process is repeated for all the other parameters in turn. This method assumes that there is no inter-dependence between the parameters. The application of Simplex optimisation or some other chemometric procedure (e.g. Doehlert matrix or factorial design [69]) facilitates optimisation of all of the parameters simultaneously, and therefore offers an elegant and rapid method of system optimisation. If more than one analyte is to be determined, then the best compromise conditions may also be identified. Other chemometric methods may be used to identify the source of noise and drift [70, 71]. Once the cause of the problem is identified, measures can then be taken to alleviate or remove the problem.

Chemometrics may also be used to manipulate and interrogate the data obtained from the ICP-AES instruments, so that maximal information may be obtained. Such approaches are common in environmental analysis and food science [72–74].

3.7 Conclusions

The technique of ICP-AES has long since reached maturity. The instrumentation is relatively inexpensive (in comparison with ICP-MS), and is therefore the technique of choice for very many applications. It has a sensitivity that, depending on the analyte, is at least equivalent to that of flame atomic absorption, but also has a multi-element capability. Conventional analysis by ICP-AES is relatively rapid, with both simultaneous instrumentation or rapid sequential scanning spectrometers being available. The linear range extends to at least 5 orders of magnitude and a wide range of analyte lines may be utilised to increase flexibility. The concentration of most analytes is likely to fall within the range of at least one of the analytical lines. If, however, pre-concentration is required, this may often be achieved relatively simply using either pre-concentration on resins or by solvent extraction, with the added advantage that often the matrix, and hence potential interferences, are removed simultaneously. There are a variety of different sample introduction methods available that may be used for both liquid and solid samples. Many of these alternative sample introduction systems lead to improved sensitivity and lower LoDs when compared with conventional nebulisation. Since a majority of commercial instruments now link to powerful computing facilities and an autosampler, large volumes of data may be obtained without the need for constant attended operation. In addition, the coupling of chromatographic methods on-line with ICP-AES detection has enabled speciation data to be obtained. Hence, ICP-AES continues to play a major role in many analytical laboratories.

References

[1] Greenfield, S. & Foulkes, M. 'Historical development of ICP-AES', in *Inductively Coupled Plasma Spectrometry and its Applications*, S.J. Hill (ed.), Sheffield Academic Press Ltd, Sheffield, 1999.

[2] Mermet, J.M. 'Fundamental principles of inductively coupled plasmas', in *Inductively Coupled Plasma Spectrometry and its Applications*, S.J. Hill (ed.), Sheffield Academic Press Ltd, Sheffield, 1999.

[3] Zander, A. 'The problem of spectral interferences and line selection in plasma emission Spectrometry', in *Inductively Coupled Plasma in Analytical Atomic Spectrometry*, A. Montaser & D.W. Golightly (eds), VCH, 1987, New York.

[4] Greenfield, S., 'Common R.F. generators, torches and sample introduction systems', in *Inductively Coupled Plasmas in Analytical Atomic Spectrometry*, A. Montaser & D.W. Golightly (eds), VCH, 1987, New York.

[5] Boss, C.B. & Fredeen, K.J., *Concepts, Instrumentation and Techniques in Inductively Coupled Plasma Optical Emission Spectrometry*, Perkin Elmer Corp., 1997, USA.

[6] Pomeroy, R.S. *CTD Detectors in Atomic Emission Spectroscopy*, VCH, New York, 1994.

[7] Bilhorn, R.B., Epperson, P.M., Sweedler, J.V. & Denton, M.B., *Appl. Spectrosc.* 1987, **41**, 1125.

[8] Dubuisson, C., Poussel, E. & Mermet, J.M., *J. Anal. At. Spectrom.*, 1997, **12**, 281–286.

[9] Sharp, B., *J. Anal. At. Spectrom.*, 1988, **3**, 613–652.

[10] Sharp, B., *J. Anal. At. Spectrom.*, 1988, **3**, 939–963.

[11] Luan, S., Pang, H.M., Shum, S.C.K. & Houk, R.S., *J. Anal. At. Spectrom.*, 1992, **7**, 799–805.

[12] McLean, J.A., Minnich, M.G., Iacone, L.A., Liu, H.Y. & Montaser, A., *J. Anal. At. Spectrom.*, 1998, **13**, 829–842.

[13] Browner, R.F. 'Aerosol generation and sample transport in plasma spectrometry', in *Inductively Coupled Plasma Spectrometry and its Applications*, S.J. Hill (ed.), Sheffield Academic Press Ltd, Sheffield, 1999.

[14] Layman, L.R. & Lichte, F.E., *Anal. Chem.*, 1982, **54**, 638–642.

[15] Botto, R.I. & Zhu, J.J., *J. Anal. At. Spectrom.*, 1994, **9**, 905–912.

[16] Wuilloud, R.G., Wuilloud, J.C., Olsina, R.A. & Martinez, L.D., *Analyst*, 2001, **126**, 715–719.

[17] Vaughan, B. & Claassen, L., *Commun. Soil Sci. Plant Anal.*, 1996, **27**, 819–827.

[18] Hettipathirana, T.D. & Davey, D.E., *J. Anal. At. Spectrom.*, 1998, **13**, 483–488.

[19] Todoli, J.L. & Mermet, J.M., *J. Anal. At. Spectrom.*, 2001, **16**, 514–520.

[20] Wang, R.J., McLaughlin, R.L. & Brindle, I.D., *J. Anal. At. Spectrom.*, 2000, **15**, 1303–1312.

[21] Zhang, X.H., Chen, D., Marquardt, R. & Koropchak, J.A., *Microchem. J.*, 2000, **66**, 17–53.

[22] Gotz, R., Elgersma, J.W., Kraak, J.C. & Poppe, H., *Spectrochim. Acta*, 1994, **49B**, 761–768.

[23] Ding, L., Liang, F., Huan Y., Cao, Y., Zhang, H. & Jin, Q., *J. Anal. At. Spectrom.*, 2000, **15**, 293–296.

[24] Boorn, A.W. & Browner, R.F., 'Applications: organics', in *Inductively Coupled Plasma Emission Spectroscopy, Part II Applications and Fundamentals*, P.W.J.M. Boumans (ed.), John Wiley and Sons, New York, 1987.

[25] Browner, R.F. & Boorn, A., *Anal. Chem.*, 1982, **54**, 1402–1410.

[26] Maestre, S., Mora, J., Gras, L. & Todoli, J.L., *Can. J. Anal. Sci. Spectrosc.*, 2000, **45**, 124–132.

[27] Savage, R.N. & Hieftje, G.M., *Anal. Chim. Acta*, 1981, **123**, 319–324.

[28] Weiss, A.D., Savage, R.N. & Hieftje, G.M., *Anal. Chim. Acta*, 1981, **124**, 245.

[29] Gervais, L.S. & Salin, E.D., *J. Anal. At. Spectrom.*, 1991, **6**, 493–495.

[30] Temprano, M.C.V.H., De La Campa, M.R.F. & Sanz-Medel, A., *Anal. Chim. Acta*, 1995, **309**, 369–378.

[31] Temprano, M.C.V.H., De La Campa, M.R.F. & Sanz-Medel, A., *J. Anal. At. Spectrom.*, 1994, **9**, 231–236.

[32] Dedina, J. & Tsalev, D.J., *Hydride Generation and Atomic Absorption Spectrometry*, Wiley and Sons, Chichester, 1995.

[33] Darke, S.A. & Tyson, J.F., *Microchem. J.*, 1994, **50**, 310–336.
[34] Kantor, T., *Spectrochim. Acta B*, 2001, **56**, 1523–1563.
[35] Skinner, C.D. & Salin, E.E., *J. Anal. At. Spectrom.*, 1997, 12, 725–732.
[36] Hill, S.J., Dawson, S., Evans, E.H., Fisher, A., Price, W.J., Smith, C.M. & Tyson, J., *J. Anal. At. Spectrom.*, 2001, **16**, 672–711.
[37] Sutton, K.L. & Caruso, J.A., 'Environmental applications of plasma spectrometry', in *Inductively Coupled Plasma Spectrometry and its Applications*, S.J. Hill (ed.), Sheffield Academic Press Ltd, Sheffield, 1999.
[38] Miles, D.L. 'Geological applications of plasma spectrometry', in *Inductively Coupled Plasma Spectrometry and its Applications*, S.J. Hill (ed.), Sheffield Academic Press Ltd, Sheffield, 1999.
[39] Robb, P., Crews, H.M. & Baxter, M.J., 'Applications of plasma spectrometry in food science', in *Inductively Coupled Plasma Spectrometry and its Applications*, S.J. Hill (ed.), Sheffield Academic Press Ltd, Sheffield, 1999.
[40] Ebdon, L., Foulkes, M. & Sutton, K., *J. Anal. At. Spectrom.*, 1997, **12**, 213–229.
[41] Ebdon, L. & Goodall, P., *J. Anal. At. Spectrom.*, 1992, **7**, 1111–1116.
[42] Peng, T.Y., Chang, G., Sheng, X.H., Jiang, Z.C. & Hu, B., *Anal. Chim. Acta*, 2001, **433**, 255–262.
[43] Ouddane, B., Skiker, M., Fischer, J.C. & Wartel, M., *Analysis*, 1997, **25**, 308–313.
[44] Hashemi, P. & Olin, A., *Talanta*, 1997, **44**, 1037–1053.
[45] Quevauviller, P., *Method Performance Studies for Speciation Analysis*, RSC pub. Cambridge, 1998.
[46] Caruso, J.A., Sutton, K.L. & Ackley, K.L., *Elemental Speciation, New Approaches For Trace Element Analysis*, Volume XXXIII *Wilson and Wilson's Comprehensive Analytical Chemistry*, D. Barcelo (ed.), Elsevier, Amsterdam, 2000.
[47] Cairns, W.R.L., Ebdon, L. & Hill, S.J., *Fresenius J. Anal. Chem.*, 1996, **355**, 202–208.
[48] Rivas, C., Ebdon, L. & Hill, S.J., *J. Anal. At. Spectrom.*, 1996, **11**, 1147–1150.
[49] Hill, S.J., Bloxham, M.J. & Worsfold, P.J., *J. Anal. At. Spectrom.*, 1993, **8**, 499–515.
[50] Mitrovic, B. & Milacic, R., *Sci. Tot. Environ.*, 2000, **258**, 183–194.
[51] Aguerre, S., Pecheyran, C., Lespes, G., Krupp, E., Donard, O.F.X. & Potin-Gautier, M., *J. Anal. At. Spectrom.*, 2001, **16**, 1429–1433.
[52] Jerrell, L.J., Dunn, M.R., Anderson, J.E. & Fannin, H.B., *Appl. Spectrosc.*, 1999, **53**, 245–248.
[53] Deng, B.Y. & Chan, W.T., *Electrophoresis*, 2001, **22**, 2186–2191.
[54] Rivaro, P., Zaratin, L., Frache, R. & Mazzucotelli, A., *Analyst*, 1995, **120**, 1937–1939.
[55] Rubio, R., Alberti, J., Padro, A. & Rauret, G., *TrAC-Trends in Analytical Chemistry*, 1995, **14**, 274–279.
[56] Lafuente, J.M.G., Sanchez, M.L.F. & Sanz-Medel, A., *J. Anal. At. Spectrom.*, 1996, **11**, 1163–1169.
[57] Telgheder, U. & Khvostikov, V.A., *J. Anal. At. Spectrom.*, 1997, **12**, 1–6.
[58] Edelson, M.C., DeKalb, E.L., Winge, R.K. & Fassel, V.A., *Spectrochim. Acta*, 1986, **41B**, 475–486.
[59] Broekaert, J.A.C., Leis, F. & Laqua, K., *Spectrochim. Acta*, 1979, **34B**, 73–84.
[60] Marathe, S.M., Biswas, S.S., Patil, P.B. & Murty, P.S., *Mikrochim. Acta*, 1992, **109**, 261–268.
[61] Edelson, M.C., 'High resolution plasma spectrometry', in *Inductively Coupled Plasmas in Analytical Atomic Spectrometry*, A. Montaser & D.W. Golightly (eds), VCH, New York, 1987.
[62] Kabayashi, S., Wakui, Y., Kanesato, M., Matsunaga, H. & Suzuki, T.M., *Anal. Chim. Acta*, 1992, **262**, 161–166.
[63] Kumar, M., *Analyst*, 1994, **119**, 2013–2024.
[64] Montaser, A. & Van Hoven, R.L., *Crit. rev. Anal. Chem.*, 1987, **18**, 45.
[65] Ohls, K.D., Golightly, D.W. & Montaser, A., 'Mixed-gas, molecular-gas and helium inductively coupled plasmas operated at atmospheric and reduced pressure', in *Inductively Coupled*

Plasmas in Analytical Atomic Spectrometry, A. Montaser & D.W. Golightly (eds), VCH, New York, 1987.
[66] Raeymaekers, B., Graule, T., Broekaert, J.A.C., Adams, F. & Tschopel, P., *Spectrochim. Acta*, 1988, **43B**, 923–940.
[67] Yang, P. & Barnes, R.M., *Spectrochim. Acta*, 1989, **44B**, 1093.
[68] Sesi, N.N., MacKenzie, A., Shanks, K.E., Yang, P. & Hieftje, G.M., *Spectrochim. Acta*, 1994, **49B**, 1259.
[69] Zougagh, M., Rudner, P.C., de Torres, A.G. & Pavon, J.M.C., *J. Anal. At. Spectrom.*, 2000, **15**, 1589–1594.
[70] Marcos, A., Foulkes, M. & Hill, S.J., *J. Anal. At. Spectrom.*, 2001, **16**, 105–114.
[71] Marcos, A. & Hill, S.J., *Analyst*, 2000, **125**, 1015–1020.
[72] Marcos, A., Fisher, A., Rea, G. & Hill, S.J., *J. Anal. At. Spectrom.*, 1998, **13**, 521–525.
[73] Herrador, M.A. & Gonzalez, A.G., *Talanta*, 2001, **53**, 1249–1257.
[74] dos Santos, E.J. & de Oliveira, E., *J. Food Comp. Anal.*, 2001, **14**, 523–531.

4 Analytical glow discharges

Norbert Jakubowski, Annemie Bogaerts and Volker Hoffmann

4.1 Introduction

Glow discharges (GD) are used in a variety of applications, ranging from plasma etching and deposition of thin films (e.g. in the microelectronics industry) to applications in lasers and as spectroscopic sources in analytical chemistry. This chapter will concentrate on selected aspects of GD used for analytical purposes. The fundamental processes will be described in detail, because even nowadays confusion exists about the dominant processes taking place in GD. Only a brief description will be given of applications, with the intention of highlighting the many different areas of analytical GD spectroscopy and its applications in the material sciences. Other aspects – e.g. the use of GD in environmental applications – are described elsewhere [1]. More details are given in [2–5].

4.2 Basic principles of the analytical GD

4.2.1 Introduction

A GD is a kind of plasma, i.e. a partially ionized gas consisting of a large number of neutral species as well as electrons and positive ions. In its simplest form, it is created by applying a potential difference (in the order of 1 kV) between two electrodes which are inserted in a cell filled with a gas (or which form the cell walls). For typical analytical applications (such as mass spectrometry (MS) or optical emission spectrometry (OES)), the gas is usually Ar (sometimes Ne), and the gas pressure is in the range 50–700 Pa.

If no potential difference were applied between both electrodes, a few electrons would be emitted from the electrodes by cosmic radiation, but these electrons would not be able to sustain the discharge. However, when a potential difference is applied, the electrons are accelerated away from the negative electrode (or cathode) and collide with the Ar gas atoms. The most important collisions are electron impact ionization and excitation of Ar. The ionization collisions create argon ions and new electrons. The Ar ions are accelerated by the potential difference (or electric field) toward the cathode, where they release new electrons (i.e. *secondary electron emission*). These electrons can again give rise to ionization collisions with the argon gas, creating new electrons and Ar ions. In this way, the potential difference between both electrodes causes the creation of sufficient electrons and Ar ions to carry electrical current from one electrode to the other, and the glow discharge is called a *self-sustaining plasma*.

The other important collision process, as mentioned above, is electron impact excitation of the Ar atoms. This gives rise to Ar atoms in excited levels, which will decay to lower levels by the emission of radiation. This explains the characteristic name of the *glow* discharge (e.g. in Ar, it emits a blue glow). These elementary processes are schematically represented in Fig. 4.1.

The use of the GD in analytical chemistry (i.e. for the trace analysis of (mainly solid) materials) is based on the phenomenon of *sputtering*, which is also illustrated in Fig. 4.1. The material to be analyzed is then used as the cathode of the GD, which is bombarded by the positive (Ar) ions accelerated by the electric field. These ions release not only secondary electrons at the cathode, but also atoms of the cathode material, which is called sputtering (i.e. *sandblasting* at the atomic level). The sputtered, analytically important atoms arrive in the plasma, where they are subject to collisions. The ionization collisions (by electrons or other plasma species; see below) create ions of the cathode material, which can be measured in a mass spectrometer – glow discharge mass spectrometry (GDMS). The excitation collisions yield excited atoms, which release photons, characteristic of the

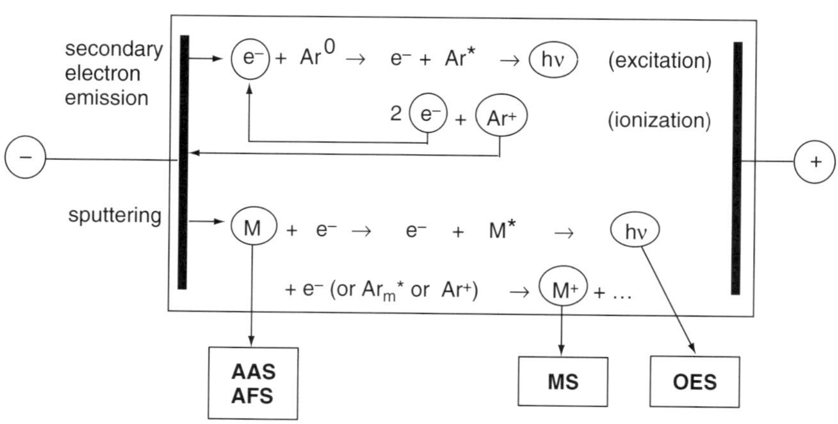

Fig. 4.1 Schematic diagram of the basic processes in an analytical argon glow discharge (electron impact excitation and ionization, secondary electron emission and sputtering), including the analytical applications (atomic absorption and fluorescence spectrometry, AAS and AFS; mass spectrometry, MS; and optical emission spectrometry, OES).

elements present in the material to be analyzed. These photons can be detected with an optical emission spectrometer – glow discharge optical emission spectrometry (GD-OES). Moreover, the sputtered atoms can also directly be detected in the plasma, using an external light source (e.g. a laser) – glow discharge atomic absorption spectrometry (GD-AAS) or glow discharge atomic fluorescence spectrometry (GD-AFS). These analytical applications are also schematically represented in Fig. 4.1.

Besides analytical chemistry, GD are also used in many other applications, i.e. for surface modification (e.g. etching of surfaces, deposition of thin films) in the semiconductor industry and for materials engineering, such as lamps (e.g. fluorescent lamps), plasma display panels (for the new types of flat TV-screens) or lasers, for environmental purposes (e.g. the destruction of volatile compounds), and for biomedical applications (e.g. sterilization of materials) [6]. However, in this chapter, we will limit ourselves to the analytical GD plasmas.

After outlining the typical discharge conditions of analytical GD (including the different electrical operation modes), the different spatial regions that can be distinguished in analytical GD will be discussed, in terms of potential and electric field distribution, charged species densities and luminosity. Next, information will be given about the role of the various species present in the plasma (e.g. their typical number density). Finally, an overview of the fundamental processes occurring in the plasma and at the cell walls will be given. Most of this information is based on the results of numerical modeling for GD developed by Bogaerts *et al.* (e.g. [7–9]) as well as on experimental verifications of this model (e.g. [10, 11]).

4.2.2 Typical discharge conditions of analytical GD

Analytical GD, in the direct current (d.c.) mode, typically operate at a voltage of 500–1400 V, an Ar gas pressure of 50–700 Pa, a gas temperature ranging from room temperature to 1000 K, and an electrical current of 1–100 mA. Although a strict subdivision cannot be made, GD used as ion source for MS generally operate at *softer* conditions (i.e. lower pressure, gas temperature and electrical current; but similar voltage) than GD used for OES. This is particularly true for the commercial GD instruments currently available. Indeed, the VG 9000 GDMS (ThermoElemental, Winsford, UK) operates at a pressure of 100 Pa and less, yielding a current of 1–10 mA, whereas the Grimm-type GD (which is the classical source for OES) usually operates at pressures above 100 Pa, yielding a current of 10–100 mA. As mentioned above, this subdivision is rather artificial, and mainly caused by instrumental considerations (e.g. pressure limitation when coupled to a mass spectrometer). For example, Grimm-type sources can also be used as ion source for MS [12, 13], operating at similar conditions as when used for OES. Moreover, there also exist home-built

GD sources, typically based on a so-called *six-way-cross* geometry (e.g. [10]), which have been used already for GDMS, GD-OES, as well as GD-AAS and GD-AFS, and which operate for example at 100 Pa and a few mA [10]. Figure 4.2 presents some typical d.c. current–voltage–pressure characteristics, as obtained from modeling calculations (but verified with experiments), both for the VG 9000 GDMS system (50–100 Pa; full circles) and for the Grimm-type GD (300–600 Pa; full triangles) [9, 14].

Besides the d.c. mode, analytical GD can also operate in radio-frequency (Rf) mode. The typical operating conditions of an analytical Rf GD are quite similar to that of a d.c. GD; however, the applied voltage and current are not constant, but they vary as a function of time. Hence, the Rf-powered electrode (which is made out of the material to be analyzed) acts alternately as cathode and anode. This makes the analysis of non-conducting materials possible, because the charge built up during positive ion bombardment (as would be the case in the d.c. GD) will be neutralized by electron bombardment during the other half of the Rf-cycle. The Rf typically used is 13.56 MHz, although lower frequencies (e.g. 3.5 MHz) have been used as well (e.g. [15]). Because of the large difference in size between the Rf-powered electrode (i.e. sample to be analyzed) and the grounded electrode (i.e. cell walls), a large negative d.c.-bias voltage is developed in front of the Rf-electrode, enabling sufficient ion bombardment and hence sputtering [15]. This large d.c.-bias voltage makes the analytical Rf GD very similar to the d.c. GD, as far as plasma species densities and processes are concerned.

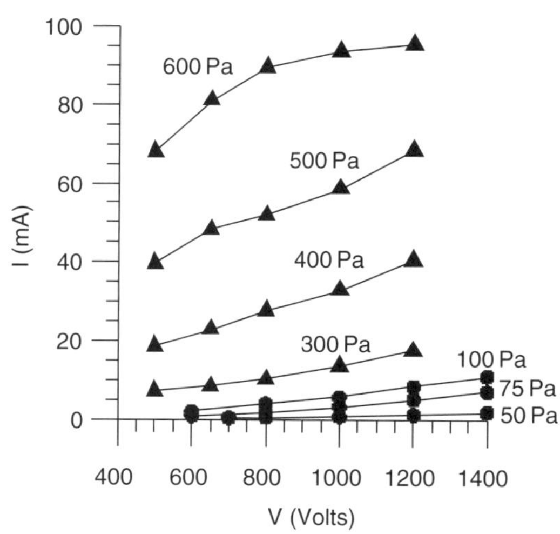

Fig. 4.2 Electrical current as a function of voltage at different pressures, as obtained from modelling calculations [14–23], illustrating the typical d.c. discharge conditions of the VG 9000 and Grimm-type glow discharge cells (full circles and full triangles, respectively).

It should be mentioned that there exist a few different commercial Rf-GD-OES instruments, but there is no commercial Rf-GDMS instrument yet.

Both d.c. and Rf discharges are normally used in a continuous mode but pulsed operation is also possible. Here, voltage and current are applied for short pulses (milli- or microsecond range), which is followed by a (generally longer) afterglow, in which the GD burns out, before the next pulse starts. Because higher pulse voltages and currents can be applied (e.g. 2 kV at 1 A in μs-pulsed d.c.) for the same average power as in a d.c. GD [16], higher ionization and excitation rates can be obtained, yielding higher analyte signals, and hence lower detection limits are expected than in continuous d.c. discharges. In GDMS, this operation mode is very useful in combination with Time of Flight instruments, which can operate synchronously with the pulsed erosion [17]. By changing the delay time between the pulse of erosion and the pulse of ion acceleration in the flight tube, it is possible to distinguish between sputtered ions from the sample and ions of the discharge gas and thus to avoid interferences in the measured mass spectrum. The different energies existing in the μs-pulsed GD between power introduction and afterglow can be exploited to perform speciation analysis [18]. Another interesting kind of pulsed discharges is the application of double pulses [19]. Up to now, no μs-pulsed Rf discharges have been realized and investigated. This is most probably caused by technical difficulties at the Rf generators.

However, apart from these somewhat different operating conditions, the ms-pulsed discharge resembles very much a short d.c. GD, with similar plasma characteristics, followed by a long afterglow. They are easy to handle in d.c. and Rf modes and commercial instruments already benefit from the advantages of reduced power introduction for the analysis of heat sensitive material. Commercial Rf generators, for example, work up to 30 kHz in pulsed mode, but the synchronized measurement of light intensity is still missing; however, it is not a major problem in this timescale. By using such synchronized detection it is also possible in the ms-pulsed GD to take advantage of the different excitations and ionizations during the pulse and in the afterglow.

Because analytical Rf and pulsed GD are not fundamentally different from d.c. discharges, we will focus in the following discussion on d.c. GD. The differences between d.c., Rf and pulsed GD can, for example, be found in refs [20, 21].

4.2.3 *Spatial regions in the analytical GD*

The analytical GD can be subdivided in three spatial regions (see Fig. 4.3a), which differ from each other by the potential difference or electric field strength (Fig. 4.3b), the charged species densities (Fig. 4.3c), and the luminosity (related to Fig. 4.3d). It should be mentioned that Figs 4.3b–d are the result of modeling calculations at a pressure of 75 Pa, a voltage of 1000 V and a current of 3 mA (e.g. [7, 8]).

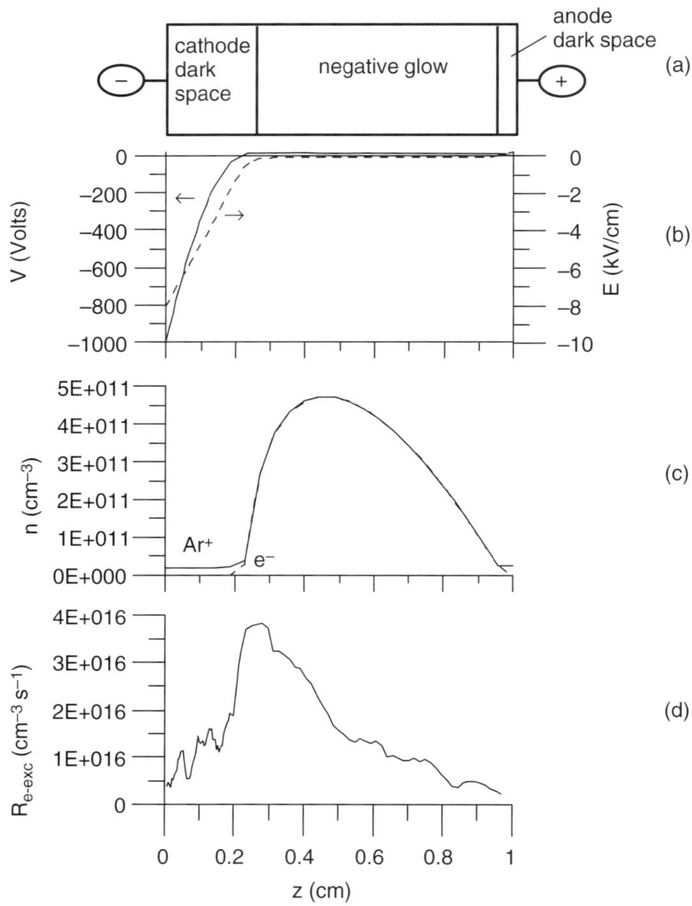

Fig. 4.3 Schematic diagram of the different spatial regions in a glow discharge (a), and the corresponding potential and electric field distributions (b), charged particle density profiles (c) and electron impact excitation rate to give an idea about the luminosity; (d), as obtained from modelling calculations at 1000 V, 75 Pa and 3 mA.

The potential difference applied between anode and cathode drops off completely in the first few millimeters in front of the cathode, i.e. in the region called *cathode dark space* (CDS). The reason for this large potential drop is the positive space charge in this region, i.e. the Ar ion density is nearly constant and rather low, but the electron density is roughly zero (Fig. 4.3c). This region is characterized by a strong negative electric field (Fig. 4.3b). The latter accelerates the electrons away from the cathode in the forward direction, so that they traverse this region without many collisions (excitation, ionization…). The low number of excitation collisions (Fig. 4.3d), and the corresponding de-excitation with emission of light explain why this region is called *cathode dark space*.

Within a few millimeters from the cathode, the potential crosses zero and becomes slightly positive. This positive value (in the order of several volts) is nearly constant till the anode and is called the *plasma potential* (Fig. 4.3b). This nearly constant potential difference gives rise to a very low electric field (Fig. 4.3b). Hence, the electrons are not accelerated to a great extent anymore, and they wander around in this region, giving rise to many collisions. The large number of excitation collisions (Fig. 4.3d) and the subsequent radiative decay explain why this region is very bright. It is therefore called the *negative glow* (NG). Ion and electron densities here are typically rather high and nearly equal to each other (Fig. 4.3c). This gives rise to a more or less electrically neutral region, which explains the nearly constant potential mentioned above.

Although the NG extends almost till the anode, there is a small region adjacent to the anode where the potential returns to zero, and the electric field is slightly positive (Fig. 4.3b). This so-called *anode dark space* (ADS) is again characterized by a low and nearly constant positive ion density and a nearly zero electron density (Fig. 4.3c), giving rise to a small positive space charge (which is in correspondence to the small potential drop).

As mentioned above, the CDS is typically a few millimeters thick, and the NG fills most of the discharge. The ADS is very thin (typically much less than 1 mm). As appears from Fig. 4.4 (which is also obtained from modeling calculations [9, 14]), the length of the CDS decreases slightly with rising voltage, and more strongly with rising pressure. As predicted by the calculations, it

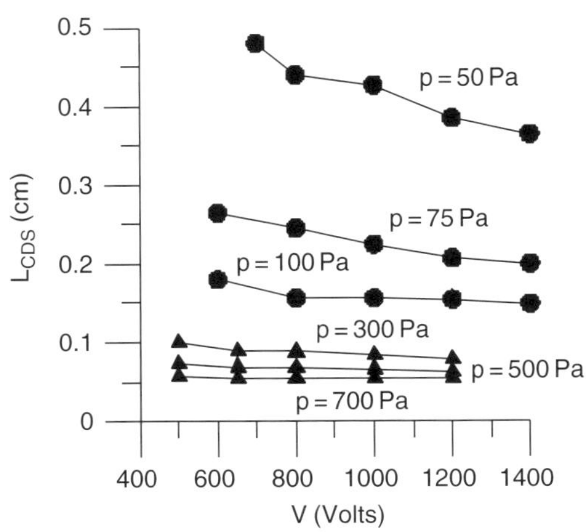

Fig. 4.4 Length of the cathode dark space (L_{CDS}) as a function of voltage at different pressures, as obtained from modeling calculations [14–23], for typical VG 9000 and Grimm-type conditions (full circles and full triangles, respectively).

ranges from about 0.5 mm (at a pressure of 700 Pa) to about 5 mm (at a pressure of 50 Pa).

4.2.4 *Species present in the analytical GD*

As mentioned above, the analytical GD is usually maintained in Ar. The Ar gas atoms are the dominant plasma species, with typical densities, as deduced from the ideal gas law ($n = N/V = p/kT$), in the order of 10^{16}–10^{17} cm^{-3}, at the typical analytical discharge conditions mentioned above. The gas atom density is more or less constant throughout the discharge, although it can be somewhat depleted near the cathode as a result of gas heating (i.e. a higher gas temperature yields a lower density, see above). According to model calculations [22], the gas heating resulting from energy transfer of energetic plasma species into the Ar gas is of minor importance at moderate discharge conditions (e.g. 600 V, 100 Pa or below, a few mA); however, it becomes significant at the strong discharge conditions of, for example, 1000 V, several hundred Pa and a few tens of mA. For the latter conditions, the gas temperature was calculated to rise to 1000 K and more, as is illustrated in Fig. 4.5 (which shows the calculated gas temperature at the maximum of its profile, as a function of voltage and pressure). The gas heating is mainly caused by elastic collisions of fast Ar ions, Ar atoms and sputtered atoms with the Ar gas atoms, resulting in energy transfer into the Ar gas [22]. This occurs mainly in the region adjacent to the cathode, which results in the highest gas temperature, and hence the lowest gas density, near the cathode. Figure 4.6 shows a typical calculated gas

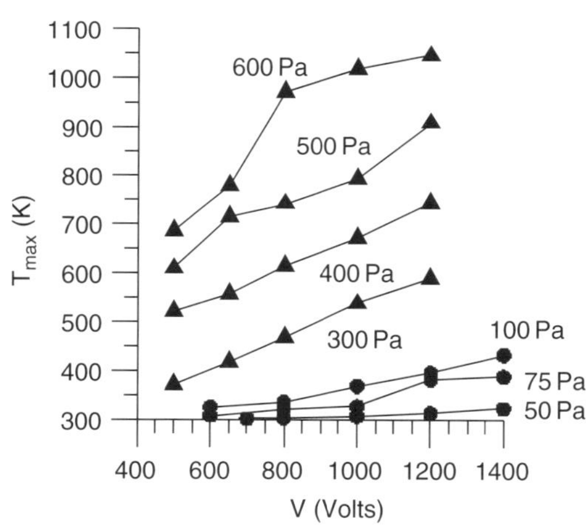

Fig. 4.5 Gas temperature at the maximum of its profile, as a function of voltage at different pressures, as obtained from modelling calculations [32], for typical VG 9000 and Grimm-type conditions (full circles and full triangles, respectively).

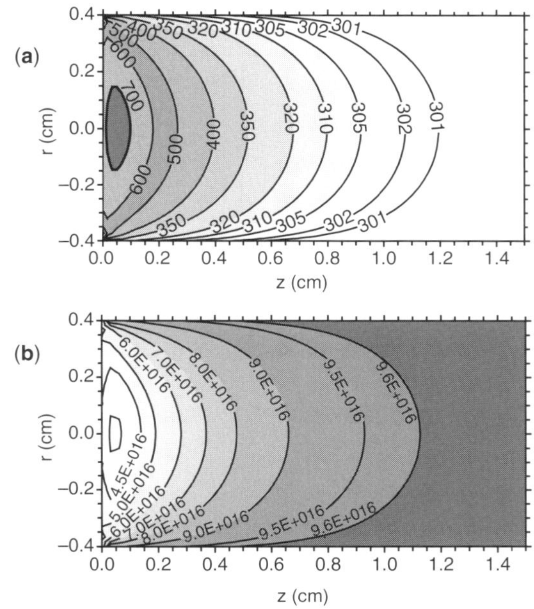

Fig. 4.6 Gas temperature distribution (a) and corresponding argon gas atom density distribution (b), as obtained from modelling calculations [32], for a Grimm-type source, at 800 V, 500 Pa and about 50 mA. Note that only the first 1.5 cm near the cathode of the Grimm-type cell is shown.

temperature profile, and the corresponding Ar gas atom density distribution, at 800 V, 500 Pa and about 50 mA, for a simplified Grimm-type geometry (only the first 1.5 cm is shown, because the gas temperature and Ar gas density are constant at longer distances from the cathode) assuming no gas flow at all. For these conditions, the gas temperature increases from room temperature (in most of the discharge region, starting at about 1 cm from the cathode) to about 700 K near the cathode. Correspondingly, the Ar gas atom density is nearly constant (almost 10^{17} cm^{-3}), starting at about 1 cm from the cathode to the end of the discharge (several centimetres long), but it drops to a minimum of about 4×10^{16} cm^{-3} near the cathode. It should be mentioned that at *softer* discharge conditions (i.e. lower pressure, voltage and current), the temperature does not rise to such a large extent, and the gas atom density remains more constant throughout the discharge.

Although most of the Ar gas is in atomic form, a fraction, in the order of 10^{-5}–10^{-3} (increasing with pressure, voltage and current), is also ionized. Hence, the analytical GD can be considered as a weakly ionized plasma. Figure 4.7 shows a typical calculated argon (Ar^+) ion density distribution, for the same conditions as in Fig. 4.6. It is clear that the Ar^+ ion density is rather low in the CDS; it reaches a maximum at a few mm from the cathode,

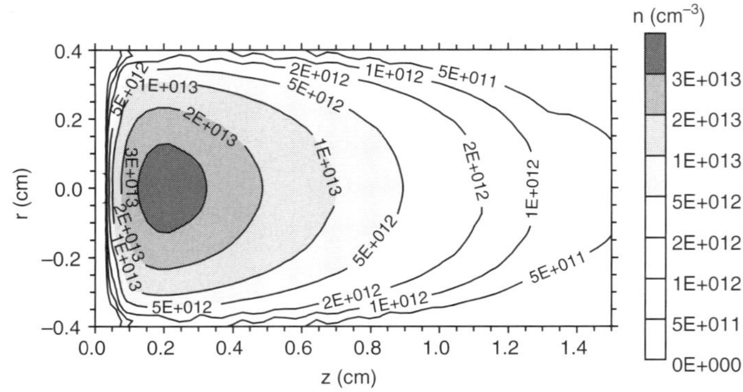

Fig. 4.7 Argon (Ar^+) ion density distribution as obtained from modelling calculations [14], for a Grimm-type source, at the same conditions as in Fig. 4.6.

and then decreases to lower values further away from the cathode, and toward the cell walls. It is worth mentioning that the electron density is characterized by nearly the same profile, except that it is zero in the CDS (see also above, Fig. 4.3). The Ar ion and electron densities typically range from 10^{11} to 10^{14} cm^{-3} at the typical analytical GD conditions, increasing with voltage, pressure and current. Figure 4.8 shows a comparison between calculated electron densities (obtained at the maximum of their profile, i.e. at ca. 2 mm from the cathode; see above, Fig. 4.7) and values measured with a Langmuir probe (at a distance of about 7 mm from the cathode) for similar albeit not exactly the same discharge conditions [23]. The qualitative agreement is quite satisfactory, with rising densities as a function of voltage and pressure. As far as the absolute numbers are concerned, a difference of a factor of 2 is observed (note the different scales in the left and right axes), which is, however, considered to be well below the expected uncertainties in both the model calculations and the experiment.

The Ar^+ ions are usually the dominant ionic species in the GD plasma. The multiply charged ions and the dimers have typically much lower densities; for example, the densities of Ar^{2+} and $Ar_2{}^+$ ions have been calculated to be typically two orders of magnitude lower than the Ar^+ ion density [24]. Further it should be mentioned that for small traces of impurities in the Ar gas, other dimers can also be formed in the plasma. Recent modelling calculations, for 1000 V and 75 Pa, have shown that when 1% H_2 is added to the Ar gas, the ArH^+ ion density is about 2.5×10^{10} cm^{-3} (i.e. only a factor of 4 lower than the Ar^+ ion density, which was calculated to be almost 10^{11} cm^{-3}) [25]. At 0.1% H_2 addition, the ArH^+ and Ar^+ ion densities were calculated to be about 1.5×10^{10} cm^{-3} and 2.2×10^{11} cm^{-3}, respectively; hence, the ratio of ArH^+/Ar^+ ion densities dropped to a factor 1/15 [25].

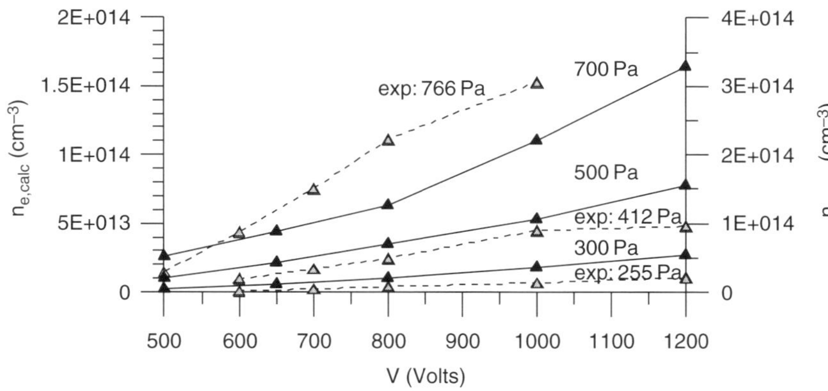

Fig. 4.8 Electron densities as a function of voltage at three different pressures, as obtained from modelling calculations at the maximum of their profile (left axis, solid lines), and from Langmuir probe measurements at 7 mm from the cathode (right axis, dashed lines), both obtained for the Grimm-type source.

Besides argon atoms and ions in the ground state, excited species are also present in the plasma. The most important are the argon atoms in the two 4s metastable levels (i.e. the lowest excited levels, at 11.55 and 11.72 eV, which cannot decay to the ground state by emission of radiation). Indeed, the Ar metastable atoms play an important role in ionization of the sputtered atoms (see below). Besides the two 4s metastable levels, there are also two other 4s levels (i.e. at 11.62 and 11.83 eV), which can decay to the ground state by emission of radiation. However, this radiation is easily reabsorbed by the Ar ground state atoms, because of their high density. At the typical analytical GD conditions, the fraction of radiation which is not reabsorbed (so-called *escape factor*) is only in the order of 10^{-3}–10^{-4} [26]. Hence, the 4s non-metastable excited levels (also called *resonant levels*) are also characterized by rather high densities. Figure 4.9 illustrates the calculated densities of these four 4s levels as a function of distance from the cathode, in a Grimm-type source (only the first 1.5 cm are shown again), at the same conditions as in Fig. 4.6. It appears that the 4s $[3/2]_2$ metastable level has the highest density, followed by the 4s′ $[1/2]_0$ metastable level. The calculated densities of the two resonant levels (i.e. 4s $[3/2]_1$ and 4s′ $[1/2]_1$) are, however, only slightly lower. Further, the densities of all four 4s levels appear to reach a pronounced maximum adjacent to the cathode, which is attributed to population of these levels by fast Ar ion and atom impact excitation [26]. Indeed, these processes were found to be more important as a production mechanism than electron impact excitation, at the typical analytical discharge conditions [26]. Such a pronounced maximum adjacent to the cathode, followed by a rapid drop as a function of distance, was also measured with atomic absorption spectrometry (AAS), by Ferreira *et al.*, for the metastable level densities in a Grimm-type source [27].

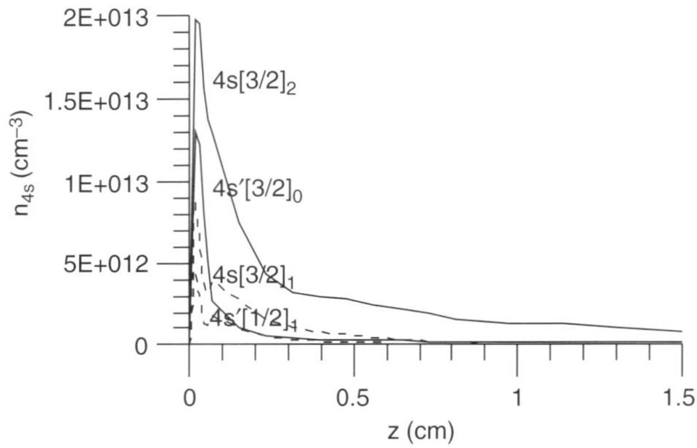

Fig. 4.9 Level populations of the argon atoms in the four 4s (metastable and resonant) levels, as a function of distance from the cathode, as obtained from modelling calculations [14], for a Grimm-type source, at the same conditions as in Fig. 4.6.

The higher excited Ar atomic levels are characterized by much lower densities, as is clear from Fig. 4.10 which shows the calculated level populations at the maximum of their profile, divided by the statistical weight, as a function of energy of the excited levels, at the same conditions as in Fig. 4.6. The Ar* ($3p^5$) 4p, 5s, 3d and 5p levels have still a considerable population density (albeit already several orders of magnitude lower than the 4s metastable and resonant levels), but the higher excited levels have much lower population densities [26].

Besides argon species, there are also sputtered species present in the plasma, which are, of course, of special importance in the analytical GD. The sputtered atom ground state density in the plasma is typically 3–5 orders of magnitude lower than the Ar atom density, and it is non-uniformly distributed, with a maximum near the cathode (as a result of sputtering from the cathode). The sputtered Cu atom density distribution, as calculated for the same conditions as in Fig. 4.6, is illustrated in Fig. 4.11a. This calculation result has been checked by laser-induced fluorescence and atomic absorption measurements [10], albeit for a different cathode material (tantalum; because of the availability of suitable laser lines), and the agreement was highly satisfactory [10].

The corresponding Cu^+ ion ground state density is still a few orders of magnitude lower than the Cu atom density, as appears from the calculation results in Fig. 4.11b, for the same discharge conditions. It is also clear that the Cu^+ ion density reaches a maximum at the same position as the Ar^+ ion density (i.e. in the beginning of the NG; see also above), as applies in fact to all ionic species. The ionization degree of Cu was calculated in the range of 10^{-4}–0.04, for typical analytical GD conditions and it rises with voltage, pressure and current. This is significantly higher than the ionization degree of Ar (see above). Indeed, the sputtered atoms are much more efficiently ionized

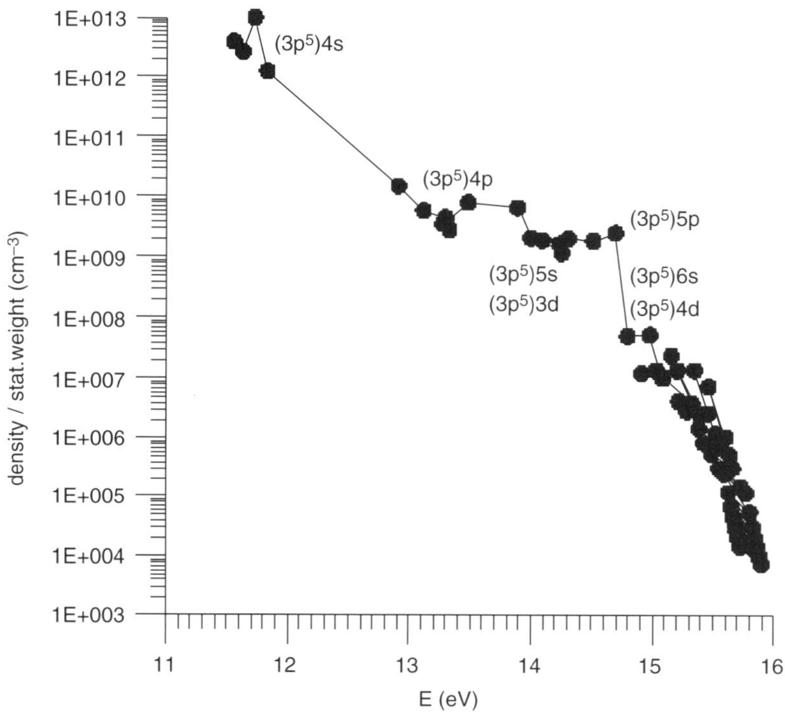

Fig. 4.10 Level populations, divided by the statistical weight, of the argon atoms in various excited levels, at the maximum of their profiles, as a function of their excitation energy, as obtained from modelling calculations [14], for a Grimm-type source, at the same conditions as in Fig. 4.6.

in the GD plasma than the Ar atoms, due to Penning ionization by Ar metastable atoms and asymmetric charge transfer with Ar ions, as will be explained below.

The Cu atoms and ions in excited levels have again a much lower population density in comparison to the ground state atomic and ionic levels, as is illustrated in Fig. 4.12, which shows the level populations at the maximum of their profiles, divided by the statistical weight, as a function of energy of the excited levels, at the same conditions as in Fig. 4.6 [28]. Further, it is also clear from this figure that the doubly ionized Cu^{2+} ions have again a much lower density in the plasma than the Cu^+ ions. The ionization degree of Cu^+ ions (i.e. the ratio of Cu^{2+} to Cu^+ ion density) was calculated to be about 3×10^{-3} at the conditions shown in Fig. 4.12, which is slightly lower than the ionization degree of Cu atoms (i.e. Cu^+/Cu^0 density ratio, in the order of 10^{-2} at the same conditions).

In conclusion, it can be stated that the Ar gas atoms, Ar^+ ions and electrons, Ar metastable atoms, sputtered (e.g. Cu) atoms and corresponding Cu^+ ions are the most important plasma species, not only in terms of their density, but

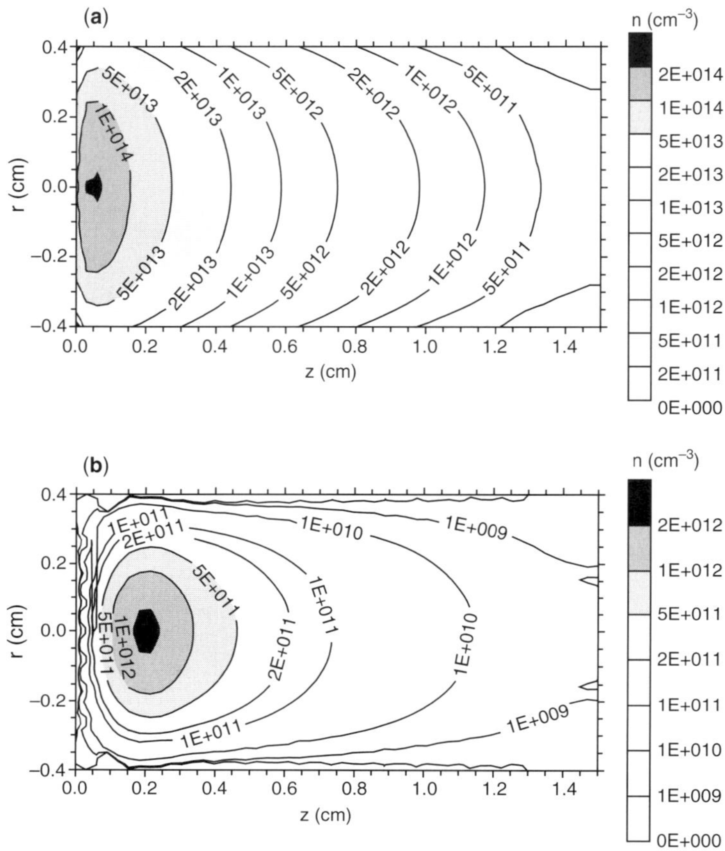

Fig. 4.11 Sputtered copper atom (a) and corresponding Cu^+ ion (b) density distributions as obtained from modelling calculations [14], for a Grimm-type source, at the same conditions as in Fig. 4.6.

also in terms of their role in the plasma. Table 4.1 gives an overview of these species, and the range of their densities, as calculated for the typical range of analytical GD conditions (see above) [7–9].

4.2.5 Most important collision processes in the plasma

As mentioned above, the GD plasma is a complicated environment, consisting of many different species, which can all interact with each other. It is therefore not feasible to give a complete overview of all possible phenomena. Probably not all processes are even known in the literature. Hence, we will restrict ourselves to the most relevant processes in analytical GD, i.e. ionization (and recombination) and excitation (and de-excitation), and to

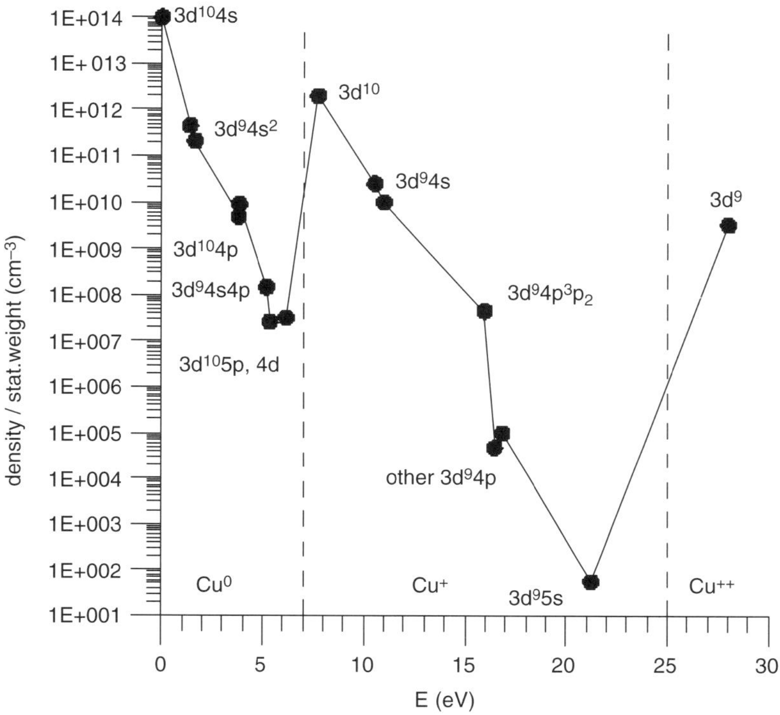

Fig. 4.12 Level populations, divided by the statistical weight, of the copper atoms and Cu^+ ions in various excited levels, as well as the Cu^{2+} ions, at the maximum of their profiles, as a function of their excitation energy, as obtained from modelling calculations [37], for a Grimm-type source, at the same conditions as in Fig. 4.6.

Table 4.1 Overview of the most important plasma species and their densities obtained from modelling calculations, in the typical range of analytical glow discharge conditions (see Fig. 4.2; i.e. voltage of 500–1400 V, pressure of 50–700 Pa, gas temperature of 300–1000 K, electrical current of 1–100 mA)

Species	Calculated range of densities (cm^{-3})
Argon gas atoms	10^{16}–10^{17}
Ar^+ ions	10^{11}–10^{14}
Electrons	10^{11}–10^{14}
Argon atoms in 4s metastable levels	10^{11}–10^{13}
Sputtered (Cu) atoms	10^{12}–10^{14}
Cu^+ ions	10^{9}–10^{12}

the plasma species playing a role in these processes, i.e. electrons, Ar atoms, Ar^+ ions, Ar atoms in various excited levels, with special focus on the metastable levels, and atoms and ions of the cathode material. Collision processes involving multiply charged particles, clusters and negative ions will not be considered here, since these species are clearly of minor importance in the argon GD.

4.2.5.1 Elastic collisions

It is worth briefly mentioning this type of collision first, because of the high rate at which they occur in the GD. Elastic collisions do not result in internal changes of the energy of the collision partners; their effect is only to redistribute the kinetic energy of the particles. This redistribution is considerable for collisions between species with comparable masses (e.g. two atoms), but it is negligible for particles with very different masses (e.g. atoms + electrons).

4.2.5.2 Ionization and excitation of argon atoms

Electron impact ionization and excitation

$$Ar^0 + e^- \rightarrow Ar^+ + 2e^-$$

$$Ar^0 + e^- \rightarrow Ar^* + e^-$$

Electron impact ionization is one of the most important and best-known processes in the GD. It is the essential process in a self-sustaining plasma, since the electrons formed in this way can again give rise to ionization, leading to electron multiplication (see Section 4.1). It can occur by the collision of electrons with argon atoms in the ground state (direct electron impact ionization) and also with argon atoms in the metastable levels at 11.55 or 11.72 eV (two-step electron impact ionization). The minimum electron energy required for the first process is 15.76 eV (i.e. the ionization potential of argon), whereas the second process can start occurring at electron energies above 4 eV. Nevertheless, the first process is much more important in the analytical GD at relatively high voltages (order of 1 kV), due to the much higher argon ground state atom density compared to the argon metastable atom density (see Figs 4.6 and 4.9; and Table 4.1). The cross sections of both processes increase with the electron energy, reach a maximum of about 3×10^{-16} and 8×10^{-16} cm^2 at about 80 and 10 eV, respectively, after which they decrease again, because the residence time of the electron around the argon atom becomes too short for efficient ionization (see Fig. 4.13, curves 1 and 2) [29, 30].

The mechanism of electron impact excitation is the same as for ionization, but less energy is transferred to the atom, so that no ionization can occur. Indeed, the electron cannot be ejected, it can only jump to a higher energy level within the atom. The total cross section of electron impact excitation as a function of the electron energy shows the same behavior as for electron impact ionization. The minimum energy required is 11.55 eV (i.e. the energy of the

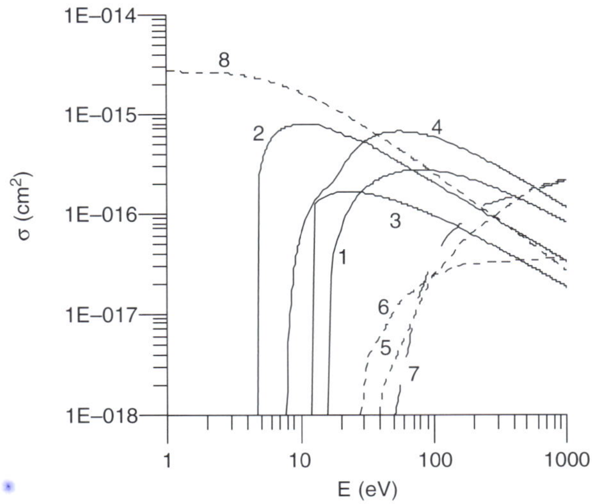

Fig. 4.13 Collision cross sections as a function of energy: solid lines: electron collisions; small dashed lines: argon ion collisions; wide dashed lines: fast argon atom collisions. (1) electron impact ionization of argon ground state atoms; (2) electron impact ionization of argon metastable atoms; (3) electron impact excitation of argon ground state atoms; (4) electron impact ionization of copper atoms; (5) fast argon ion impact ionization of argon ground state atoms; (6) fast argon ion impact excitation of argon ground state atoms; (7) fast argon atom impact ionization and excitation of argon ground state atoms;. (8) symmetric charge transfer between argon ions and argon atoms (i.e. elastic scattering in the backward direction).

lowest excited level). A maximum of about 1.6×10^{-16} cm^2 is reached at about 20 eV [29] (see curve 3 in Fig. 4.13).

Fast argon ion and argon atom impact ionization and excitation

$$Ar^0 + Ar^+_f \rightarrow Ar^+ + Ar^+_f + e^-, \quad Ar^0 + Ar^0_f \rightarrow Ar^+ + Ar^0_f + e^-$$

$$Ar^0 + Ar^+_f \rightarrow Ar^* + Ar^+_f, \quad Ar^0 + Ar^0_f \rightarrow Ar^* + Ar^0_f$$

In analogy to electron impact ionization and excitation, argon ions and atoms can also cause the ionization and excitation of argon atoms if their energy is sufficiently high. The cross sections of these processes are similar to the ones for electron impact ionization and excitation, i.e. rising with increasing impact energy, until a maximum is reached, and falling off at high energies [31]. However, the cross section curves are shifted towards higher energies, due to the larger masses of atoms and ions. Indeed, the process becomes important only at ion and atom energies of more than 100 eV and the maximum is reached above 1000 eV (see curves 5, 6 and 7 in Fig. 4.13). The cross section values at the maximum are comparable to the ones for electron impact ionization and excitation (cf. curves 1 and 3). In the GD, energetic argon ions and atoms are only found close to the cathode, where they have gained much energy from the electric field in the CDS (at least for the ions). Therefore, argon ion and atom impact ionization and excitation are only significant close to the cathode, and the importance of these processes in the GD increases with increasing voltages. At voltages of about 1 kV which are commonly used in analytical GD, these processes are clearly not negligible (see, e.g., [14]). The importance of fast argon ion and atom impact excitation as population mechanisms for the Ar (4s and other) excited levels was already demonstrated above. Furthermore, although the relative contributions of fast argon ion and atom impact ionization to the total ionization of argon were calculated to be only in the order of 1–4% (for the Ar^+ ions) and 2–14% for the fast Ar atoms [9, 14], these processes had to be included in the modelling work to enable the reproduction of the experimental current–voltage characteristics [14].

Argon metastable atom collisions leading to the ionization of one of the atoms

$$Ar^*_m + Ar^*_m \rightarrow Ar^+ + Ar^0 + e^-$$

When two argon metastable atoms collide with each other, together they have sufficient energy (2 times 11.55 or 11.72 eV) to knock off one electron and to yield the ionization of one of the atoms. The rate constant of this process is about $6.4 \times 10^{-10}\ cm^3\ s^{-1}$ [32]. This process is, however, clearly of minor importance in the analytical GD compared to electron impact ionization [33].

Symmetric charge transfer

$$Ar^+_f + Ar^0_s \rightarrow Ar^0_f + Ar^+_s$$

When a fast argon ion collides with a slow argon atom, an electron can be transferred from the atom to the ion without changes in the kinetic energy of the two colliding particles. In this way a fast argon atom and a slow argon ion are formed. Hence, it is not a real ionization process since there is no increase in the number of ions; only a fast argon ion has disappeared and a new slow argon ion has been created. However, the process is included in this overview since it is of major importance in the GD. Indeed, this process is responsible for the creation of a large number of fast argon atoms, which play an important role in the ionization and excitation of argon (see above) and especially for sputtering at the cathode (see below). The cross section of this process is in the order of $2–3 \times 10^{-15}\ cm^2$ at energies of a few eV and decreases gradually towards higher energies [34] (see curve 8 in Fig. 4.13). It should be mentioned that this process is actually a form of elastic collision, because there is no change in the kinetic energy of the collision partners [34].

Photo-ionization/excitation. In principle, the argon atoms can be ionized or excited by any suitable energy input. Therefore, photo-ionization and -excitation should also be mentioned here [35]. The cross section of photo-ionization shows a maximum of about 3.7×10^{-17} cm^2 (i.e. about seven times lower than the maximum in the electron impact ionization cross section) at the threshold photon energy (i.e. 15.8 eV, corresponding to about 800 Å) and decreases very rapidly at higher energies [35]. The cross section of photo-excitation is comparable to that of photo-ionization [35]. In ref [27] it is, however, demonstrated that in a GD, photo-ionization and -excitation are clearly of minor importance compared to electron (and fast ion and atom) impact ionization and excitation.

4.2.5.3 Ionization and excitation of sputtered (analyte) atoms

In principle, the same processes that cause the ionization and excitation of argon atoms apply also to the ionization and excitation of analyte atoms. However, very little is known of the above mentioned processes for the analyte atoms. Only electron impact ionization data (e.g. [36]) and some very limited electron impact excitation data (e.g. [37]) are available in the literature, and only for some specific sputtered elements. In addition to the above mentioned processes, two other collision types seem to be of special importance for the ionization (and possibly simultaneous excitation) of analyte atoms, i.e. Penning ionization by Ar metastable atoms and asymmetric charge transfer with Ar ions. Because these two processes are of higher importance than electron impact ionization (i.e. the dominant ionization mechanism for argon) and because they are absent for argon, the sputtered atoms are more efficiently ionized in the GD than the argon atoms, as was already discussed above (cf. ionization degree).

Electron impact ionization and excitation. The mechanisms are the same as for the corresponding processes of argon atoms. The cross section curve of electron impact ionization as a function of the electron energy is comparable for all elements [36]. Therefore, this process can be considered as rather unselective. The electron impact ionization cross section for copper is presented in Fig. 4.13 (curve 4). The cross sections of electron impact excitation of the analyte atoms exhibit similar curves.

Penning ionization

$$M^0 + Ar^*_m \rightarrow M^+ + Ar^0 + e^-$$

If an argon atom in a 4s metastable level collides with an analyte atom, the energy of the metastable level (i.e. 11.55 or 11.72 eV) can be used to ionize the analyte atom if the ionization potential of the latter is lower than the metastable energy. Since most of the atoms of the periodic table have an ionization potential lower than this value, Penning ionization is also more or less unselective. Cross sections of this process are rather difficult to find in the literature for all elements. However, some empirical formulas are available, expressing a relation between the cross section and the mass and radius or polarizability of the colliding particles [38, 39]. Generally, the Penning ionization cross sections between argon metastable atoms and analyte atoms are in the order of $5 \times 10^{-15}\,\mathrm{cm}^2$ [38, 40]. This process is suggested to be the dominant ionization mechanism of sputtered atoms, certainly in low pressure GD (e.g. [41–43]). The only elements which cannot be ionized by Penning ionization in argon are H, N, O, F, Cl, Br.

Asymmetric charge transfer

$$M^0 + Ar^+ \rightarrow (M^+)^* + Ar^0$$

The collision between an analyte atom and an argon ion can lead to the transfer of an electron from the atom to the ion if the energy difference between the argon ion ground state or metastable level and the energy levels of the resulting analyte ion is sufficiently small; the efficiency of this process generally decreases with growing energy difference between the levels. Asymmetric charge transfer is therefore a rather selective process, unlike Penning ionization which occurs unselectively for all elements having an ionization potential below the argon metastable energy levels, independently of the relative position of the levels (see above).

Cross section data of asymmetric charge transfers are very difficult to find in the literature, at least for the typical (relatively low) ion energies in a GD, and between argon ions and typical analyte elements (for a discussion, see [44]). Hence, the relative importance of this process in the GD is still a bit controversial. Steers and co-workers, as well as Wagatsuma and Hirokawa, have clearly demonstrated the occurrence of asymmetric charge transfer in GD for Ar^+/Cu, Ne^+/Cu, Ne^+/Al, Ar^+/Fe, Ar^+/Ti and Ne^+/Fe (see e.g. [45–48]). The process has also been shown to be important in hollow cathode discharges for Ne^+/Cu, Ar^+/Fe, Ne^+/Fe, Ar^+/Ti, He^+/Cu and Ar^+/Cu (e.g. [49–54]). In the early investigations of Coburn and Kay [41], Penning ionization was considered to be the most important ionization process and charge transfer was neglected. The authors, however, considered only charge transfer in which the ground state of the analyte ions is formed; and since the energy difference is then far too large, they ruled out that possibility. However, as was demonstrated by Steers and co-workers [45–47], the resulting analyte ion can also be formed in an excited state so that the energy difference is much smaller, and asymmetric charge transfer becomes, therefore, more probable. In refs [55] and [56], cross section data are available for He^+/Cd and He^+/Zn respectively. In these cases, a good energy overlap is found and, moreover, the Penning ionization and asymmetric charge transfer cross sections are found to be of comparable magnitude. From the argon ion number densities typically calculated for analytical GD (see Fig. 4.7 and Table 4.1), we expect that asymmetric charge transfer is a quite significant ionization mechanism for specific elements.

The energy levels of argon ions and of ions for three elements (Fe, Cu and Ag) are shown schematically in Fig. 4.14, to illustrate for which elements asymmetric charge transfer with argon is possible [44]. It is clear that Fe II possesses many levels suitable for asymmetric charge transfer; Cu II has no levels lying close to the Ar II ground state and only one level showing close overlap with the Ar II metastable state; Ag II has no levels at all that could account for asymmetric charge transfer with argon.

4.2.5.4 Positive ion–electron recombination

Electron-ion recombination is the reverse process of ionization, i.e. an electron coalesces with a positive ion to form a neutral atom. From the conservation laws of momentum and energy, it follows that a simple two-body coalescence is not allowed [35]. However, some alternative recombination processes can occur. To our knowledge, the recombination processes apply to both argon and analyte ions, and they will therefore be discussed simultaneously (A^+ is an arbitrary ion).

Three-body recombination

$$A^+ + e^- + B \rightarrow A^0 + B$$

A third body takes part in the collision process, taking away the excess energy and allowing it to satisfy the conservation laws. The third body B can be every particle present in the plasma, or it can be the cell wall. If the third body is a heavy particle, the electron can lose only a small fraction (i.e. $2\,m_e/M_B$) of its energy to the third body. Therefore, this reaction is very slow [57]. Massey and Burhop [58] estimated that the coefficient of this three-body recombination process is about $10^{-11} \times p\ \mathrm{cm^3\,s^{-1}}$, where p is the gas pressure in Torr. Therefore, for a gas pressure of about 1 Torr, and for electron and argon ion densities in the order of $10^{12}\ \mathrm{cm^{-3}}$ (see Table 4.1), the rate of this recombination process is of the order of $10^{13}\ \mathrm{cm^{-3}\,s^{-1}}$. When comparing this with typical ionization rates in the order of 10^{16}–$10^{17}\ \mathrm{cm^{-3}\,s^{-1}}$ (see e.g. [7–9]), it can be concluded that this kind of recombination is insignificant for the typical analytical GD conditions. However, when the third body is an electron, the three-body recombination process is clearly more efficient, since the electrons can take away the excess energy. Rate constants for this process are reported in the order of $10^{-24}\ \mathrm{cm^6\,s^{-1}}$ for typical electron temperatures expected for analytical GD [59]. For electron and argon ion densities in the order of $10^{12}\ \mathrm{cm^{-3}}$, the rate of this recombination is about $10^{12}\ \mathrm{cm^{-3}\,s^{-1}}$.

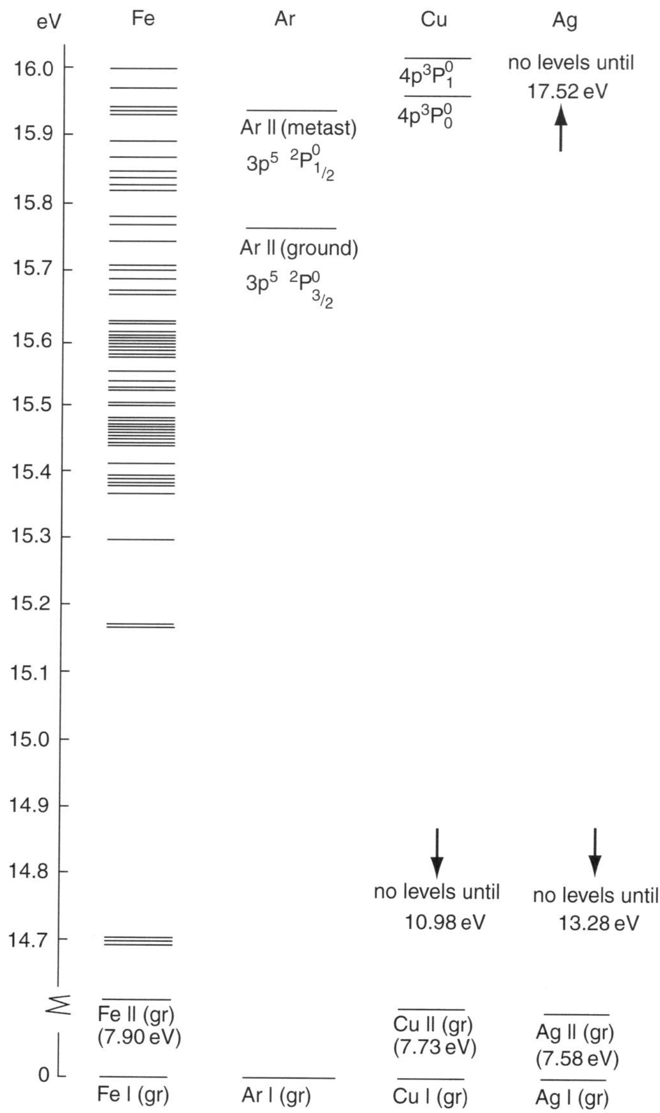

Fig. 4.14 Schematic representation of the energy levels of the element ions, that could account for asymmetric charge transfer with argon ions, for three elements: Fe, Cu and Ag. The zero level is taken at the ground state of the atoms. The Ar^+ ion ground state and metastable level are situated at 15.76 eV and 15.94 eV, respectively. Only the region of interest is shown here, i.e. 14.7–16.0 eV, in the assumption that asymmetric charge transfer occurs only for levels lying from 1 eV below till 0.02 eV above these Ar^+ ion levels. Reprinted from ref. [53] with permission of the Royal Society of Chemistry.

Hence, this process is also negligible compared to ionization in the analytical GD.

Radiative recombination

$$A^+ + e^- \rightarrow A^* + h\nu$$

The excess energy is carried away by a photon. The rate constant of this process is about 10^{-11} cm^3 s^{-1} in the case of argon [60], yielding a recombination rate of about 10^{13} cm^{-3} s^{-1} at typical electron and argon ion densities, which is again negligible compared to the ionization rate.

Dissociative recombination

$$AB^+ + e^- \rightarrow (AB)^* \rightarrow A^* + B$$

When the ion is molecular, a two-body recombination process is possible since the collision product can dissociate and the recombination energy converted into kinetic and potential energy of the dissociation products. When AB^+ is equal to $Ar_2{}^+$, the rate constant is in the range of 10^{-7}–10^{-6} cm^3 s^{-1} [60]. Hence, the process is clearly more efficient than radiative recombination. Nevertheless, this process will not be of major importance in the GD since $Ar_2{}^+$ ions are not considered to be dominant species in the GD plasma (see above).

Two-stage recombination

$$A^0 + e^- \rightarrow A^-$$

$$A^- + A^+ \rightarrow A^0 + A^0$$

The electron attaches to a neutral atom to form a negative ion. The negative ion collides with a positive ion, an electron is transferred, and two neutral atoms are formed. The probability of forming a negative ion (step 1) depends on the electronegativity of the atom. For argon, this probability is very low, so that this kind of recombination is also negligible in an argon discharge.

In general, it can be concluded that electron–ion recombination is significant only at high electron and ion densities and will be of minor importance in the analytical argon GD discussed here.

4.2.5.5 De-excitation

De-excitation is the inverse of excitation. Indeed, except for the metastable levels, the excited levels of the atoms and ions are only short-lived, and the electron configuration soon returns to the ground state in one or several transitions. Each transition is accompanied by the emission of a photon with specific energy. If such a photon has an energy between 1.7 and 3.0 eV (corresponding to 720 and 410 nm respectively), it is detected by the human eye. Hence, the de-excitation processes produce a glow and are therefore responsible for the characteristic name of the *glow* discharge. As mentioned above, if the lower levels have a high population density (as is the case for the ground state atoms), a substantial fraction of the emitted radiation will be reabsorbed by the lower level, leading to *radiation trapping*.

4.2.6 Processes occurring at the walls

When a particle collides with the walls of the GD cell, different phenomena may occur, depending on the kind of particle. An electron may be reflected, absorbed or cause the emission of secondary electrons. An ion or atom may be reflected (possibly in another form), implanted (probably with structural rearrangements), cause secondary electron emission or eject one of the wall atoms (called sputtering). The latter two processes are of special importance for the analytical GD, and will therefore be discussed in more detail. Secondary electron emission can occur at all walls, whereas sputtering is more or less restricted to the cathode, since high particle bombarding energies are required for this process.

4.2.6.1 Secondary electron emission

When a particle strikes a surface, an electron can be emitted. This process is necessary to maintain the GD, i.e. new electrons can be supplied to compensate for the electron losses at the walls. Secondary electron emission can be caused by the bombardment of electrons, ions, neutrals and photons. The number of electrons emitted per incident particle is called the secondary electron emission coefficient. It depends on the kind of bombarding particles and their energy, and on the kind of wall material.

By electron bombardment. This process is only important at the anode walls, and is negligible at the cathode since the strong electric field in front of it prevents the electrons from bombarding the cathode. When electrons strike a surface, three types of electrons are emitted [35], (i) the elastically reflected primaries, with energies equal to the energy of the incident electron, (ii) inelastically reflected primaries, with energies lower than the incident electron energy, and (iii) true secondary electrons, with energies of a few eV. The latter group is often the major one [35]. The secondary electron emission coefficient by electron bombardment, δ, is in the order of 1, but depends on the electron energy and the kind of surface material. It typically shows a maximum at electron energies of about 600–800 eV [35].

By atom and positive ion bombardment. This process can occur at both cathode and anode walls. For clean surfaces, the secondary electron emission coefficient, γ, is nearly independent of the ion or atom kinetic energy at energies below about 1000 eV. It is typically in the order of 0.1 for positive argon ions and nearly zero for neutral ground state argon atoms. At higher energies, γ starts to increase with the ion or atom kinetic energy [35]. This suggests that secondary electron emission results from a constant potential energy component (ca. 0.1 for the argon ions and almost zero for the argon atoms), and a kinetic energy component (similar for both) which plays a role only at energies below about 1000 eV. The mechanism of *potential energy ejection* is explained by Auger neutralization. The potential energy component of the ions is of the order of the ionization potential, which seems amply sufficient for secondary electron emission. Neutral ground state atoms do not possess such a potential energy component, which explains why γ is negligible at energies below about 1000 eV. However, metastable atoms do possess a potential energy (i.e. the energy of the excited metastable level), and hence will be able to give rise to secondary electron emission. Oechsner measured secondary electron emission coefficients for Ar^+ ions of 1 keV on 11 different polycrystalline surfaces [61]. He proposed a theoretical formula, which showed good agreement with the experimentally obtained values:

$$\gamma = \frac{0.2}{\epsilon_F}(0.8\,E_i - 2\Phi)$$

where ϵ_F is the Fermi energy measured from the bottom of the valence band, E_i is the effective potential energy of the incoming ion, and Φ is the work function of the metal. If this formula is applicable to species other than Ar^+ ions, the secondary electron emission coefficient of argon metastable atoms (E_i = 11.55 or 11.72 eV) on metal surfaces can be calculated and typical values of about 0.005–0.05 would be obtained.

It should be noticed that the value of the secondary electron emission coefficient by ion and atom bombardment is largely dependent on the surface conditions (crystal phase, contaminations). Indeed, it appears that for contaminated (or *gas-covered*) surfaces the secondary electron emission coefficients of both argon ions and atoms can vary over many orders of magnitude as a function of the bombarding energies [29]. Moreover, the values characteristic for pure metals can differ significantly from the ones of alloys or nonconducting materials [35].

By photon bombardment. The ejection of electrons due to photon bombardment is called photo-emission. The photoelectric yield γ_p for most pure metals is only 10^{-4}–10^{-3} electrons per photon in the visible and near UV wavelength range, since the photons are usually more efficiently reflected. It increases however toward shorter wavelengths (i.e. of the order of 0.1 at 100 nm [35]). The effects of photoelectric emission in GD are, however, not well understood.

Field emission. At extremely high electric fields (ca. 10^7 V/cm), electrons can be emitted by the mere action of the electric field. However, since such high electric fields do not occur in analytical GD, this effect can be considered negligible.

4.2.6.2 Sputtering

When energetic particles (i.e. gas ions, gas atoms and also ions of the cathode material) bombard the cathode surface, they can penetrate the surface and set up a series of collisions between atoms of the cathode material, until they have lost their energy. In this three-dimensional collision cascade, which lasts for a very short period of time only (ca. 10^{-12} s [62]), atoms lying at the surface can obtain some energy larger than their surface binding energy so that they can escape from the surface. This is called sputtering. It is believed that the majority of the sputtered particles are neutral atoms. Ions can also be sputtered, but positive ions will immediately return to the cathode, by the strong electric field in front of it. The energy of the sputtered atoms is in the order of a few eV [63].

An important quantity with respect to sputtering is the sputtering yield, Y. It is defined as the number of sputtered atoms per incident particle. Analytical expressions or estimates for sputtering yield data have been developed by many authors from theories [64], empirical relations [65, 66] and computer simulations [67] of sputtering processes. Sigmund has developed a comprehensive theory describing the effects of different ion–target combinations and energies, based on the assumption that sputtering is closely related to other ion-bombardment phenomena, and solving the linearized Boltzmann equation [64]. The calculation consists of four different steps: (i) the determination of the amount of energy deposited by energetic particles near the surface, (ii) the conversion of this energy into a number of low energy recoil atoms, (iii) the determination of the number of these atoms that can reach the surface, and (iv) the selection of those atoms that have sufficient energy to overcome the surface binding forces. However, systematic derivations from the original Sigmund formula have been pointed out for some cases, such as light-ion and low-energy sputtering [65]. A few modifications to this original formula have been proposed, but some discrepancies still remain. On the other hand, a number of empirical formulae have been developed [65, 66], in which the discrepancies of the Sigmund formula have been solved, and which can hence be applied to all ion–target combinations [66]. One such formula, which is typically used in modelling calculations, is the so-called second Matsunami formula for normal incidence [65, 66]:

$$Y(E) = 0.42\frac{\alpha QK\sigma_n(\epsilon)}{U_s[1 + 0.35U_s\sigma_e(\epsilon)]}\left[1 - \sqrt{\frac{E_{th}}{E}}\right]^{2.8}$$

where $Y(E)$ is the sputtering yield for incident ions (and atoms) of energy E; U_S is the sublimation energy of the cathode material (in eV); Q is an empirical parameter, dependent on the cathode material; $\sigma_e(\epsilon)$ and $\sigma_n(\epsilon)$ are Lindhard's elastic and inelastic reduced stopping cross sections.

$$\sigma_e(\epsilon) = 0.079\frac{(M_1 + M_2)^{3/2}}{M_1^{3/2}M_2^{1/2}}\frac{Z_1^{2/3}Z_2^{1/2}}{\left(Z_1^{2/3}+Z_2^{2/3}\right)^{3/4}}\sqrt{\epsilon}$$

$$\sigma_n(\epsilon) = \frac{3.441\sqrt{\epsilon}\ln(\epsilon + 2.718)}{1 + 6.355\sqrt{\epsilon}+\epsilon\left(-1.708 + 6.882\sqrt{\epsilon}\right)}$$

ϵ is the reduced energy:

$$\epsilon = \frac{0.03255}{Z_1Z_2\left(Z_1^{2/3}+Z_2^{2/3}\right)^{1/2}}\frac{M_2}{M_1 + M_2}E$$

where Z_1 and Z_2 are the atomic numbers of the bombarding and sputtered particles, respectively, and M_1 and M_2 are their mass numbers. K is the conversion factor from $\sigma_n(\epsilon)$ to $S_n(E)$ (reduced energy to energy in eV):

$$K = \frac{S_n(E)}{\sigma_n(\epsilon)} = 8.478\frac{Z_1Z_2}{\left(Z_1^{2/3}+Z_2^{2/3}\right)^{1/2}}\frac{M_1}{M_1 + M_2}$$

α is an empirical parameter, function of M_1/M_2:

$$\alpha = 0.08 + 0.164\left(\frac{M_2}{M_1}\right)^{0.4} + 0.0145\left(\frac{M_2}{M_1}\right)^{1.29}$$

E_{th} is an empirical parameter, function of M_1/M_2 and U_S. It represents the threshold energy required for sputtering:

$$E_{th} = U_s\left[1.9 + 3.8\left(\frac{M_1}{M_2}\right) + 0.134\left(\frac{M_2}{M_1}\right)^{1.24}\right]$$

This empirical formula is in principle valid for all combinations of bombarding ions/atoms and cathode materials [66].

From this formula it can be concluded that the sputtering yield is a complex function of the incident energy and the masses and atomic numbers of the bombarding particles and surface target. In general, the sputtering yield increases less than linearly with the mass of the incident particles. Hence, the sputtering yield does not reach a maximum when the masses of incident particles and surface target are close to each other (e.g. [66]), which has sometimes been suggested from simplified sputtering yield expressions.

As far as the energy of incident particles is concerned, a threshold energy is required to give sputtering. Indeed, the atoms at the cathode surface must obtain sufficient energy to overcome their surface binding energy. A good measure of the surface binding energy is the heat of sublimation. It is suggested that the minimum energy for sputtering must be about four times the heat of sublimation of the cathode material [62]. Above this minimum energy, the sputtering yield increases with the energy of the bombarding particles. It reaches a broad maximum at energies in the order of several keV, after which it decreases again, as ion implantation becomes important [35].

The kind of cathode material is also a determining factor for sputtering. In general, the sputtering yield increases with the atomic number of the cathode material, within each row of the periodic table [68]. Copper, silver and gold have the highest sputtering yield among transition metals. This trend is explained as follows [69]: if the penetration depth in the surface increases, a larger fraction of the energy is expended in collisions that do not lead to sputtering. The increase in sputtering yield is associated with the filling of the electron shells, especially the d shells. As the shells fill, the target looks more opaque, and the incident particles are not able to penetrate very deeply and hence give rise to more sputtering. It should, however, be noticed that the sputtering yields of the different elements seldom differ by more than one order of magnitude among each other, as opposed to the rates of evaporation which differ by several orders of magnitude [62]. Owing to these rather uniform sputtering yields, the GD is relatively free from matrix effects.

Besides the kind of cathode material, the cathode surface also has influence on the sputtering yield, i.e. the sputtering yield decreases by surface contaminations, or by the formation of oxide layers or adsorbed gas layers on the cathode surface. Moreover, the sputtering can result in the formation of a regular pattern of submicroscopic cones on the surface. These are due to some components in the cathode material with a lower sputtering yield. These components can migrate over the cathode surface and form small clusters, thereby protecting the underlying material from sputtering. This effect also results in a decrease of the sputtering yield.

Finally, the Matsunami formula given above applies to normal incidence. The angle of the incident particles is, however, also an important factor for sputtering. Oechsner has found that for rare gas ions bombarding a polycrystalline copper target with energies of 0.5–2 keV, the sputtering yield reaches a maximum at incident angles of about 60–80° relative to the surface normal [70]. Indeed, at low incident angles, the sputtering yield increases with rising angle due to the increased probability of the collision cascade to propagate back to the cathode surface and hence to result in sputtering. At incident angles higher than 60–80°, the sputtering yield decreases, since the incoming particles are more likely to reflect off the surface without any penetration or momentum transfer, so that sputtering becomes unlikely. It should, however, be mentioned that these sputter angle studies were performed with ion guns in a vacuum setting where perfect control of the sputtering angles is possible. This is, however, not the situation in a GD. It is expected that the majority of the ions from the GD plasma bombard the cathode with normal incidence, because they are directed by the electric field in front of the cathode. The argon atoms, on the other hand, which play a dominant role in the sputtering process (see below), are subject to a large number of scattering collisions and can in principle bombard the cathode from all directions.

Figure 4.15 shows the relative contributions of fast argon ions, fast argon atoms and cathode (copper) ions to the sputtering process, calculated for a range of different pressures and voltages. It appears that the fast argon atoms (solid lines, full circles) play a dominant role in sputtering, at least in the low pressure range or at low voltages. The reason is that a large number of fast argon atoms

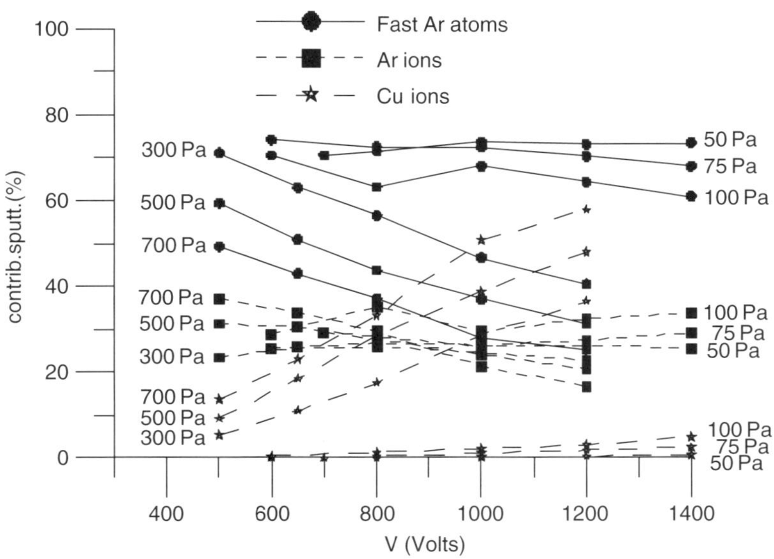

Fig. 4.15 Relative contributions to sputtering of the fast argon atoms (solid lines, full circles), Ar^+ ions (small dashed lines, full rectangles) and Cu^+ ions (wide dashed lines, stars), as a function of voltage at different pressures, as obtained from modelling calculations [14], [23], for typical VG 9000 and Grimm-type conditions.

are created near the cathode, due to collisions of argon ions with argon gas atoms. Indeed, the flux of fast argon atoms bombarding the cathode was calculated to be typically two orders of magnitude larger than the flux of fast argon ions (e.g. [7]). The contribution of fast argon atoms is calculated in the order of 70% at low pressure and voltage, but this value drops to roughly 25% at the highest voltage and pressure investigated, due to the increasing role of sputtered copper ions (see below). The contribution of argon ions to sputtering (small dashed lines, full rectangles) is calculated to be in the order of 20–30% at all pressures and voltages under study. Finally, as just mentioned above, the copper ions also play a non-negligible role in sputtering, with a relative contribution of less than 1% at the lowest voltage and pressure investigated, and increasing up to about 60% at the highest voltage and pressure under study (wide dashed lines, star symbols). Indeed, although the copper ions have a lower flux when bombarding the cathode (cfr. the densities), they are characterized by much higher energies, and because the sputtering efficiency increases with rising energy (see above), the role of copper ions (i.e. so-called *self-sputtering*) is not negligible, and appears to become dominant at high voltage and pressure.

4.3 Glow discharge optical emission spectroscopy

4.3.1 Introduction

Glow discharge optical emission spectroscopy (GD-OES) today is a well-tried analytical technique for the multi-element determination of concentrations in bulk and surface of flat solid samples. The range of concentration goes from μg/g up to main components. In the past, conducting materials were very rapidly analyzed, typically up to a depth of many micrometers. The quantification procedure for these direct current GD-OES (DC-GD-OES) depth profiles is in-built in the commercial software of all modern GD-OES instruments, and standardization for Zn-based coatings on steel has commenced [71].

The application of Rf discharge to GD-OES extended the field of application to non-conducting materials. Efforts are going on to include Rf mode also in the standardization but at the present stage of development, Rf-GD-OES turned out to be too complex to be included from the beginning.

Much research effort has gone into applying GD-OES for liquid and gaseous samples, e.g. for environmental analysis [1]. Pulsing in the ms range was studied in d.c. and Rf mode [72] and was mainly used to reduce the thermal stress of the sample at sputtering. The application of μs pulsed GD, previously realized only in the d.c. mode, led to a number of new features. The strong increase of emission line intensities, e.g. is known for a long time from hollow cathode lamps [73].

Using the fast data acquisition technique now available for the light measured by the photo multipliers (PMTs) in modern instrumentation, it has become clear, that today GD-OES is also very suitable for the analysis of very thin layers in the nanometer range. Fast stability of the high voltage of the Rf-GD-OES system is needed for the analysis of such very thin non-conducting layer systems.

4.3.2 Source and instrumental development

Even if GD have been well-known for many decades, the breakthrough in analytical chemistry was achieved by the invention of a commercially suitable source for flat samples by Grimm [74]. This source is characterized by a tube-like anode and a distance between the front side of the tube and the flat sample at a cathode potential of about 0.2 mm. A special feature is that this source is evacuated between the sample and the outside of the anode tube. Only by this is the discharge confined inside the tube at high pressures and power densities. Other source types are used for special applications and investigations and are summarized in [4].

There are three main producers of GD-OES instruments in the market at the moment: Jobin Yvon (France), LECO (USA) and Spectruma (Germany). The d.c. sources of LECO and Spectruma are, due to historical reasons, very similar and are offered together with an optional Rf source. Jobin Yvon offer only Rf sources and systems. All Rf sources are also based on the Grimm design, but the three companies use different approaches for the coupling of the Rf voltage. At Jobin Yvon the hot electrode is positioned at the backside of the sample, LECO couples the Rf voltage direct from the front side and Spectruma inserts an insulator between the hot electrode and the sample. The Rf power supply of all producers is very similar and based on 13.56 MHz generators with matching systems to adapt the low output impedance of the generators with the high impedance of the plasma.

The emitted light of the discharge is focused either on the entrance slit or on the grating of the spectrometer. For a long time, only evacuated polychromators with photomultiplier detection were used in the spectral range between 110 and 500 nm. Sometimes one additional channel was provided by a monochromator.

Very recently charged coupled devices (CCD) technology was introduced by Spectruma and LECO in GD-OES instrumentation; this allows a lot of flexibility in GD-OES analysis and also makes it less expensive. This technique covers a wide spectral range and gives information comparable to that given by the old photographic plates, e.g. about the existence of molecular band emission and other spectral background. With respect to speed, spectral resolution and sensitivity this technique still cannot compete with the normally used polychromators and detection by PMTs, but a fast development of CCD technology is guaranteed.

4.3.3 Quantitative depth profile analysis

The quantification of depth profiles is based on the concept of constant emission yields, which means that the light emitted by a certain mass of an element sputtered per time interval only depends on the electrical discharge parameters (in the d.c. mode: voltage and current). The influence of pressure and matrix is assumed to be negligible, which was found to be true in most known applications. Some exceptions are discussed later.

The emission yields are determined by analyzing standard material with known concentrations and measured sputtering rates (sputtered mass per time interval). If the electrical parameters for both calibration and measurement are the same (e.g. by adjusting the pressure) the eroded mass of an element per time follows directly from the measured intensities. After calculation of the sputtering rate by normalization of all concentrations to 1, the concentration of an element can be calculated from the eroded mass. Finally the actual depth is calculated with the help of a density model. If it is impossible to maintain constant electrical discharge conditions (e.g. in the Rf mode) the emission yield must be determined taking into account the electrical parameters and a corresponding model must be used [75, 76].

Because of the high amount of steel produced and used all over the world, the Zn coating process to reduce corrosion is of high economical interest and shall serve as an example here. The coatings are produced by different process types, like hot dipping, electroplating and even by polymer coatings. Because these layers mostly are deposited in the μm range, GD-OES is an important and unique technique for their production control.

The change of the matrix from Zn to Fe requires the application of a multimatrix calibration. Because these Zn layers have a thickness of some μm, constant sputtering conditions can be ensured by regulation of the pressure to constant voltage and current. In the quantification model used, therefore, no corrections for changing plasma parameters must be done, because the influence of the pressure on the emission yields is considered to be negligible.

Special effort was used by the company LECO to produce a novel zinc standard material for this application [77]. Figure 4.16 shows the calibration curves of Mn and Cr in zinc (CZ 2009–14) and iron (NIST 1761–67) at 700 V and 20 mA using a 4 mm source, measured with a GDA 750 from Spectruma.

These curves demonstrate the principle of multi-matrix calibration. The essential prerequisite for this procedure is the determination and inclusion of the sputtering rate of the standards into the quantification. After this correction the emission yield (per mass and time emitted light) is expected to be matrix independent. In most cases this model works very well [78], but in the case of Zn standards, Weiss found that there are additional effects existing, which should be found and included in the models.

The application of this kind of calibration curves in the commercial software allows the GD-OES user to transform an intensity time curve (see Fig. 4.17)

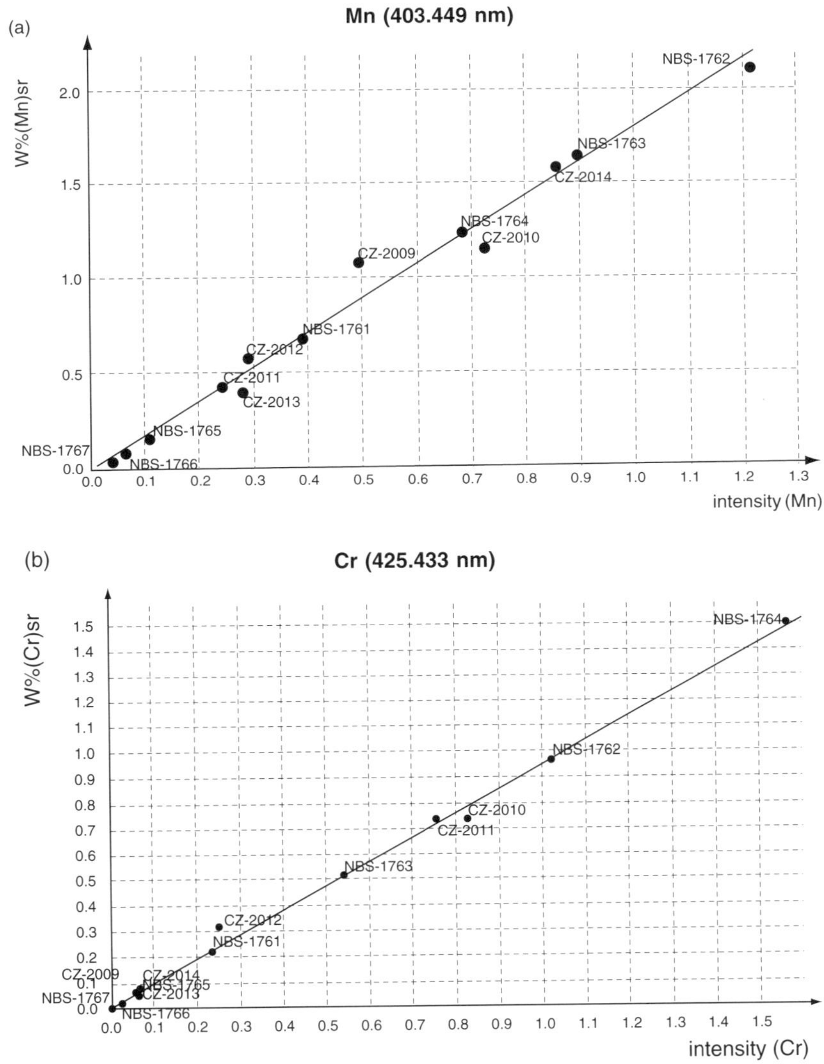

Fig. 4.16 (a) Calibration curve of Mn in zinc and iron at 700 V and 20 mA using a 4 mm source; (b) calibration curve of Cr in zinc and iron at 700 V and 20 mA using a 4 mm source.

easily into a quantified concentration depth profile (see Fig. 4.18). In the intensity time profile some additional information about the discharge and the sample is included as in the quantified one. In this figure, for example, the variation of pressure from 2.4 to 2.8 hPa is obvious and caused by the change of the matrix over about 100 s. By it the discharge parameters, voltage and current, are maintained constant. This qualitative picture includes also the elements

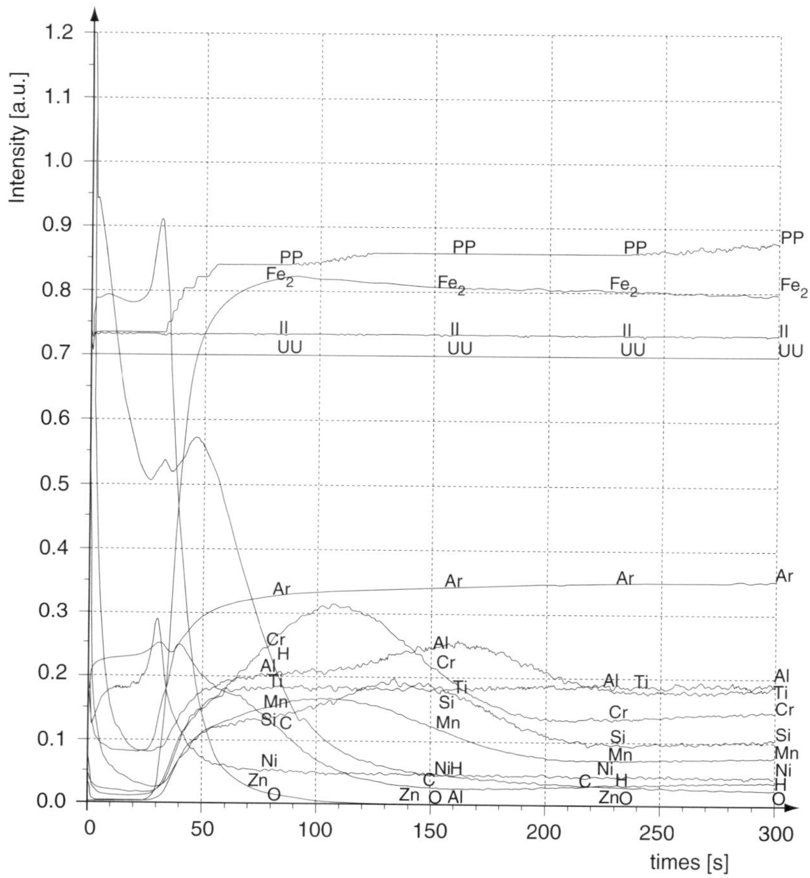

Fig. 4.17 GD-OES intensity time profile of a μm thick electroplated Zn layer on steel, measured at constant voltage (700 V) and current (20 mA), 4 mm anode diameter.

Ar and H, which are not quantified but deliver additional information for the interpretation of the results. The Ar signal increases in spite of the constant discharge conditions, because the pressure and therefore the Ar density are increased. In some cases it is also observed that the Ar line intensity changes, even at constant pressure and constant electrical discharge conditions. One reason for such changes are the different reflectivities of the matrices (see interferences in Section 4.3.5) and in this case it is also to be expected that the emission yields of the analytical lines depend on the matrix.

Other essential information is included in the line intensity of the light elements. In this case the intensity of the H and C lines proves that there is a high amount of C and H in the layer as well as in the interface. This effect

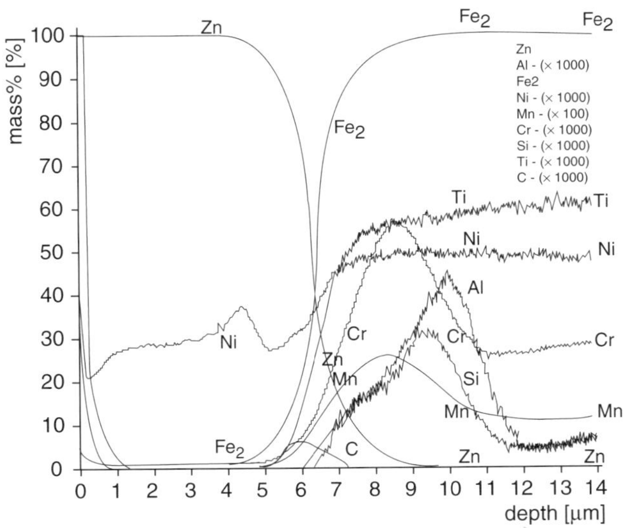

Fig. 4.18 Quantified concentration depth profile of Fig. 4.17.

is well known from the electroplating process and also from the steel industry. Because of the lack of suitable standard material, H cannot be included in the quantification and there are also more standard materials necessary for a reliable multi-matrix calibration of high concentrations of C, N and O. Nevertheless, the information about the presence of H is essential for the characterization of the material and very difficult to measure with other methods. Moreover the influence of H on the light emission of lines of other elements is well known and discussed in more detail in Section 4.3.4.

A strong emission of the N line mostly indicates a leakage between source and sample and a repetitive measurement is required.

Unfortunately, the exponential decrease of all light element line intensities at the beginning of the discharge caused by impurities of sample and source is difficult to distinguish from real impurities of light elements in the sample; this is much more significant in thin layer analysis (see Fig. 4.21). The quantified depth profile in Fig. 4.18 reveals some additional information. At first the Zn layer thickness is obtained and secondly the enrichments of Cr and Mn followed by those of Al and Si appear very clear.

4.3.4 Application of a CCD detector

LECO as well as Spectruma offer at present GD-OES systems with detection by CCDs [79]. The GDA 150 from Spectruma represents a sytem which uses exclusively a CCD for detection and has a time resolution of

0.1 s, a spectral range of 160–500 nm at a resolution of about 0.3 nm and detection limits of about 100 μg/g. The spectral range of the CCD system in the GDA 750, which uses the CCD in combination with the usual polychromator and PMTs, is usually 200–730 nm.

Pb/Sn alloys are widely used in many technical applications, e.g. in solders in electronic and microelectronic devices and as material for organ pipes. The analysis of the Pb and Sn concentration in binary alloys of these elements is an interesting application in GD-OES. Because of the existence of Pb and Sn as separate phases in the alloy, determination is difficult by methods with higher lateral resolution like Electro probe Micro-Analyzer (EPMA), and the determination of an average concentration of the area in the sputtered crater is an advantage of GD-OES. The concentration range of interest is between 10 and 90% Sn and therefore detection limits are of less importance. This is essential, because due to the low melting point of this alloy, only low sputtering conditions can be used. The selection of suitable analytical lines is possible and the accuracy is about 5%. Figure 4.19 shows a calibration curve of the CCD system in the GDA 750 by Spectruma using 10–90% Sn containing calibration samples. It must also be pointed out that in the spectral range above 450 nm, where the normal polychromators have no access, very strong and useful analytical lines, like H at 656.3 nm, exist.

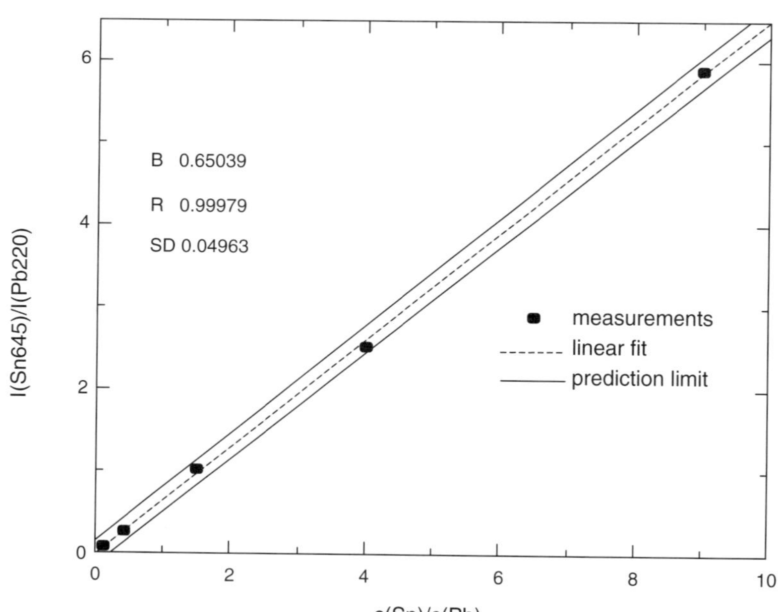

Fig. 4.19 Pb/Sn calibration graph, 700 V, 6 mA, 2.3 hPa, 2.5 mm diameter anode.

Molecular bands can be identified, which provide additional information about the species beside the element identification. They also give valuable hints about disturbances of spectrometer channels at fixed wavelengths in the polychromator systems. However, the real advantage of such a system becomes obvious if more than single line intensities are observed. This information can be seen in Fig. 4.20, where a contaminated surface of Cu was sputtered. In fact a finger print of Mr Arne Bengtson was analyzed.

The first spectrum when sputtering the contaminated surface (Fig. 4.20a) clearly shows the emission of the OH-band at 308.9 nm, which indicates moisture in the plasma. This area of the spectrum is free of band emission at the end of the analysis (Fig. 4.20b) when the surface is cleaned by the sputtering itself. The existence of H in the plasma means the disturbance of the lines by the molecular bands and the appearance of a continuous background with its maximum at about 300 nm. This effect was investigated in detail by Hodoroaba *et al.* [80–82] and could be explained by the decay of the H molecule by Penning excitation with Ar metastables. This continuous background disturbs all weak lines in a wide spectral range.

The strong atomic Cu resonance lines at 324 and 327 nm exist in both spectra and are only a little influenced by the contamination. In fact these lines increase little with higher H concentration in the plasma. In contrast, the ionic Cu lines in the range of 210–230 nm are nearly vanished at the beginning, where a high amount of H exists in the discharge. Similar differences between the atomic and ionic Ar lines but in the opposite direction were found when sputtering Cu. In this case the atomic lines decrease and the ionic lines increase with the H concentration.

These results mean of course that no simple correction of the H-effect can be expected. The models must include the intensity and shape of the continuous background as a function of the discharge parameters as well as the specific behaviour of the different lines. Models which include Ar to define the electrical parameters have to also consider the influence of H on these lines.

The situation to find suitable corrections for this effect is even more complicated, because no standard materials with high H concentration exist in the market. It was found that a mixture of the sputtering gas with molecular H can simulate the influence of H from the sample quite well, most probably because the equilibrium between atomic and molecular H in the plasma is quickly established [83]. Using this approach the dependence of the line intensities on the H concentration in the sputter gas was studied [84] and a model for its correction was published [85].

4.3.5 *Radio frequency GD and interference effect*

The application of Rf discharge to GD-OES extended the field of application to non-conducting materials. The rapid attainment of stable electrical parameters and the

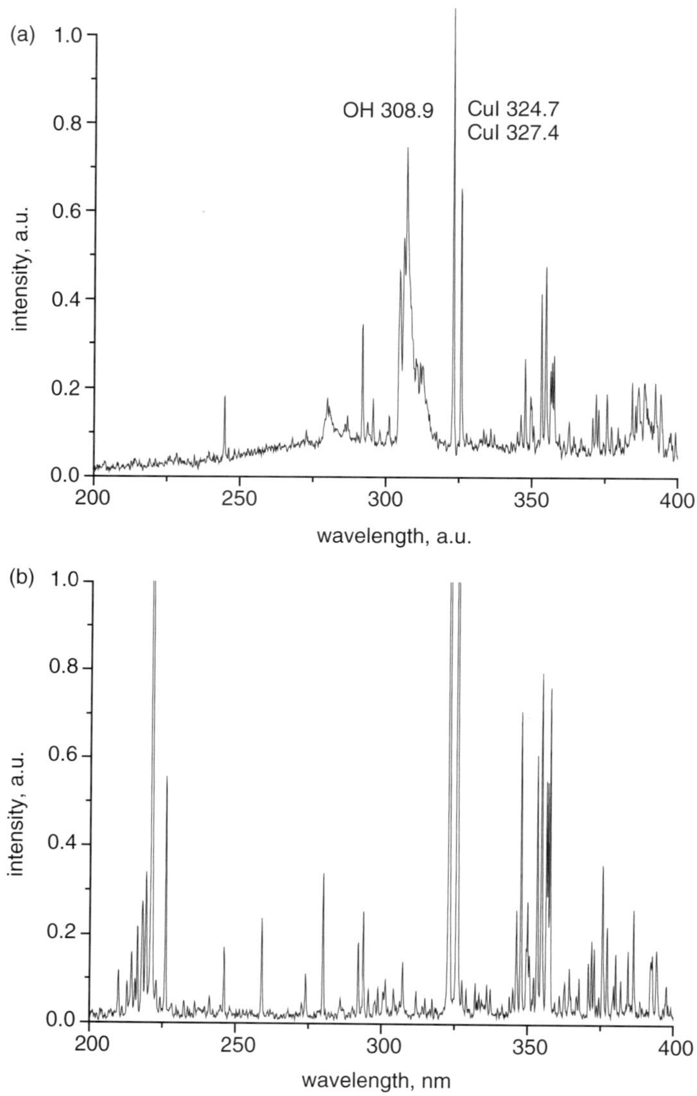

Fig. 4.20 Comparison of the spectra of the contaminated surface of a copper sample (a) and of pure copper (b) measured by a CCD spectrometer.

high voltage available in the Rf-GD-OES system recently developed by IFW and LECO [86] allow the analysis of very thin non-conducting layer systems as well as the analysis of thick non-conductors. Much progress was also made in commercially available matched Rf generators, where the power losses in the matching system, cables and connectors were reduced. However,

Rf-GD-OES still is mostly applied to thin layer analysis. Also, if the sample under investigation is expected to be conductive, often the Rf mode is used [87, 88]. Even if these investigations were done by Rf-only instruments, there are still principal advantages, e.g. avoiding sparks through very thin oxides at the surface. Another reason for the application of Rf-GD-OES is in the use of substrates for the layers, which also sometimes include non-conducting layers like SiO_2. Such examples can be seen in Fig. 4.21, where the Cu diffusion in modern microelectronic circuits through thin barrier layers was investigated [89].

The layer system in the original as deposited state is well resolved and a study of the influence of treatment was performed by GD-OES in combination with other techniques [90–92]. These kinds of depth profiles prove the excellent depth profile capabilities of GD-OES for very thin layers, e.g. the problem of mixing, existing in other techniques like secondary ion mass spectrometry (SIMS), atomic emission spectrometry (AES) and X-ray photoelectron spectroscopy (XPS) is less important in GD-OES.

It must be pointed out that this kind of investigation requires a very fast and stable technique. No mechanical matching in a matchbox or regulations of the pressure are allowed. Therefore, the pressure is kept constant in these analyses and quantification must take into account the different electrical discharge conditions in the different matrices. The intensity oscillations, which exist in the SiO_2 layer, were investigated in more detail and led to new conclusions [93, 94]. Due to the dependence of these oscillations on the wavelength, it became clear that they are caused by an interference effect at transparent layers on substrates with high reflectivity like silicon. The emitted light from the plasma in

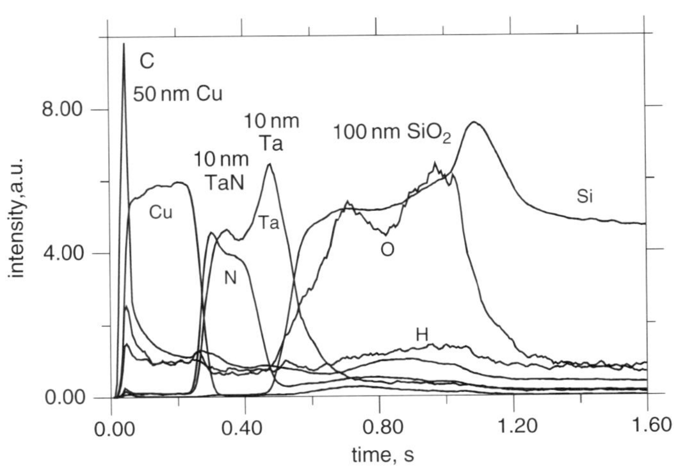

Fig. 4.21 Rf-GD-OES depth profile of a microelectronic multi-layer system, C 156.1, Cu 220.3, N 174.2, Ta 362.6, O 130.2, H 121.5, Si 288.1.

the GD is reflected at the surface of the transparent layer as well as at the interface between layer and substrate (see Fig. 4.22). The absorption of light in the layer itself restricts this effect mostly to wavelengths above 200 nm. Because the length of the light path through the layer reduces during the sputtering process, the intensity of the interfering waves changes (see Fig. 4.23). These oscillations complicate and disturb of course the quantification of the concentration but allow at the other side a very accurate determination of layer thickness, if the refractive index is known.

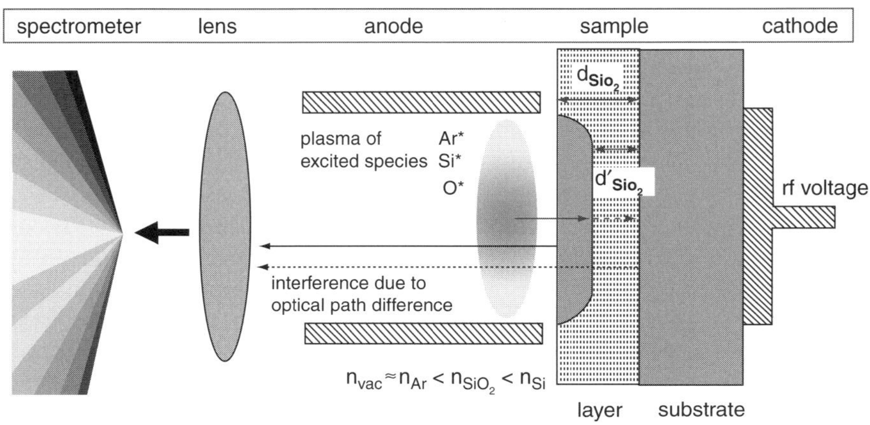

Fig. 4.22 Interference of the reflected light from the sample in the glow discharge source [94].

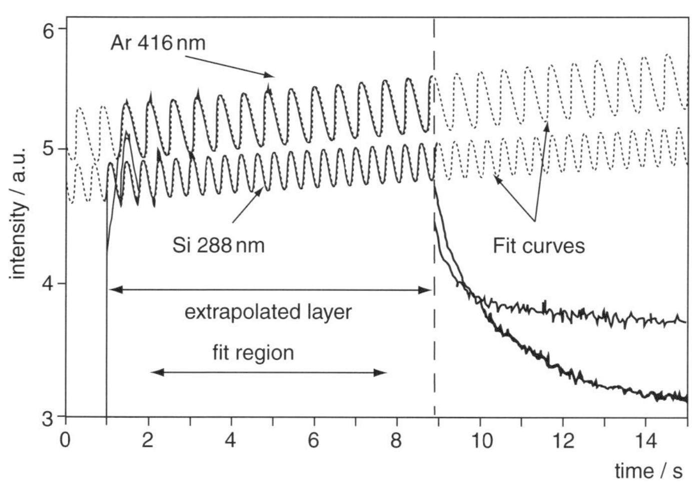

Fig. 4.23 Oscillations of Ar and Si line analyzing an about 1950 nm thick SiO_2 layer on Si with fitted curves source [94].

The understanding of this effect however made undoubtedly clear that sample reflectivity is able to influence the light intensity and thus the analytical result. The number of oscillations also shows that the amount of reflected light can reach higher amounts, in the worst case of a mirror up to 100%. Therefore, sample reflectivity must be counted as one of the factors which are able to cause matrix dependence.

Whereas the interference effect is mostly observed in Rf-GD-OES analyses, because transparent layers are mostly non-conductive, the effect of reflectivity is a general one.

4.3.6 Quantification approaches and measurement of electrical parameters

As already mentioned in Section 4.3.3, in the quantification of DC-GD-OES depth profiles for multi-matrix quantification of Rf-GD-OES depth profiles, it is also necessary to determine the electrical parameters of the discharge. Special hard- and software developments in this field were done and still are going on.

The hardware developments [95, 96] are focused on the direct determination of the consumed power in the plasma together with determination of impedance, which is matrix-dependent, if the pressure is kept constant. The measurement of these electrical parameters is possible in conducting samples and seems to be possible also in non-conductors (see Figs 4.24 and 4.25).

In Fig. 4.24 the dependence of voltage and current on time in an Rf discharge is presented. The superimposed capacitive current is subtracted and only the data of interest for the behaviour of the plasma is shown. These curves were compared with the result of model calculations and a principal agreement was found [15].

The so-called current/voltage characteristics are presented in Fig. 4.25 for three different voltages and a copper sample at the left side, three different pressures and a copper sample in the middle and a copper and

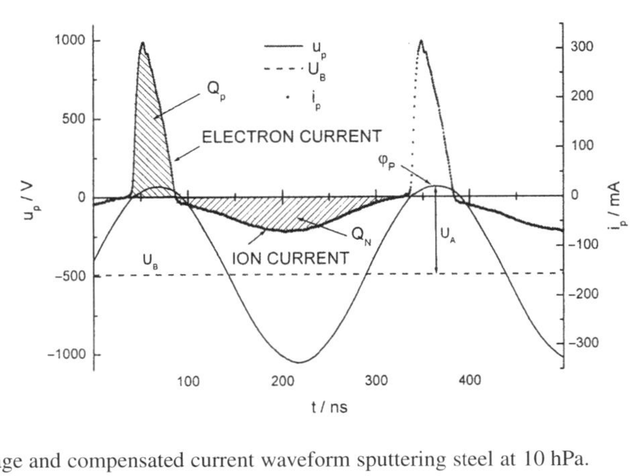

Fig. 4.24 Voltage and compensated current waveform sputtering steel at 10 hPa.

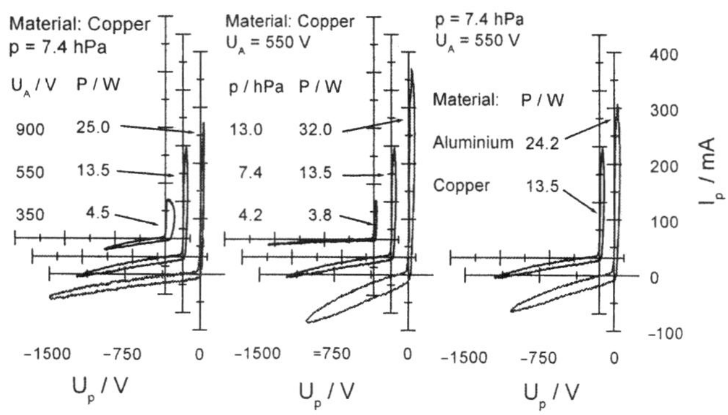

Fig. 4.25 I(U) characteristics of the GD dependent on the voltage (left), pressure (middle) and sample material (right).

aluminium sample at same voltage and pressure at the right side. It is clear that the slope of these curves in the negative voltage part only changes for the middle and right curves, which proves that the impedance changes for the ion current, when pressure or matrix is changed – well known from the d.c. mode. By the inclusion of electrical equivalent circuits into the evaluation, this approach can be used hopefully in future also universally for multi-matrix quantification of non-conducting samples.

However, further research and development is needed in this area and therefore other approaches were used up to now for the implementation in software and commercial instruments, e.g. the so-called subtraction method was used. It is based on the determination of the delivered power from the Rf generator, if no plasma is on (at no pressure in the source), and the subtraction of this amount from the power, when a plasma is burning. This approach delivers generally better values for the power, but has no access to the plasma impedance at all. Another approach done by Bengtson and co-workers [97, 98] takes instead the changing Ar emission intensity in dependence on the electrical parameters into account and calculates thereby the electrical parameters. Danger in this approach are the changing Ar emission intensities not caused by the electrical parameters, but, e.g., by varying reflectivity of the sample or hydrogen in the plasma. The experience gained from this approach however can be directly used when other techniques for the determination of the discharge parameters will be available.

4.4 Glow discharge mass spectrometry

In comparison with emission spectroscopy, mass spectrometry has several advantages. Nearly all elements of the periodic table can be determined simultaneously with their isotopic composition, spectra are much easier to interpret and detection limits at ng/g levels and below can be achieved with a dynamic range of more than 9 orders of magnitude. Although the advantages of GDMS are convincing, the analytical community has shown only a hesitant acceptance due to the fact that only one instrument at an elevated price level is commercially available.

4.4.1 Source and instrumental development

Many factors have had an important influence on ion source and instrumental development in GDMS, such as the strengths and weaknesses of spark source mass spectrometry (SSMS); the pioneering work of Coburn and Harrison; the success of sources in GD-OES; the success of *inductively coupled plasma mass spectrometry* (ICP-MS) in elemental mass spectrometry and the fact that only one instrument, the VG 9000 (Thermo Elemental), was commercially successful in the past two decades.

GD as a universal atomization and ionization source for mass spectrometry had been first used in the early days of mass spectrometry by the pioneers Aston, Thomson and Dempster [99–101]. Later on, due to the development of the electron impact ionization source and vacuum arc and spark discharges used as ionization sources in mass spectrometry, the analytical interest in these discharges was lost for a long period of time. For at least 25 years solid mass spectrometry was dominated by SSMS beginning in the early 1960s. This method was convincing by its extremely low detection limits at sub ng/g levels and its true multi-element capability [102]. Its disadvantage has been the insufficient precision in the range above 10% coming from a discontinuous operation of the sparks. For many applications the accuracy was not sufficient and the use of photo-plates for detection was very laborious and time-consuming.

Due to the main limitations, researchers started to look for continuous discharges, such as arcs and GD, and the revival of the latter as an ion source for mass spectrometry was initiated by Coburn using a planar diode sputtering system (the GD) coupled to a quadrupole mass filter [103]. Coburn and his co-workers later extended their work to Rf-discharges for the direct analysis of conducting and non-conducting solids too, and even Teflon had been analyzed in their first work [104].

A hollow cathode DC-GD ion source had first been coupled to an SSMS instrument by Harrison and Magee [105] and a couple of months later a second publication on the same topic followed from Colby and Evans [106]. Only one year later Harrison and co-workers started to explore Rf GD using again the SSMS instrument [107]. Glass samples as well as liquid residues were analyzed.

Only one year after their first GDMS publication, Harrison and co-workers demonstrated that d.c. sources could also be coupled to the less expensive quadrupole mass analyzers instead of the sector field devices used for SSMS [108]. The fundamental work of this group and the leading applications had stimulated not only the development of new sources but had also opened the door to use alternative mass analyzers.

The first instrument, the VG 9000 (Thermo Elemental, Winsford, UK), was launched in the market in 1985 and the first analytical results appeared in 1986 [109]. Many attempts have been made to launch additional GD instruments in the market. A second sector field device was announced by *Kratos*, and quadrupole-based instruments were launched by VG Micro Trace (Winsford, UK), Extrel (Pittsburgh, USA) and by Turner Scientific (Appleton, UK). Also Finnigan MAT (now Thermo Finnigan, Bremen, Germany) had offered an add-on GD source for their ICP-MS instrument (Element 1), but, with the exception of the VG 9000, all instrumental approaches disappeared leaving room for further instrumental developments.

Parallel to the early developments in GDMS, GD had gained considerable reputation as excitation sources in optical emission spectroscopy, as discussed in the previous chapter, for a variety of applications due to their stability, robustness and simple construction. This was the main reason for the development of ion sources using design criteria well accepted in AES among which the Grimm-type design was the most important.

The main advantage of the Grimm-type was that the flat sample itself served as the vacuum sealing part thus enabling fast sample changing and source cleaning. Due to effective cooling, higher powers could be used and due to a restricted discharge area, this source turned out to be superior in terms of stability in comparison to pin type sources. A source of this design was coupled to a quadrupole mass analyzer [12]. The analytical performance of the system was characterized in the bulk analysis of steel standard reference materials using operational parameters selected from extensive optimization measurements [110]. The analytical results showed for 30 elements, satisfactory agreement with their certified values. For multi-element analysis, the detection limits were about 0.1 μmol/mol while single element determination was possible with five times lower detection limits. The relative standard deviation (RSD) was in the range from 2 to 15%. A special and important feature of the Grimm-type design is that depending on the working conditions chosen, this source can also be applied for in-depth analysis [111].

Later the same Grimm-type source was also coupled to an ICP-MS instrument based on a double focusing sector field device (ICP-SFMS) with high mass resolution capabilities (prototype of the Element 1, Thermo Finnigan, Bremen, Germany) [13], so as to compete with the VG 9000. The high mass resolution, which could be changed continuously from $R = 400$ up to 8000 with the prototype instrument, was particularly helpful in elucidating and overcoming contributions from spectral interferences. Overall, detection limits at ng/g levels and below were achieved, but at least the dynamic range of only 6 orders of magnitude of a secondary electron multiplier (SEM) in counting mode was by far not sufficient for solid analysis. Some years later, the commercial version of this instrument (Element 1) was used again in combination with a home-built Rf-GD source by Becker and co-workers for the analysis of non-conducting samples [112]. They achieved detection limits in the low mass resolution mode of 10 ng/g for B in high-ohmic GaAs using both an analogue and a counting mode of the commercial instrument, but they had been limited by memory effects of deposited material in the lens system of the mass spectrometer [113].

In both examples the mass spectrometers used had been originally developed for ICP-MS with the idea of demonstrating that GDMS systems at a lower price scale can be successfully used to solve actual analytical problems. Other research groups have also coupled GD sources to their commercial ICP-MS instruments [114, 115] and a commercial version of such a hybrid ICP/GD-MS instrument was commercially available for a couple of years (TS Sola, Turner Scientific, Appleton, UK).

The early work of the Coburn and Harrison group has given a strong impetus to the use of Rf discharges in GDMS. In the following Rf GD sources had been coupled to various instruments for instance an SSMS instrument [116]: to the Element 1, to quadrupole instruments and to the VG 9000. Over the years, Rf sources had been strongly promoted by Marcus and co-workers, but this will be discussed in more detail in Section 4.4.6 [117].

The Grimm-type design is very well accepted now in mass spectrometry. A pulsed version of a modified Grimm-type source with a relatively high gas flow rate was recently coupled to a modified ICP-TOFMS by Yang *et al.* [118]. The source was operated with a pulse width of 20 μs, pulse voltages of up to 1600 V and pulse currents of up to 200 mA. Pulsing of a GD has also been described previously to enhance the sputtering rate, atomization, excitation and ionization by application of a short but high discharge power. Due to short pulse width, the thermal load to the sample is very low, so that even water cooling of the sample was not required. Full mass spectra were taken with increasing delay times of up to 200 μs and are shown in Fig. 4.26. The main result of this figure is that all argon-related ions showed a faster fall-off in comparison to the analyte ions, which could be used analytically to overcome limitations by spectral interferences.

In emission spectroscopy Grimm-type sources were developed that could be operated in a d.c. as well as in an Rf mode [119]. Such a source had been coupled to a quadrupole ICP-MS instrument previously [119]. The peculiarity of this source was that it operates with high argon gas flow rates of up to 100 ml/min. One idea of this concept is that the speed of the gas flow should be high enough to increase the overall transport efficiency of ions towards the MS. This idea, to use high gas flow rates for various reasons, was first discussed by Mason [120] and later also by Miller *et al.* [121]. They observed a dramatic clean-up of the mass spectrum compared to a more static

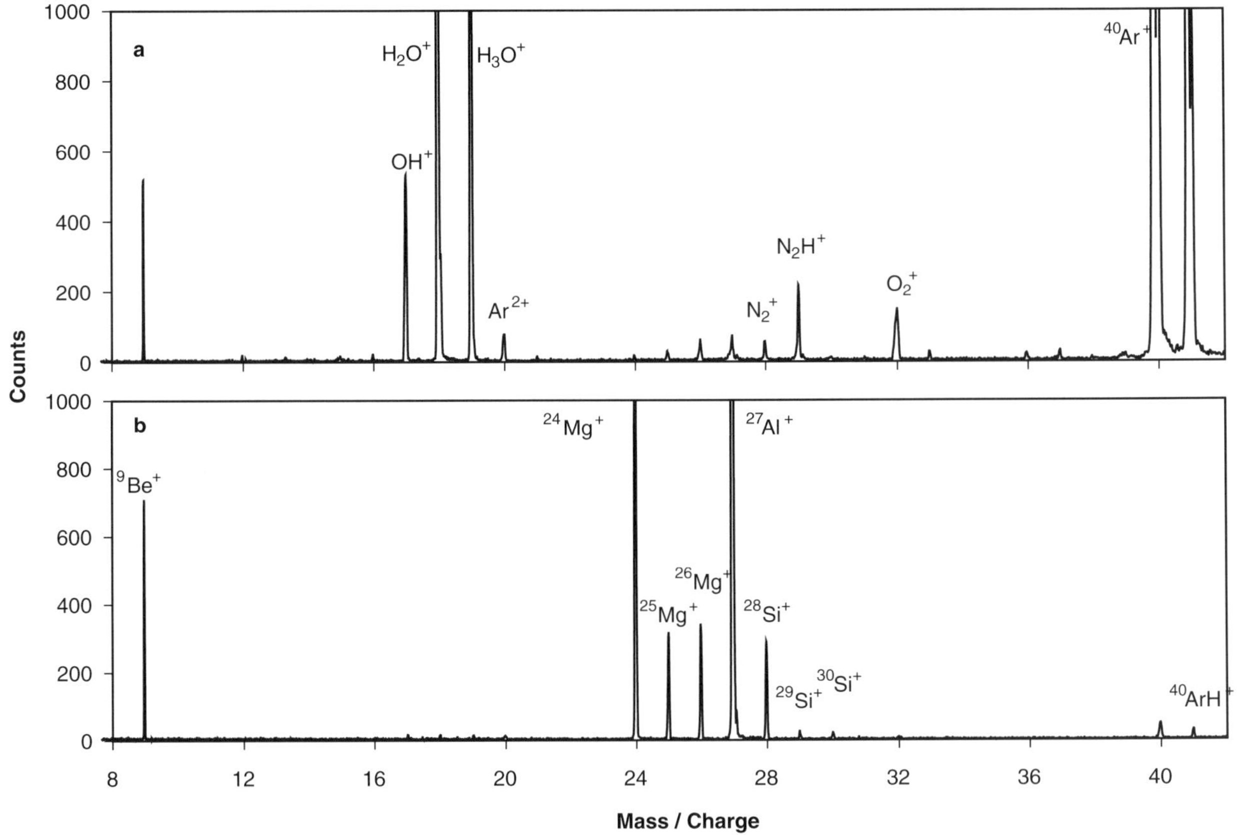

Fig. 4.26 Effect of temporal delays on removal of spectral interferences in NIST 1122 Brass: (a) 100 μs delay, (b) 240 μs delay.

flow of conventional source. A Grimm-type source operated with high gas flow rates and high discharge power of up to 70 W had been coupled by Beyer *et al.* to an inductively coupled plasma sector field mass spectrometry (ICP-SFMS) instrument (Axiom, ThermoElemental, Winsford, UK) [122]. Figure 4.27 shows the dependence of the gas flow rate on the ion intensity of argon and iron. From this figure it can be seen that the iron ion intensity exceeds the one of argon by about a factor of 10, whereas for a comparable source operated with a gas flow rate of less than 1 ml/min, the ratio was less than 0.1. Again here, the spatial decay of the GD plasma can be used to overcome problems caused by spectral interferences, whereas in the pulsed GD a temporal decay showed the same effect.

A very recent approach makes use of the fact that the Mattauch-Herzog geometry often used in SSMS instruments allows a simultaneous detection of all ions. Instead of photo-plates Hieftje and co-workers used electronic devices such as array detectors for simultaneous detection of all ions. This approach could give a new impetus to the instrumental development in GD- as well as in ICP-MS [123].

4.4.2 Instrumentation

From a technical point of view GD ion sources can be coupled to any MS instrument, for instance, a single, double or triple quadrupole arrangement, quadrupole ion trap MS, time of flight MS (ToF MS), sector field devices with

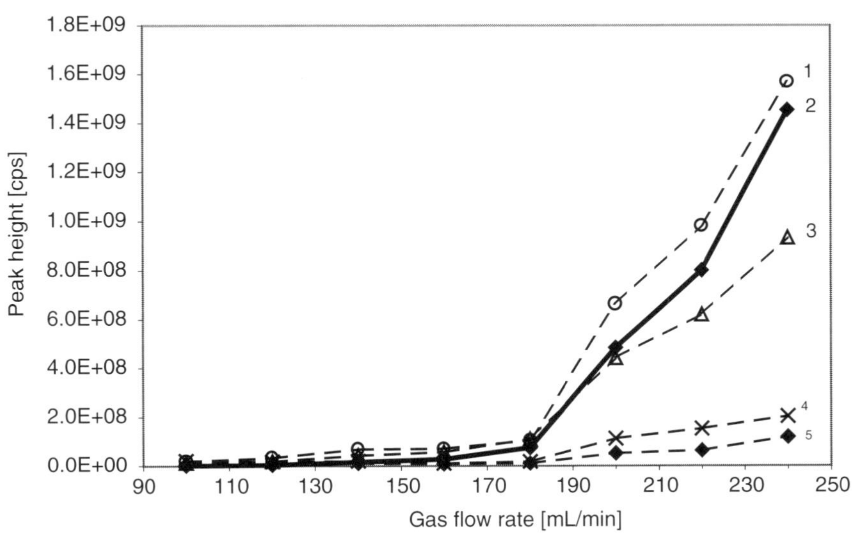

Fig. 4.27 Effect of gas flow rate on removal of spectral interferences in NIST 1263 A Steel. 1, $^{56}Fe^{+}$(/10); 2, $^{40}Ar^{+}$; 3, $^{40}Ar^{1}H^{+}$; 4, Ar^{2+}(*100); 5, $^{16}O^{1}H^{+}$(*1000).

low and high mass resolution capabilities, Fourier transform ion cyclotron resonance MS. More details concerning the different approaches of mass analyzers can be found in [5, 124].

Due to the differences of working pressure in the source and mass analyzer most often an interface in between is required for differential pumping. Quadrupole mass analyzers have often been used for home-built instrumentation, because they are cheap and easy to operate. Concerning their resolution they are typically operated with unit mass resolution, which is in principle sufficient for elemental analysis because each isotope needs only to be separated from its neighbours, but in praxis it is not sufficient to overcome spectral interferences, which can become the main limitation for ultra-trace analysis.

Spectral interferences may be subdivided into several groups, as they can be attributed to the presence of isobaric atomic ions, multiply charged ions and molecular ions of various origin. Isobaric ions overlap exists where isotopes of different elements coincide at the same nominal mass. For each element except indium, one isotope can be selected for quantification which is at least free from isobaric overlap, but in most cases this is not the most abundant one. Multiply charged ions will be found in the mass spectrum at a position m/z where z is the ion charge. In particular, doubly charged ions of the main matrix constituents are most often observed in GDMS whereas the discharge gas, for instance Ar, can also be observed in higher ionization states. Molecular ions may consist of atoms of the discharge gas and its contaminants (water, oxygen and nitrogen), and/or components of the matrix (cluster ions). Many approaches to overcome these spectral interferences have been described in the literature and most often they are only useful to solve specific problems and therefore the most straightforward approach to handle this problem is the application of high mass resolution.

Whether an interfering ion signal can be separated from an analytical signal, the isotope of interest, depends only on the mass difference of the peak maximum of these ion signals and the resolution of the instrument. The mass resolution R is defined as

$$R = \frac{m}{\Delta m}$$

where Δm is the mass difference necessary to achieve a valley of 10% between two neighbouring peaks of identical intensity at mass m and mass $m \pm \Delta m$. Due to the fact that most often the intensities of neighbouring signals are not identical, a more practical definition is used to measure or define the mass resolution. In this definition the peak width is measured at 5% of the maximal peak height of a single interference free peak. This definition will lead to the same mass resolution as in the 10% definition if the two neighbouring peaks are equally high and symmetric.

In Table 4.2 some examples of typical interferences observed in GDMS are compiled. For instance, in the case to separate $^{40}Ar^{+}$ from $^{40}Ca^{+}$ a very high mass resolution of about 190 000 must be applied to separate these isobaric interferences, but this is far above the upper limit of commercial instrumentation.

Table 4.2 Mass resolution required to separate a interference from isotope of interest

Molecular ion	Nuclide	Resolution
R < 3000		
$^{12}C_2^+$	$^{24}Mg^+$	1605
$^{14}N_2^+$	$^{28}Si^+$	958
$^{12}C^{16}O^+$	$^{28}Si^+$	1557
$^{14}N^{16}O^{1}H^+$	$^{31}P^+$	968
$^{15}N^{16}O^+$	$^{31}P^+$	1458
$^{16}O^{2+}$	$^{32}S^+$	1801
$^{12}C^{16}O_2^+$	$^{44}Ca^+$	1281
$^{40}Ar^{12}C^+$	$^{52}Cr^+$	2375
$^{40}Ar^{14}N^+$	$^{54}Fe^+$	2088
$^{40}Ar^{16}O^+$	$^{56}Fe^+$	2502
R: 3000–10 000		
$^{40}Ar_2^+$	$^{80}Se^+$	9688
R > 10 000		
Mo^{2+}	$^{48}Ti^+$	10 920
$^{40}Ar^{56}Fe^+$	$^{96}Mo^+$	13 041
$^{40}Ar^+$	$^{40}Ca^+$	1 92 497
$^{58}Fe^+$	$^{58}Ni^+$	27 998

It should be mentioned that the required resolution of about 1 90 000 to separate both isotopes is far above the upper limit of commercial instrumentation. On the one hand most examples given in this table demonstrate that a resolution R of less than 5000 is sufficient to overcome most interferences in practical work, but on the other hand a number of poly-atomic molecular ions, for instance hydrides, argides and oxides of the matrix components, in a mass range above 100 require a resolution close or above the upper end of the achievable range of commercial instrumentation. Additionally the price to be paid for high mass resolution is high concerning the price of sector field devices and one should have in mind that higher mass resolution is always associated with a loss in sensitivity which is inversely proportional to R.

As has been mentioned previously, up to now the only commercially available instrument is the VG 9000, a high mass resolution sector field device. A sketch of the VG 9000 is shown in Fig. 4.28. The instrument is a double focusing sector field device and it consists of a magnetic sector, an electric sector, an entrance and an exit slit, a detector device and the ion source itself. The geometry in which the magnetic sector is followed by an electric sector is well known as the reverse Nier-Johnson geometry.

In the magnetic sector the ions are separated due to their mass-to-charge ratio. The electric sector is needed to overcome differences in ion energy, which can otherwise worsen the achievable mass resolution. By lowering the slit-width, the instrument can be operated with higher mass resolution which is a prerequisite to overcome spectral interferences by interfering ions. A high mass resolution of up to 10 000 is in principle possible with this instrument, but typically $R = 5000$ is used. The instrument is equipped with two separate

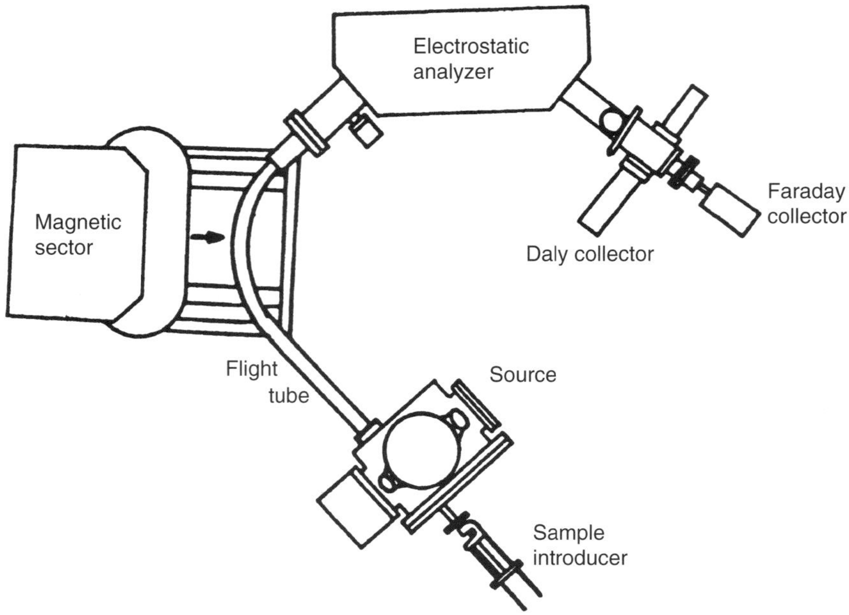

Fig. 4.28 Schematics of the commercial GDMS instrument (VG 9000) [with courtesy of Thermo Elemental].

detectors. For high ion intensities of the matrix, a Faraday cup is used, whereas for trace and ultra-traces, a Daly detector operated in an ion counting mode is applied. Automatic switching between both detectors has been provided so that a dynamic range of more than 9 orders of magnitude could be guaranteed in a single scan.

A sketch of the standard ion source for pin samples is shown in Fig. 4.29. The source consists of three main parts: two electrodes, the conducting sample used as cathode and the anode with the exit aperture. The pin type sample is fixed in a special holder and introduced into the ion source by a solid probe arrangement. The sample is operated as the cathode and is isolated from the anode source body by a ceramic cover. The anode body is cryogenically cooled, by which residual gases can be reduced significantly. The whole source floats at the acceleration voltage of the sector field device. For flat samples a so-called *mega* source is also commercially available.

4.4.3 Sample preparation

As has been mentioned previously, the strength of GD-OES and GDMS is the direct analysis of solid samples. Therefore most strategies used for sample preparation are identical for both methods. As mentioned previously, the d.c.

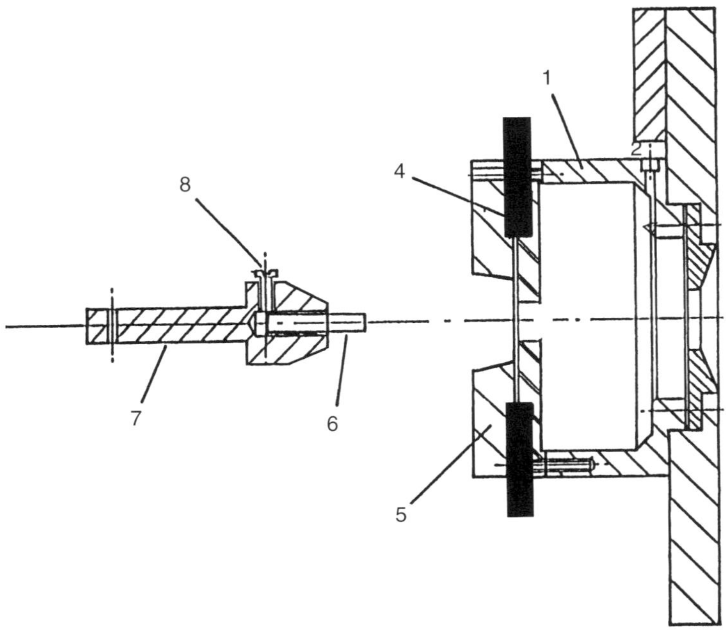

Fig. 4.29 Glow discharge ion source for pin type samples: 1, anode; 2, Ar inlet; 3, Ion exit slit; 4, isolation ring (boron nitride or quartz); 5, cathode back plate; 6, sample; 7, sample holder; 8, fixing screw (with permission of Cornel Venzago, Infracor, Hanau, Germany).

GD demands a conducting sample in pin or flat form. The simplest way to proceed is by machining to the final shape, but additionally the final surface is polished by grinding. Most often a surface cleaning by etching with a mineral acid or the removal of organic contaminations by use of an organic solvent is applied afterwards.

Powder samples can be pressed with a die in a suitable form, and even non-conducting powdered samples can be analyzed either directly [125] or after mixing with a conducting host matrix, for instance a powder or a very pure ductile solid such as gallium or indium [3]. For completeness it should be mentioned that liquid residues have also been analyzed by GDMS, but as we want to restrict ourselves to applications of solid materials; more details about these applications are given elsewhere [3].

4.4.4 Quantification

Similar to GD-OES, GDMS also is not an absolute method and therefore calibration is a prerequisite for an accurate analysis. In comparison to AES, quantification in GDMS is simpler due to a relatively low number of *lines* in the mass spectrum. The latter is nothing other than the well-known stable isotopes with also well-known natural abundances. In our environment we have only 92 elements of the periodic table with at least 287 stable isotopes in the mass range of up to 238.

The most common approach for quantification has been introduced by the manufacturer of the commercial instrument as the so-called *ion beam ratio* (IBR) procedure. It can be applied for the analysis of pure materials, where one element represents the matrix m and can be estimated to have a concentration of 100%. In this case the assumption is made that the ratio of the intensity for any analytically relevant isotope with respect to the total ion current is representative of the ratio of the number of atoms of this isotope relative to the atoms of the matrix. In this case, the concentration c of an isotope x is obtained from

$$C_x = (I_x/I_m).$$

This is a semi-quantitative approach, because here it is assumed that the sensitivity of the element x comes close to that of the matrix m. Usually, the differences in sensitivity can be as high as a factor of 3 so that quantified concentration is inaccurate by this factor. Nevertheless, this accuracy is sufficient for most ultra-trace analyses so that this approach is most often applied for the commercial instrument.

For a quantitative analysis, a calibration procedure with relevant standards or certified reference materials (CRM) is essential. The slope of the calibration curve is the sensitivity $S_{x,m}$ of the element x in the matrix m. We define here the elemental sensitivities S_x of an element x in the sample as the slope of the calibration line of the measured intensities I_x of an element x according to their certified content $c_{\text{cert},x}$ in μg/g, and corrected for the isotope abundance. The relative sensitivity factor (RSF) of an element x is calculated from the division of the elemental sensitivity and the sensitivity S_{St} of the internal standard:

$$\text{RSF}_x = \frac{S_x}{S_{St}}$$

The content of an element x can then be calculated relative to the concentration of the internal standard:

$$c_x/c_{St} = \frac{(I_x/I_{St})}{\text{RSF}_x}$$

For ultra-trace analysis in a pure matrix m, where m is identical to the internal standard, the concentration of the internal standard is assumed to be $c_{St} = 1$. Due to the fact that the number of atoms counted at the detector is proportional to the number of atoms in the sample, the RSF and the calculated content are expressed in atomic concentrations [110, 126].

A prerequisite for this procedure is that a sputter equilibrium is reached and the intensity ratio remains constant during the time needed for analysis. The main limitation for calibration of course is the lack of CRM at trace and ultra-trace levels for as many elements as possible. As has been discussed earlier in this chapter, many different physical processes, such as sputtering, ionization, extraction and transport of ions, contribute to the RSF and therefore they are valid only for a selected set of working conditions of the discharge and the instrument used. To a certain extent they can be predicted by theoretical models [9, 14] and, surprisingly in practice, the range of RSFs is relatively limited from about 0.1 to 4, and it was found that they are only slightly dependent on working conditions for a given source operated for optimized sensitivities. It even has been shown that RSFs are very similar for different matrices if identical internal standards are chosen. This is the reason why for the commercial instrument, a library of RSF values is available, which often is used for quantification to improve the accuracy for those matrices for which no reference materials or standards are at hand (see for instance ref [127]). Again, due to the very limited range of RSFs, a semi-quantitative approach by using an RSF = 1 is also possible, especially at ultra-trace levels when the demand for highest accuracy is not a great deal any more [128].

4.4.5 Analytical figures of merit

Obviously, the analytical figures of merit that can be achieved by GDMS strongly depend on the method applied for analysis. Therefore they depend on the matrix, the sample preparation, operational conditions of the source and the instrument, etc. They shall be discussed in more detail in the next section, but nevertheless some typical values shall be given here to get a feel of the method and to discuss some problems which are related to the technique.

4.4.5.1 Sensitivity

The sensitivity in GDMS depends on the geometry of source and sample, the matrix, the chosen discharge and instrumental conditions. For the matrices Al, Cu or Au, the total ion matrix currents of 1×10^{-10} A are achieved for the VG 9000 measured with $R = 4000$ and hence sensitivities of several thousand counts per second (cps) and per μg/g are obtained. The noise level is typically less than 0.1 cps.

4.4.5.2 Limit of detection

In GDMS, it is difficult to apply the standard definitions for the limit of detection (LoD) using the 3 σ definition, because for this definition a low blank is required which should come from the sample investigated. Therefore, ultra-pure samples are required to calculate the LoD, but most often these samples are not available and therefore a so-called instrumental blank (background noise) in combination with the sensitivity of an element is used to calculate the detection limit, thus giving a more or less theoretical LoD. For various matrices the following LoDs have been achieved with the VG 9000:

$$\text{H, C, N, O:} \quad < 1\ \mu g/g$$

$$\text{F, Cl, Br, J:} \quad 1 - 0.1\ ng/g$$

$$\text{all others except noble gases:} \quad < 0.1\ ng/g$$

With an integration time of 1 min per isotope, LoD of less than 1 ng/g can be achieved, so that elements that are relevant for the electronic industry such as Th and U can be measured down to 0.01 ng/g. For the latter, integration time was increased to 5 min, so that a total analysis time of an hour or more is needed for a multi-element coverage.

4.4.5.3 Reproducibility and precision

In GDMS, we differentiate between an external reproducibility, which includes sample preparation and sample changing, and an internal reproducibility which is calculated from repeated scans. Both are expressed by the precision calculated from the standard deviations of intensity measurements. Both internal and external precision depend on the homogeneity and content of the element in the sample. Therefore, most often in the literature, data for relative standard deviations (RSD) are given which depend less on concentration. Internal precision measured on CRM SS-487/1 making three repeat measurements range from $<3\%$ RSD for major matrix components to $<10\%$ for traces and $>10\%$ for ultra-traces. The latter are limited by the counting statistics only.

4.4.5.4 Robustness

Robustness is a measure of the dependence of sensitivities on discharge or instrumental conditions. Although the absolute intensities strongly depend on discharge conditions, the RSFs are generally less dependent.

4.4.5.5 Accuracy

A prerequisite for a sufficient accuracy is of course that GDMS has a high robustness and is free of any influence of the concentration of matrix constituents on the signal intensity, which is known as a non-spectroscopic interference (matrix effect). On the one hand, it has been mentioned earlier in this chapter and has often been claimed in literature too, that matrix effects are not a limiting factor in GDMS, because the steps of atomization and ionization are separated in time and space so that the matrix should not have any influence on the ionization process as it is the case, for instance, in SIMS. On the other hand, from a physical point of view many physical processes and parameters can have an influence on the sensitivity S as well as on the RSFs, and thus might influence the accuracy. For instance, the sample is an electrode of the discharge and therefore the physical properties such as temperature (and cooling of the sample), surface structure, cleanliness, composition, or the chosen discharge parameters can be expected to have an influence on sputtering and plasma conditions and therefore on the sputtering, transport of material, ionization and extraction efficiency.

From a pragmatic point of view, up to now there has been little evidence for matrix effects and this is demonstrated in the next example, where the accuracy of GDMS has been assessed in a round robin analysis. For this purpose, Venzago *et al.* performed a round robin analysis of aluminium samples with a purity of 4N (99.99%) [129]. Ten European labs participated and contributed their results including reference values from rapid neutron activation analysis (RNAA) and wet chemistry. Figure 4.30 shows the bias between GDMS and the reference values in μg/g for 31 elements. From this figure it can be seen that for 23 elements no significant bias between GDMS and accepted reference values can be detected.

4.4.6 Applications

There are many advantages in applying GDMS for direct analysis of solid samples, which are mainly caused by the non-thermal process of atomization by physical sputtering. Furthermore, the steps of atomization and ionization are separated in time and space and therefore, fractionation and matrix effects are strongly reduced in comparison, e.g., with SSMS or SIMS, and matrix-independent RSFs can be used and thus multi matrix quantification became possible.

From the analytical point of view, GDMS is convincing due to the multi-element capability, the very high sensitivity, the low and stable background, and thus the low detection limits. Of course, alternative and powerful spectroscopic methods are available if no ultimate detection limits are required. For instance, if for high contents a high precision is required, X-ray fluorescence (XRF) is still superior or, if no standards are available, still methods based on dissolution of the solid material are applied such as ICP-AES and ICP-MS. The latter are also very promising for a direct analysis of a solid if, for instance, laser ablation is used for sample introduction.

The strength of the latter methods is that at least for all solid samples, procedures are known how to dissolve them. Simple quantification is possible by the use of internal standards added to a matrix-matched solution or more directly by standard addition procedures. On the other hand, dissolution of solids is time consuming and laborious and may become crucial due to the risks of contamination. By the dilution of the solid, the detection limits are worsened and often the matrix load can cause drift effects or sensitivity

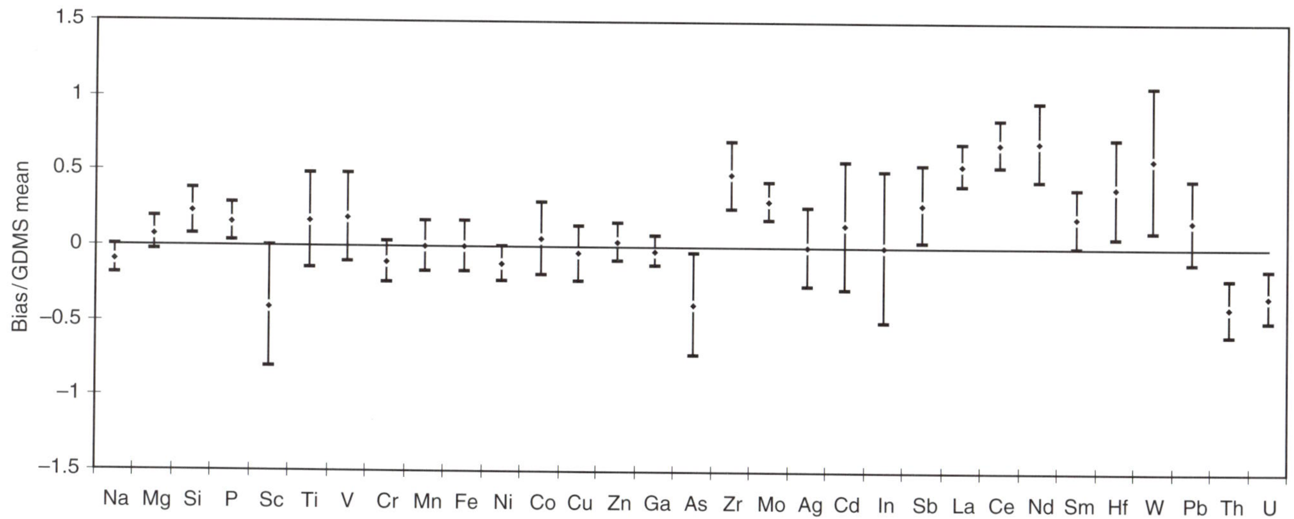

Fig. 4.30 Significance of bias [μg/g] between reference value (by rapid neutron activation analysis, wet chemistry) (from ref. 129 with permission of JAAS).

changes of elements (matrix effects), so that further dilution of the sample is required. Contributions from a variety of spectral interferences introduced by the chemicals in the preparation procedure are severe limitations in lower resolution instruments for a number of elements.

It is a basic advantage of GDMS for the direct analysis of solids to avoid most of these limitations. Additionally, sample preparation is much simpler, a fast analysis is possible and any risk of contamination from sample pre-treatment is avoided. Finally, direct analysis avoids a sensitivity loss due to dissolution, and therefore considerably lower detection limits can be achieved.

Nevertheless, some limitations should also be mentioned. First of all, the high price of the commercial instrument is counter-productive to a broader acceptance, the instrument is not automated and the sample throughput is limited. Finally, a lack of standards needed for calibration is a definite handicap. In the following, some selected applications from the material sciences shall be presented which best demonstrate the analytical capabilities of GDMS.

4.4.6.1 Bulk analysis of conducting and semi-conducting samples

The domain of DC-GDMS for a very long time has been and still is the bulk analysis of conducting and semi-conducting materials. For this application the discharge parameters, discharge pressure and discharge voltage or current are optimized for highest sensitivity. A certain pre-burn phase is required to get stable sensitivities and to clean the sample surface by sputtering.

The VG 9000 instrument operated with high mass resolution is, in particular, convincing by extremely low detection limits for a great variety of elements in bulk analysis of pure metals (Al [129–131], Ti [132], Ga [133], Pt [134] and Mo [135]), alloys (Fe, Al and Cu [136] and AlSi [137]) and semiconductor materials (Si-wafers [138]).

An interesting report from Mykytiuk *et al.* covered the GDMS analysis of CdTe, GaAs, Ga and As [139]. The authors claimed that no standards had been applied for calibration and they pointed out the extremely high dynamic range achieved for these materials of more than 9 orders of magnitude. Even elements such as C, N and O had been quantified in the matrices mentioned. In the work of Wilhartitz *et al.*, the matrices Mo, W and Cr had been investigated with different methods including GDMS and it was shown that inhomogeneities can have an influence on the accuracy, because the amount of material sputtered (at mg levels) is relatively low in comparison to the amount used for solution analysis [140]. Vieth *et al.* analyzed 56 elements in a number of standard reference materials based on Al, Ti, Fe, Ni, Cu and Fe and from the similarity of the RSFs in different matrices, they concluded the absence of matrix effects [141].

It is a vicious circle. For many advanced materials, the knowledge of the non-metals C, N, O, H, S and P is even nowadays a challenge, because the detection limits achieved with traditional methods usually applied, such as photometry techniques or elementary analyzers, are not sufficient any more for their characterization and therefore it is also not possible to produce new standard materials at much lower levels needed for calibration of GDMS.

4.4.6.2 Bulk analysis of non-conducting samples

Non-conducting materials cannot be analyzed directly by DC-GDMS, because at least a semi-conducting electrode has to be used as the cathode of the discharge arrangement. Various techniques are described in the literature for d.c. discharges to overcome this limitation. Some of them, for instance the secondary cathode approach [142], are limited to specific application, for instance the analysis of glass samples only [143], whereas the mixing of non-conducting powders with conducting powders, a technique well known from SSMS, used as a binder is a more universal approach. Mixed samples of aluminium oxide powder with copper powder in the ratio 1:5 were prepared as compacted pellets [144]. With detection limits in the region of a few μg/g, these experiments ascertained that with GDMS satisfactory analysis of non-conductive samples was also possible.

A more universal approach for analysis of non-conducting materials is the operation of the GD source with Rf voltages but it should be mentioned that until now, no commercial Rf source is available for the VG 9000. Nevertheless, Duckworth *et al.* used a slightly modified original source of the VG 9000 and coupled an Rf potential via a direct insertion probe to the non-conducting sample, for instance a glass sample [145]. They could demonstrate the potential of this approach but had been limited by a broadening of the ion energy distribution, which resulted in a worsening of the ion transmission.

Main applications for Rf discharges can be seen in the semiconductor industry, for the analysis of glasses and ceramics as well as for polymers. For instance, a home-built Rf-GD source was developed by Becker and co-workers for the analysis of non-conducting samples. They coupled this source to an ICP-SFMS operated with low mass resolution and had achieved detection limits of 10 ng/g for B in high-ohmic GaAs [113]. Other home-built instruments had been used to analyze Teflon [104], transition metal oxides [146], glass samples [107, 145, 146], and ceramics [147]. In all these examples the composition of the materials investigated had been of main interest. A completely different approach with important analytical potentialities is the use of GDMS for fingerprinting of different polymers [148]. An example of such a spectral fingerprinting is given by Shick *et al.* showing that information of both atomic and molecular composition of the sample can be obtained from the mass spectra [149].

4.4.6.3 In-depth analysis of surface layers

The fundamental idea of applying GDMS for in-depth analysis comes from the fact that a well-defined fraction of atoms sputtered from the sample surface is converted into ions and reaches the detector nearly simultaneously, so that in principle the ions detected are representative of the surface composition if the scanning time of the MS is chosen appropriately. Of course, this concept requires fast scanning mass spectrometers such as quadrupoles and TOFs or alternatively thick layers. By performing an in-depth analysis the intensity time profile is measured. From the knowledge of the sensitivity the ion intensity can be converted into the concentration and from the time needed to sputter a well-known crater depth, the whole concentration-depth-profile can be reconstructed. But this approach can be applied only if the layer or the bulk material provides an internal standard for quantification and if sensitivities, or better, RSFs do not change for the bulk and layer material. It is also assumed that the sputter rates do not change severely during the sputtering of the layer, so that the depth at each time can be easily calculated from the total crater depth measured after the analysis. However, even in the case when the matrix changes between layer and substrate, either the concept of useful yields or normalization to total ion intensities (for more details, see ref [150]) can be used for quantification.

Although for emission spectroscopy more than 80% of all instruments sold worldwide are used mainly for surface analysis, the acceptance of GDMS for this application is not forthcoming. One reason, of course, is that the commercial source is not optimized for this purpose. Nevertheless, it has been shown by Hall and Sanderson [151] that the commercial instrument operated with a GD source with flat-cell geometry can also be applied for the analytical characterization of surfaces and near-to-surface layers.

Due to the slow scanning speed of the sector field device, most examples given in the literature are limited to thick surface layers only, but could be extended to thin layers too when a quadrupole based instrument would become commercially available for a limited period of time [152].

Similar to GD-OES, sources of the Grimm-type geometry operated in a d.c. mode are very well suited in GDMS for in-depth analysis [153] of conducting layers if adequate discharge parameters are chosen to obtain a flat sputter crater bottom, as has been mentioned previously. In Fig. 4.31 the concentration versus depth profile is shown for five double layers of Cr and Ni with a thickness of 25 nm each [154]. The data was measured with a quadrupole based instrument in 40 s only and a depth resolution of 9 nm for the first and 17 nm for the last layer interface has been calculated. This example demonstrates that these sources can be applied even for thin layers but are as well suited for technical layers with a thickness of more than 1 μm, for which other methods such as SIMS or SNMS would need many hours of analysis time. In the literature examples are given for Grimm-type sources for the analysis of galvanization layers [154], chromate/phosphate passivation layers on Al [154] and Ni coated copper [155].

In all cases satisfactory depth resolution could be obtained in a rather short analysis time, so that this technique has its strength for analysis of metallic

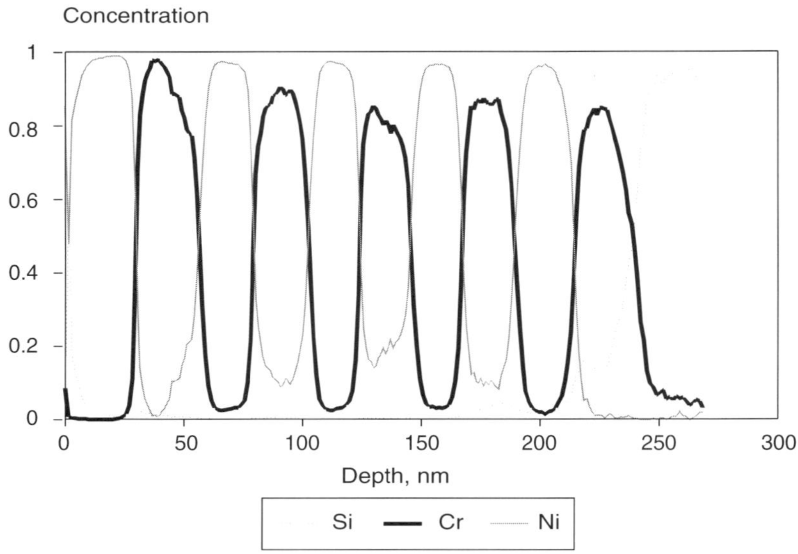

Fig. 4.31 Concentration versus depth profile calculated from a measured intensity versus time profile of five double layers Cr/Ni with a thickness of 25 nm of each single layer on a silicon substrate.

technical surface layers. Although metallic coatings on various substrates play a very important technical role for the future, more and more applications of non-conducting layers have to be analytically characterized. This holds true, for instance, for layers of hard materials or ceramics on metals to improve the physical, thermal or chemical resistance of the substrate material.

Many new materials with outstanding physical properties such as solid oxide fuel cells [156] or high temperature super-conducting materials are of significant technical interest. Understanding of the stoichiometry of the main components as well as trace contaminations in these layers is the key issue for future technical improvements, and from an analytical point of view there are no techniques better than Rf GDMS capable to solve problems related to such materials. GDMS has the strength to be at the forefront of these new technologies if adequate commercial sources become commercially available. Again, for the development of such sources, experiences from GD-OES may be useful, because only some modifications are required to operate Grimm-type source with Rf voltages, as has been mentioned previously.

Recently, the source described previously had been modified and coupled to a quadrupole based ICP-MS (TS Sola) [119]. A free running Rf generator with a frequency of 3.5 MHz and a maximum voltage of 3 kV peak to peak was used as the power supply. In Fig. 4.32, the intensity time profile of selected elements is shown for the analysis of an Al_2O_3 layer with a thickness of 5 μm

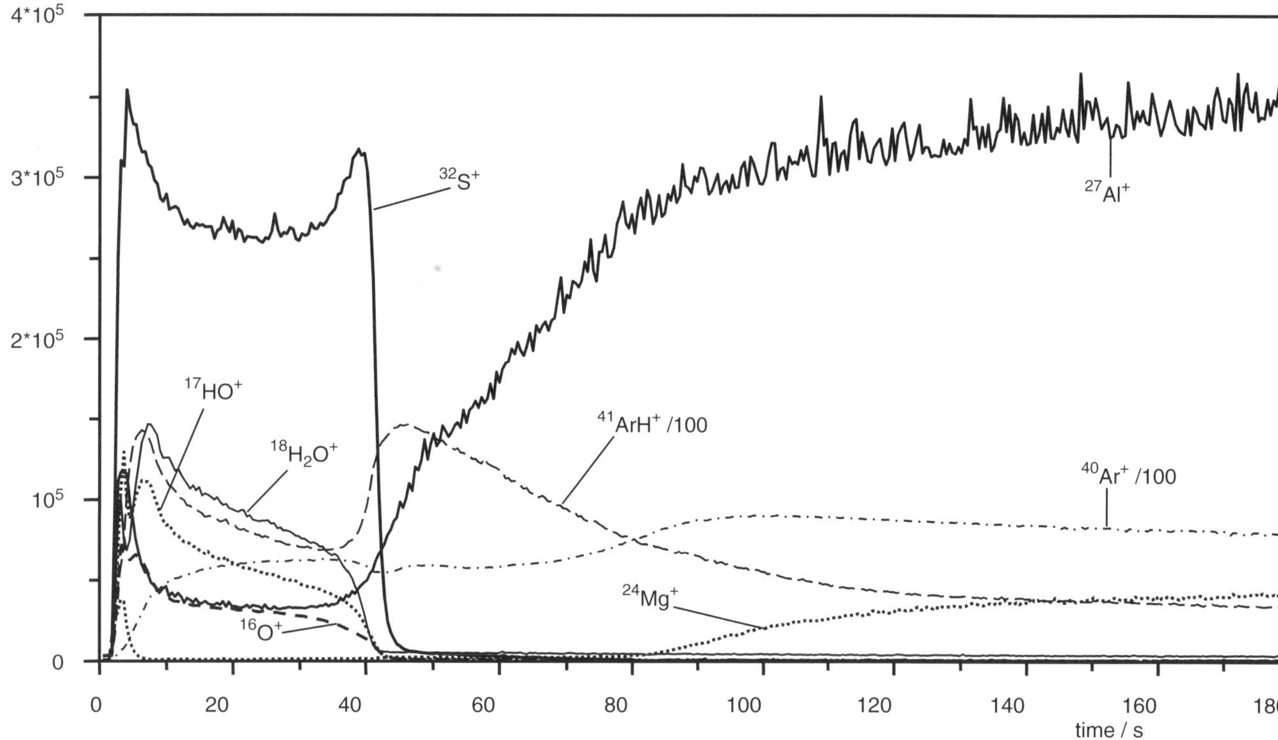

Fig. 4.32 Intensity versus time profile of a Al_2O_3 with a thickness of 5 μm measured with an Rf-Grimm-type-GD source and coupled to a quadrupole mass analyzer.

on an aluminium substrate. Although the sputter crater shows a certain curvature, the intensity time profile rises steeply which is a hint for a sharp interface between the layer and the substrate material. The layer had been produced in an electrolytic process using H_2SO_4 as the electrolyte and therefore it was interesting to see the distribution of sulphur in the layer, showing a certain enrichment at the boundary between layer and substrate. The changing intensity of Al relative to O is a hint that the stoichiometry in the layer is not constant, which for the production process is most important to know and to control.

4.5 Conclusion

Analytical GD spectroscopy is well accepted nowadays as a powerful technique for the direct analysis of all kinds of solid samples, conducting materials as well as non-conducting ones. It has the ability to fulfil the new challenges coming from the material sciences, where even more elements (full elemental coverage) at much lower concentrations in bulk and layer materials have to be analyzed. More and more d.c. sources are exchanged by Rf-driven sources, for which at least novel matching units are needed which allow the measurement of the applied discharge voltage as real current passing the plasma only. This prerequisite is needed at least for the breakthrough of these sources being used for quantitative analysis in AES as well as MS.

References

[1] S. Baude, J.A.C. Broekaert, D. Delfosse, N. Jakubowski, N.G. Orellana-Velado, R. Pereiro & A. Sanz-Medel, *J. Anal. At. Spectrom.*, 2000, **15**, 1516.

[2] J.A.C. Broekaert, *Appl. Spectrosc.*, 1995, **49**, 12A.

[3] C.M. Barshick, Glow discharge mass spectrometry, in C.M. Barshick, D.C. Duckworth & D.H. Smith (eds), *Inorganic Mass Spectrometry, Practical Spectroscopy Series*, 23, Marcel & Dekker, New York, 2000, p. 31.

[4] R. Payling, D. Jones & A. Bengtson, *Glow Discharge Optical Emission Spectrometry*, John Wiley & Sons, Chichester, 1997.

[5] R.K. Marcus, *Glow Discharge Spectroscopies*, Plenum Press, New York, 1993.

[6] A. Bogaerts, E. Neyts, R. Gijbels & J. van der Mullen, *Spectrochim. Acta, Part B*, 2002, **57**, 609.

[7] A. Bogaerts & R. Gijbels, *J. Anal. At. Spectrom.*, 1998, **13**, 945.

[8] A. Bogaerts & R. Gijbels, *Anal. Chem.*, 1997, **69**, 719A.

[9] A. Bogaerts & R. Gijbels, *Spectrochim. Acta, Part B*, 1998, **53**, 437.

[10] A. Bogaerts, E. Wagner, B.W. Smith, J.D. Winefordner, D. Pollmann, W.W. Harrison & R. Gijbels, *Spectrochim. Acta, Part B*, 1997, **52**, 205.

[11] A. Bogaerts, R.D. Guenard, B.W. Smith, J.D. Winefordner, W.W. Harrison & R. Gijbels, *Spectrochim. Acta, Part B*, 1997, **52**, 219.

[12] N. Jakubowski, D. Stüwer & G. Tölg, *Int. J. Mass Spectrom. Ion Proc.*, 1986, **71**, 183.

[13] N. Jakubowski, I. Feldmann & D. Stüwer, *J. Anal. At. Spectrom.*, 1997, **12**, 151.

[14] A. Bogaerts & R. Gijbels, *J. Appl. Phys.*, 1995, **78**, 6427.
[15] A. Bogaerts, L. Wilken, V. Hoffmann, R. Gijbels & K. Wetzig, *Spectrochim. Acta, Part B*, 2002, **57**, 109–119.
[16] C. Yang, K. Ingeneri & W.W. Harrison, *J. Anal. At. Spectrom.*, 1999, **14**, 693.
[17] C. Yang, M. Mohill & W.W. Harrison, *J. Anal. At. Spectrom.*, 1999, **15**, 1255.
[18] V. Majidi, M. Moser, C. Lewis, W. Hang & F.L. King, *J. Anal. At. Spectrom.*, 2000, **15**, 19.
[19] C. Yang, E. Oxley & W.W. Harrison, *J. Anal. At. Spectrom.*, 2001, **16**, 1131.
[20] A. Bogaerts & R. Gijbels, *J. Anal. At. Spectrom.*, 2000, **15**, 1191.
[21] A. Bogaerts & R. Gijbels, *J. Anal. At. Spectrom.*, 2000, **15**, 895.
[22] A. Bogaerts, R. Gijbels & V.V. Serikov, *J. Appl. Phys.*, 2000, **87**, 8334.
[23] A. Bogaerts, A. Quentmeier, N. Jakubowski & R. Gijbels, *Spectrochim. Acta, Part B*, 1995, **50**, 1337.
[24] A. Bogaerts & R. Gijbels, *J. Appl. Phys.*, 1999, **86**, 4124.
[25] A. Bogaerts & R. Gijbels, *J. Anal. At. Spectrom.*, in press.
[26] A. Bogaerts, R. Gijbels & J. Vlcek, *J. Appl. Phys.*, 1998, **84**, 121.
[27] N.P. Ferreira, J.A. Strauss & H.G.C. Human, *Spectrochim. Acta*, Part B, 1981, **36**, 215.
[28] A. Bogaerts, R. Gijbels & R.J. Carman, *Spectrochim. Acta, Part B*, 1998, **53**, 1679.
[29] A.V. Phelps & Z.Lj. Petrovic, *Plasma Sources Sci. Technol.*, 1999, **8**, R21.
[30] H.A. Hyman, *Phys. Rev. A*, 1979, **20**, 855.
[31] A.V. Phelps, *J. Chem. Phys. Ref. Data*, 1991, **20**, 557.
[32] C.M. Ferreira & A. Ricard, *J. Appl. Phys.*, 1983, **54**, 2261.
[33] A. Bogaerts & R. Gijbels, *Phys. Rev. A*, 1995, **52**, 3743.
[34] A.V. Phelps, *J. Appl. Phys.*, 1994, **76**, 747.
[35] B. Chapman, *Glow Discharge Processes*, Wiley, New York, 1980.
[36] L. Vriens, *Phys. Lett.*, 1964, **8**, 260.
[37] A.Z. Msezane & R.J.W. Henry, *Phys. Rev. A*, 1986, **33**, 1631.
[38] L.A. Riseberg, W.F. Parks & L.D. Schearer, *Phys. Rev. A*, 1973, **8**, 1962.
[39] M. Bourène & J. Le Calve, *J. Chem. Phys.*, 1973, **58**, 1452.
[40] S. Inaba, T. Goto & S. Hattori, *J. Phys. Soc. Jpn*, 1983, **52**, 1164.
[41] J.W. Coburn & E. Kay, *Appl. Phys. Lett.*, 1971, **18**, 435.
[42] M.K. Levy, D. Serxner, A.D. Angstadt, R.L. Smith & K.R. Hess, *Spectrochim. Acta, Part B*, 1991, **46**, 253.
[43] A. Bogaerts & R. Gijbels, *J. Appl. Phys.*, 1996, **79**, 1279.
[44] A. Bogaerts & R. Gijbels, *J. Anal. At. Spectrom.*, 1996, **11**, 841.
[45] E.B.M. Steers & R.J. Fielding, *J. Anal. At. Spectrom.*, 1987, **2**, 239.
[46] E.B.M. Steers & F. Leis, *Spectrochim. Acta, Part B*, 1991, **46**, 527.
[47] E.B.M. Steers & A.P. Thorne, *J. Anal. At. Spectrom.*, 1993, **8**, 309.
[48] K. Wagatsuma & K. Hirokawa, *Spectrochim. Acta, Part B*, 1996, **51**, 349.
[49] P.B. Farnsworth & J.P. Walters, *Spectrochim. Acta, Part B*, 1982, **37**, 773.
[50] E.M. van Veldhuizen & F.J. de Hoog, *J. Phys. D*, 1984, **17**, 953.
[51] R.S. Hudson, L.L. Skrumeda & W.J. Whaling, *J. Quant. Spectrosc. Radiat. Transfer*, 1987, **38**, 1.
[52] K. Danzmann & M. Kock, *J. Phys. B*, 1981, **14**, 2989.
[53] L. Csillag, M. Janossy, K. Rozsa & T. Salamon, *Phys. Lett.*, 1974, **50A**, 13.
[54] S.C. Rae & R.C. Tobin, *J. Appl. Phys. 1988*, **64**, 1418.
[55] P. Baltayan, J.C. Pebay-Peyroula & N. Sadeghi, *J. Phys. B*, 1985, **18**, 3618.
[56] P. Baltayan, J.C. Pebay-Peyroula & N. Sadeghi, *J. Phys. B*, 1986, **19**, 2695.
[57] E.W. McDaniel, J.B.A. Mitchell & M.E. Rudd, *Atomic Collisions: Heavy Particle Projectiles*, Wiley, New York, 1993.
[58] H.S.W. Massey & E.H.S. Burhop, *Electronic and Ionic Impact Phenomena*, Oxford University Press, Oxford, 1952.
[59] V.S. Vorob'ev, *Plasma Sources Sci. Technol.*, 1995, **4**, 163.
[60] M.A. Biondi, *Phys. Rev.*, 1963, **129**, 1181.

[61] H. Oechsner, *Phys. Rev. B*, 1978, **17**, 1052.
[62] W.W. Harrison & B.L. Bentz, *Prog. Analyt. Spectrosc.*, 1988, **11**, 53.
[63] R.V. Stuart & G.K. Wehner, *J. Appl. Phys.*, 1964, **35**, 1819.
[64] P. Sigmund, *Phys. Rev.*, 1969, **184**, 383.
[65] Y. Yamamura, N. Matsunami & N. Itoh, *Radiat. Eff.*, 1983, **71**, 65.
[66] N. Matsunami, Y. Yamamura, Y. Itikawa, N. Itoh, Y. Kazumata, S. Miyagawa, K. Morita, R. Shimizu & H. Tawara, *Atom. Data and Nucl. Data Tables*, 1984, **31**, 1.
[67] M.T. Robinson & I.M. Torrens, *Phys. Rev. B*, 1974, **9**, 5008.
[68] A. Benninghoven, F.G. Rüdenauer & H.W. Werner, *Secondary Ion Mass Spectrometry: Basic Concepts, Instrumental Aspects, Applications and Trends*, Wiley, New York, 1987.
[69] E.W. McDaniel, *Collision Phenomena in Ionized Gases*, Wiley, New York, 1964.
[70] H.A. Oechsner, *Appl. Phys.*, 1975, **8**, 185.
[71] A. Bengtson, S. Hänström, E.L. Piccolo, N. Zacchetti, R. Meilland & H. Hocquaux, bcr information, support to standardisation, Control of coated steel products setup of a standard method for quantitative depth profile analysis of Zn beased coatings by direct current glow discharge optical emission spectroscopy, Report EUR 18862 EN, ISBN-92–828–5989–4, 1999.
[72] W.W. Harrison, Ch. Yang & E. Oxley, *Anal. Chem.*, 2001, **73**, 480A.
[73] H. Prugger, R. Groβkopf & R. Torge, *Spectrochim. Acta, Part B*, 1971, **26**, 191.
[74] H. Grimm, *Spectrochim. Acta, Part B*, 1968, **23**, 443.
[75] A. Bengtson, *Spectrochim. Acta, Part B*, 1994, **49**, 411.
[76] Z. Weiss, *J. Anal. At. Spectrom.*, 1994, **9**, 351.
[77] Z. Weiss & P. Smid, *J. Anal. At. Spectrom.*, 2000, **15**, 1485.
[78] Z. Weiss, *J. Anal. At. Spectrom.*, 2001, **16**, 1275.
[79] V. Hoffmann, Y.S. Kim & M. Analytis, *MO6 at the 2002 Winter Conference on Plasma Spectrochemistry*, abstract.
[80] V.D. Hodoroaba, V. Hoffmann, E.B.M. Steers & K. Wetzig, *J. Anal. At. Spectrom.*, 2000, **15**, 951.
[81] V.D. Hodoroaba, V. Hoffmann, E.B.M. Steers & K. Wetzig, *J. Anal. At. Spectrom.*, 2000, **15**, 1075.
[82] V.D. Hodoroaba, E.B.M. Steers, V. Hoffmann & K. Wetzig, *J. Anal. At. Spectrom.*, 2001, **16**, 423.
[83] A. Bogaerts & R. Gijbels, *J. Anal. At. Spectrom.*, 2000, **15**, 441.
[84] A. Bengtson & S. Hänström, The influence of hydrogen on emission intensities in GD-OES – consequences for quantitative depth profile analysis, Fifth International Conference on Progress in Analytical Chemistry in the Steel and Metals Industries, 12–14 May 1998, Proceedings, European Commission, The European committee for the study and application of analytical work in the steel industry (CETAS), ISBN 92–828–6905–9, pp. 47–54.
[85] R. Payling, M. Aeberhard & D. Delfosse, *J. Anal. At. Spectrom.*, 2001, **16**, 50.
[86] V. Hoffmann, H.J. Uhlemann, F. Präβler, K. Wetzig & D. Birus, *Fresenius J. Anal. Chem.*, 1996, **355**, 826.
[87] K. Shimizu, H. Habazaki, P. Skeledon & G.E. Thompson, *Surf. Interface Anal.*, 2000, **29**, 887.
[88] K. Shimizu, H. Habazaki, P. Skeledon, G.E. Thompson & G.C. Wood, *Surf. Interface Anal.*, 2000, **29**, 151.
[89] V. Hoffmann, R. Dorka, L. Wilken & K. Wetzig, New developments in glow discharge optical emission and mass spectrometry, *Proceedings 8th Symposium on Laser Spectroscopy*, Taejon/Korea, 3. 4.11.00 (2000), 78–82.
[90] M. Hecker, N. Mattern, V. Hoffmann, C. Wenzel, J. Bartha, H.J. Engelmann & E. Zschech, MRS Proc. AMC 2001, in print.
[91] M. Hecker, D. Fischer, V. Hoffmann, H.-J. Engelmann, A. Voss, N. Mattern, C. Wenzel, C. Vogt & E. Zschech, Influence of N content on microstructure and thermal stability of Ta-N thin films for Cu interconnection, *Thin Solid Films*, 2002, **414** (Nr 2), 184–191.
[92] M. Hecker, R. Hübner, R. Ecke, S. Schulz, H.J. Engelmann, H. Stegmann, V. Hoffmann, N. Mattern, T. Gessner & E. Zschech, Effect of annealing on the microstructure of ultrathin tungsten nitride diffusion barriers for copper metallization, *Microelectronic Engineering*, 2002, **64** (Nr 1–4), 269–277.

[93] V. Hoffmann, R. Kurt, K. Kämmer, R. Thielsch, Th. Wirth & U. Beck, *Appl. Spectrosc.*, 1999, **53**, 987.
[94] R. Dorka, R. Kunze & V. Hoffmann, *J. Anal. At. Spectrom.*, 2000, **15**, 873.
[95] F. Lunel, Ph.D. thesis, Universite de Paris-Sud, 1997.
[96] L. Wilken, V. Hoffmann, H.J. Uhlemann, H. Siegel & K. Wetzig, *J. Vac. Sci. Technology*, submitted.
[97] P. Homman, F. Präβler, V. Hoffmann, S. Hänström & A. Bengtson, Investigation of Ar Emission Lines as Control Parameters in Glow Discharge Optical Emission Spectrometry (GD-OES), Fifth International Conference on Progress in Analytical Chemistry in the Steel and Metals Industries, 12–14 May 1998, Proceedings, European Commission, The European committee for the study and application of analytical work in the steel industry (CETAS), ISBN 92–828–6905–9, pp. 59–70.
[98] A. Bengtson, S. Hänström, E.L. Piccolo, N. Zacchetti, R. Meilland & H. Hocquaux, *Surf. Interface Anal.*, 1999, **27**, 743.
[99] F.W. Aston, *Mass Spectra and Isotopes*, Longmans, New York, 1942.
[100] J.J. Thomson & G.P. Thomson, *Conduction of Electricity through Gases*, Vol. 1, Cambridge University Press, 1928.
[101] A.J. Dempster, *Proc. Am. Phil. Soc.*, 1935, **75**, 755.
[102] G. Ramendik, J. Verlinden, R. Gijbels, Spark source mass spectrometry, in *Inorganic Mass Spectrometry* (eds) F. Adams, R. Gijbels & R. Van Grieken, John Wiley & Sons, New York, 1988, p. 17.
[103] J.W. Coburn, *Rev. Sci. Instrum.*, 1970, **41**, 1219.
[104] J.W. Coburn, E.W. Eckstein & E. Kay, *J. Vac. Sci. Technol.*, 1975, **12**, 151.
[105] W.W. Harrison & C.W. Magee, *Anal. Chem.*, 1974, **46**, 461.
[106] B.N. Colby & C.A. Evans, *Anal. Chem.*, 1974, **46**, 1236.
[107] D.L. Donohue & W.W. Harrison, *Anal. Chem.*, 1979, **51**, 673.
[108] C.G. Bruhn, B.L. Bentz & W.W. Harrison, *Anal. Chem.*, 1975, **47**, 1528.
[109] K. Robinson & R. Naylor, *Europ. Spectrosc. News*, 1986, **68**, 18.
[110] N. Jakubowski, D. Stüwer & W. Vieth, *Anal. Chem.*, 1987, **59**, 1825.
[111] N. Jakubowski & D. Stüwer, *J. Anal. At. Spectrom.*, 1992, **7**, 951.
[112] J.S. Becker, A.I. Saprykin & H.J. Dietze, *Int. J. Mass Spectrom. Ion Proc.*, 1997, **164**, 81.
[113] J.S. Becker & H.J. Dietze, *Int. J. Mass Spectrom. Ion Proc.*, 2000, **197**, 1.
[114] H.J. Kim, E.H. Piepmeier, G.L. Beck, G.G. Brumbaugh & O.T. Farmer III, *Anal. Chem.*, 1990, **62**, 1368.
[115] Y. Shao & G. Horlick, *Spectrochim. Acta, Part B*, 1991, **46**, 165.
[116] A.I. Saprykin, J.S. Becker & H.J. Dietze, *Fresenius J. Anal. Chem.*, 1996, **355**, 831.
[117] R.K. Marcus, *J. Anal. At. Spectrom.*, 1994, **9**, 1029.
[118] C. Yang, M. Mohill & W.W. Harrison, *J. Anal. At. Spectrom.*, 2000, **15**, 1255.
[119] V. Hoffmann, Poster presented at the European Winter Conference on Plasma Spectrochemistry, Hafjell, Norway 2001, p. 4–38.
[120] R.S. Mason, *Phys. Rev. E*, 1997, **55**, 7462.
[121] P.D. Miller, D. Thomas, R.S. Mason & M. Liezers, in *Recent Advances in Plasma Source Mass Spectrometry* (eds) G. Holland, BPC Wheatons Ltd, Exceter, UK, 1995, p. 91.
[122] C. Beyer, I. Feldmann, D. Gilmour, V. Hoffmann & N. Jakubowski, to be published in *Spectrochim. Acta, Part B.*
[123] D.A. Solyom, T.W. Burgoyne & G.M. Hieftje, *J. Anal. At. Spectrom.*, 1999, **14**, 1101.
[124] A. Montaser, *Inductively Coupled Plasma Mass Spectrometry*, Wiley-VCH, New York, 1998.
[125] W. Schelles, S. De Gendt, V. Muller & R. Van Grieken, *Appl. Spectrosc.*, 1995, **49**, 939.
[126] W. Vieth & C.J. Huneke, *Spectrochim. Acta, Part B*, 1991, **46**, 137.
[127] C. Vandecastele & C.B. Block, *Modern Methods for Trace Elemental Determination*, John Wiley, New York, 1993.
[128] F.L. King, J. Teng & R.E. Steiner, *J. Mass Spectrom.*, 1995, **30**, 1061.

[129] C. Venzago, L. Ohanessian-Pierrard, M. Kasik, U Collisi & S. Baude, *J. Anal. At. Spectrom.*, 1998, **13**, 189.
[130] G. Kudermann, *Fresenius J. Anal. Chem.*, 1988, **331**, 697.
[131] L.F. Vassamillet, *J. Anal. At. Spectrom.*, 1989, **4**, 451.
[132] D. Fang & P. Seegopaul, *J. Anal. At. Spectrom.*, 1992, **7**, 959.
[133] W. Vieth & J.C. Huneke, *Anal. Chem.*, 1992, **64**, 2958.
[134] M. van Straaten, K. Swenters, R. Gijbels, J. Verlinden & E. Adriaenssens, *J. Anal. At. Spectrom.*, 1994, **9**, 1389.
[135] M. Saito, F. Hirose & H. Okochi, *Anal. Sci.*, 1995, **11**, 695.
[136] M. Saito, *Fresenius J. Anal. Chem.*, 1995, **351**, 148.
[137] C. Venzago & M. Weigert, *Fresenius J. Anal. Chem.*, 1994, **350**, 303.
[138] D.M.P. Milton, R.C. Hutton & G.A. Ronan, *Fresenius J. Anal. Chem.*, 1992, **343**, 773.
[139] A.P. Mykytiuk, P. Semeniuk & S. Berman, *Spectrochim. Acta Rev.*, 1990, **13**, 1.
[140] P. Wilhartitz, H.M. Ortner, R. Krismer & H. Krabichler, *Mikrochim. Acta*, 1990, **2**, 259.
[141] W. Vieth & J.C. Huneke, *Spectrochim. Acta, Part B*, 1991, **46**, 137.
[142] W. Schelles, St. De Gendt, K. Maes & R. Van Grieken, *Fresenius J. Anal. Chem.*, 1996, **355**, 858.
[143] D.M.P. Milton & R.C. Hutton, *Spectrochim. Acta, Part B*, 1993, **48**, 39.
[144] J.C. Woo, N. Jakubowski & D. Stüwer, *J. Anal. At. Spectrom.*, 1993, **8**, 881.
[145] D.C. Duckworth, D.L. Donohue, D.H. Smith, T.A. Lewis & R.K. Marcus, *Anal. Chem.*, 1993, **65**, 2478.
[146] D.C. Duckworth & R.K. Marcus, *Anal. Chem.*, 1989, **61**, 1879.
[147] A.I. Saprykin, S. Becker & H.J. Dietze, *Fresenius J. Anal. Chem.*, 1996, **355**, 831.
[148] S.D. Hanton, *Chem. Rev.*, 2001, **101**, 527.
[149] C.R. Shick, P.A. DePalma & R.K. Marcus, *Anal. Chem.*, 1996, **68**, 2113.
[150] N. Jakubowski & D. Stüwer, *J. Anal. At. Spectrom.*, 1992, **7**, 951.
[151] D.J. Hall & N.E. Sanderson, *Surf. and Interface Anal.*, 1988, **11**, 40.
[152] A. Raith, R.C. Hutton & J.C. Huneke, *J. Anal. At. Spectrom.*, 1993, **8**, 867.
[153] N. Jakubowski, Analyse von Oberflächen und oberflächennaher metallischer Schichten mittels niederauflösender Glimmentladungsmassenspektrometrie, Thesis, University of Hohenheim, Germany, 1991.
[154] N. Jakubowski & D. Stüwer, *J. Anal. At. Spectrom.*, 1992, **7**, 951.
[155] J.C. Woo, H.J. Kim, H.B. Lim, D.W. Moon & K.W. Lee, *Bull. Korean Chem. Soc.*, 1992, **13**, 620.
[156] J.S. Becker, U. Breuer, J. Westheide, A.I. Saprykin, H. Holzbrecher, H. Nickel & H.J. Dietze, *Fresenius J. Anal. Chem.*, 1996, **355**, 626.

5 Microwave plasma atomic emission spectroscopy

Peter C. Uden

5.1 Introduction

Various microwave plasmas have been used as sources for analytical atomic spectroscopy for many years. The microwave-induced plasma (MIP) has been the most widely applied microwave plasma for elemental analysis, but others such as capacitively coupled and stabilized capacitive, surface-wave (surfatron) and microwave plasma torch have also been used. Although they are less widely applied than inductively coupled plasmas (ICPs), they have lower initial and operation costs and power requirements and have advantages in some instances, e.g. as gas chromatographic detectors, which are the key focus of this chapter. Many reviews on analytical aspects of microwave plasmas are available. Croslyn *et al.* provide an excellent critical review for the analytical chemist [1] and Winefordner *et al.* evaluate such spectrochemical sources [2]. Annual reviews are found in journals such as the *Journal of Analytical Atomic Spectroscopy – Atomic Spectroscopy Updates* [3].

Microwaves are electromagnetic waves in the 300 MHz–300 GHz frequency range and are utilized in a variety of fields such as radio and space communications, and analytically in sample preparation modes through microwave heating. At high microwave frequencies, conventional wiring and electronics are not used because of lead reactance and long energy transit times and thus they must be transmitted in a number of ways, by antennas through space and by waveguides or co-axial cables. Upon transmission to a discharge tube containing a suitable plasma gas, absorbed microwave power can generate a plasma which is a partially ionized gas medium having overall electrical neutrality. Maximum power is transmitted where the electric field component in the discharge tube is maximized. The plasma is *ignited* and sustained by collision of electrons with gas atoms, if a collision produces at least one new electron for each recombination.

5.2 Microwave-induced plasmas (MIP)

Helium or argon plasmas are maintained within a cavity, which serves to focus power from a microwave source (usually operated for analytical purposes at 2.45 GHz) into a discharge cell (often a quartz tube). The MIP may be operated at atmospheric or reduced pressure. Excitation mechanisms and thermodynamic equilibrium consideration have been compared for MIP and ICP [4]. Electron temperatures are higher in the MIP than in the ICP, but electron number density (free electron population) is lower, giving lower gas temperatures. Despite this effect, the high power densities of the helium discharges, in particular, produce high spectral emission intensities for many elements including non-metals. Low power MIP systems are less useful for liquid samples because the lower plasma energy available does not desolvate and vaporize aerosols effectively, resulting in plasma instability and extinction, but have been used as element specific gas chromatographic (GC) detectors especially for non-metals. There have been many studies on analytical figures of merit for MIPs [5, 6] and comparisons with other plasmas. Thus Sanz-Medel *et al.* compared performance of the Beenakker cavity [7], surfatron and microwave torch for chlorocarbons [8], the former proving superior. Variations in torch design have substantial effects on plasma properties. Tangential flow torches [9, 10] have advantages of improved stability while laminar flow torches have shown improved detection limits [11]. Torches can be cooled by liquid outer sheaths to give enhanced sensitivity and reduced interferences [12, 13]. While it has been suggested that the MIP may have advantages over the ICP for Mass Spectral detection notably in operational costs, Hieftje *et al.* noted obstacles regarding air entrainment and maintenance of adequate ion flux [14].

Gas phase introduction is the optimal operating mode for MIP usage. For direct sampling this may be by hydride generation or other chemical reaction, electrothermal vaporization (ETV) or graphite furnace atomization. Barnett miniaturized a hydride generation (HG) system for MIP, which allowed sub-ng determination of arsenic, antimony, tin and lead [15]. Halogens have been determined as hydrogen halides in a flow cell [16]. ETV-MIP was reviewed by Matusiewicz [17] and Dittrich *et al.* obtained a 90 ng detection limit for sulfur [18]. Abdillahi determined some non-metallic elements by ETV-MIP [29]. Graphite furnaces (GF) have also been used for MIP sample introduction. Broekaert *et al.* optimized sample introduction and obtained low picogram detection limits for some metals using GF-MIP [19, 20]. They also obtained sub-ng/ml detection for arsenic, antimony and selenium with a sequenced HG-GF-MIP system [21]. Gong *et al.* evaluated a direct-coupled hydride generation – microwave plasma torch for arsenic and antimony determination; detection limits of 8.1 and 3.2 ng/ml respectively were obtained for a 135 W plasma [22]. A schematic diagram of their instrumentation is shown in Fig. 5.1. Cold vapor mercury MIP determination using direct sampling or on line amalgamation and trapping has shown low pg detection for the element [23] and for organomercury compounds [24]. Croslyn *et al.* provide comprehensive figure of merit summary tabulations for more than 25 elements for gas phase, non-GC sample introduction and liquid phase introduction [1]. Table 5.1 presents a selection of elemental analytical figures of merit for various sample introduction modes.

Established liquid nebulization methods have been used for liquid sample introduction. Concentric glass nebulizers with Scott type spray chambers have been used [33], but pneumatic nebulization has been more popular; typical

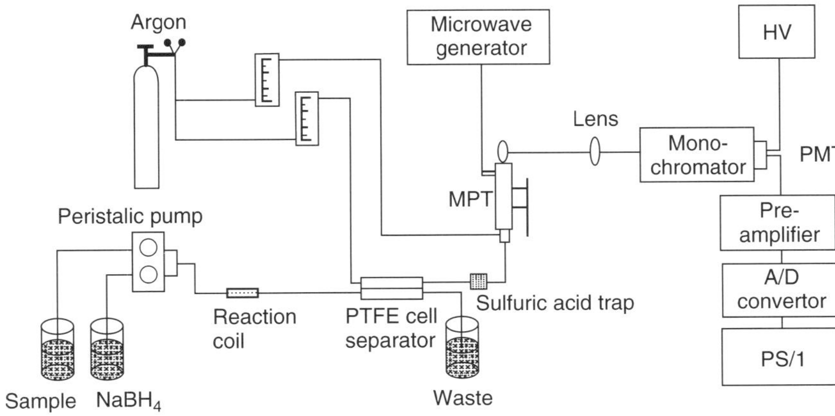

Fig. 5.1 Schematic diagram of hydride generation microwave plasma torch-AES system. Reproduced by permission from Gong *et al.* [22] Copyright 2001, Elsevier Science B.V.

detection limits being reported in the μg/ml range [34]. In general though for all but non-metals and metalloids, the establishment of ICP and ICP-MS has limited practical application of MIP except for some high performance liquid chromatography (HPLC) applications. Few elements have been determined as solids or slurries by direct introduction to the MIP and little quantitative data are available [1]. Laser ablation (LA) has been evaluated [35] but relative standard deviations (RSDs) were poor, 10–30% at the ppm level and higher at sub-ppm ranges. Spark ablation has been shown to provide improved precision but more energetic spark sources are probably needed for sub-ppm measurements [36]. Direct entrainment of solids at room temperature has been demonstrated by Carnahan *et al.* with 500 W helium MIPs [37].

5.3 MIPs as chromatographic detectors

In gas chromatography (GC), interfaced detection (hyphenation) can greatly expand its usefulness. When MIPs are used as atomic emission chromatographic detectors (AED), detection is based on monitoring the emission spectra that result from the excitation of the solute from the chromatographic column. Monitoring of emission lines that are specific to a particular analyte element enables the detector to respond only when a molecule that contains that element is eluted. While it may be arguable to denote such detection rigorously as *specific* (implying an infinitely high relative response against background) in many cases, there is sufficiently great selectivity over the background to warrant this. Specificity and selectivity are often used interchangeably in GC-AED discussions.

Table 5.1 Selected analytical figures of merit for the MIP with gaseous sample introduction

Element	λ nm	Sample Introduction Method	Plasma Gas	Power watts	LOD ng/ml	LOD ng	Linear Dynamic Range	Ref.
Ag	328.1	C	He	100	0.2	0.0006	0.002–3.0 ng	25
As	228.8	D	He	135	8.1		500	22
Br	470.5	A	He	130		1	0.005–50 μg	26
Cd	228.8	C	He	500	0.6	0.010	0.01–10 μg/ml	27
Cl	479.5	B	He	500	8	0.120	0.25–20 μg/ml	27
Co	240.7	E	Ar	90		0.305		20
Cr	425.4	E	Ar	90		0.047		20
Cu	324.7	E	Ar	90		0.024		20
Fe	238.2	E	Ar	90		0.055		20
Hg	253.6	F	He	80	0.01	0.0005	>4 orders	28
I	206.2	C	He	130		0.2	3–4 orders	30
In	410.2	C	He	75	141		>3 orders	30
Mn	259.3	E	Ar	90		0.011	>4 orders	20
Ni	352.4	E	Ar	90		0.220		20
Pb	405.7	E	He	75–135	1		>3 orders	20
S	180.7	A	He	100	0.13		4 orders	31
Sb	259.8	D	He	135	3.2		500	22
Se	196.0	D	He	100	40		>3 orders	32
Sn	317.5	D	He	100	290	2.9	1–500 μg/ml	15
Tl	377.5	E	Ar	90		0.028		20

A – Chemical generation; B – Chemical vaporization; C – Electrothermal vaporization; D – Hydride generation; E – Graphite furnace; F – Amalgamation trap.

Gas chromotography with atomic emission detection (GC-AED), was introduced by McCormack *et al.* [38] and Bache and Lisk [39] in 1965, using a reduced pressure MIP. They reported detection of P, S, F, Cl, Br, I and C with detection limits between 10^{-7} and 10^{-12} g/s, but selectivities against carbon were very poor. Research described in the early literature was conducted with laboratory-built instruments, as commercial instruments were not introduced until the late 1970s when a multi-channel instrument using a reduced pressure helium plasma, polychromator and Rowland Circle optics was introduced [40]. Advances were made, notably that helium was preferred as the plasma gas because it provided a simpler background spectrum, higher excitation energies and a wider linear response range than argon. Hagen *et al.* used chlorofluoroacetic anhydride to introduce elemental taggants (F and Cl) to facilitate GC-AED of acylated amines [41] while Zeng *et al.* optimized oxygen selective detection [42]. These studies were reported and reviewed in numerous publications [43, 44].

Another improvement was the atmospheric pressure plasma cavity; the TM_{010} resonant cavity described by Beenakker improved power transfer efficiency and allowed axial viewing of the plasma, instead of transverse through the discharge tube walls whose properties change with time [7]. The atmospheric pressure plasmas simplified and improved GC detection, but problems such as memory effects, non-specificity and low sensitivity for some elements were still present. Initial applications for packed column GC necessitated solvent venting. Early applications for packed column GC necessitated a valve for solvent venting to avoid extinguishing the plasma [45], but inlet injection volume splitting typically used for capillary GC avoids this system complexity. It is still customary however to utilize instrumental control software to bypass detector signal during solvent elution, since this causes detector overload even though the plasma remains lit.

TM_{010} cavities have been used for numerous capillary GC-MIP element-selective analyses and examples of these are discussed subsequently for metals and non-metals. In 1989, Hewlett Packard introduced the first commercially available multi-element GC-MIP system (AED the HP 5921A) [46, 47], thereby increasing the interest and availability of gas chromatographs with atomic emission detection (GC-AED). Hewlett Packard introduced a second generation GC-AED, the G2350, in 1997 [48, 49].

The ideal GC detector must monitor eluting mobile phase continuously and immediately, both qualitatively and quantitatively, to give a permanent output record. An element-selective GC detector must respond to the eluate element independently of the molecular structure of the analyte. Such structure-independent response allows the use of one or a few reference compounds to quantify any sample, using what is referred to as *compound-independent calibration* (CIC). CIC reduces costs of analysis, since many reference materials are expensive, toxic or not available at the desired purity level. It also reduces analysis time, since preparation of multi-component standard mixtures is not necessary.

The operation and applicability of different detectors can be evaluated by several performance figures of merit, which include the sensitivity, noise, minimum detectable quantity or detection limit, response time and selectivity. Sensitivity is a measure of the magnitude of the signal generated by the detector for a given amount of analyte. Ideally, it should be independent of analyte concentration, in which case a plot of detector signal as a function of analyte concentration should be linear. However, in practice, sensitivity diminishes at some value for each particular detector, and linearity is a measure of the extent to which sensitivity is directly dependent on sample size. The linear dynamic range of a detector is often taken as the range of sample amount within which the detector response is constant to within 20%. A detector is considered to be selective if its response to a certain type of analyte differs from that to another type of analyte by a factor higher than 10. A detector that is not selective in any fashion is called universal.

Plasma emission detectors have high selectivity, sensitivity and large linear dynamic range; they can be used to detect a wide variety of elements, often simultaneously and in unresolved matrixes. They are classified according to how the plasma is generated. In the ICP, the discharge results from the interaction of a radio-frequency (RF) field with argon flowing through a quartz tube set within a copper induction coil to generate a plasma that can reach 10 000K. The ICP is generally used for liquid chromatography (LC) detection, but its main drawback is that it provides poor excitation efficiency for non-metallic elements.

The direct-current plasma (DCP) in which the plasma can be generated directly by an electrical discharge through the argon gas produces a line-rich spectrum valuable for multi-element analysis. It has been used for HPLC and GC [50, 51], primarily for metals which are readily excited in argon plasmas.

Figure 5.2 shows a diagram of a GC-AED (HP 5921A) instrument system. The sample is injected through a split/splitless or an on-column injection port; the components are separated in a capillary GC column, and enter the discharge tube situated within the detector cavity that holds the plasma. Additional helium flow is added to the helium carrier gas to sustain the plasma. Small amounts of other gases, such as hydrogen, oxygen or methane, may be added to stabilize the plasma or enhance detection of a particular element of interest. These gases function as *reagent* or *scavenger gases*. The amount and composition of the reagent gas mixture can be varied to improve detection.

The MIP temperature range is between 5000 and 6000°C with sufficient energy to break the bonds and excite the atoms present in the molecules that elute from the column. The atoms formed are excited to higher electronic states and as they decay to lower energy levels, they emit light at characteristic wavelengths. The light emitted is focused on to a spectrometer and the intensities of

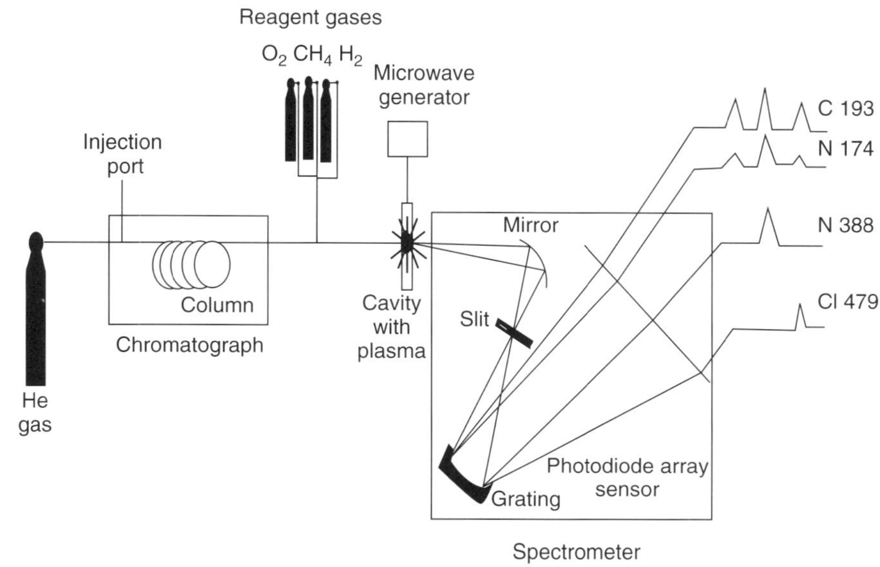

Fig. 5.2 Gas chromatograph microwave plasma atomic emission detector system.

the emission lines are measured at wavelengths that are characteristic of the atoms of interest. The intensity of the emitted light is proportional to the number of atoms in the plasma and is ideally independent of the structure of the parent compound. The HP 5921A uses a modified version of the *Beenakker* cavity [7] wherein the plasma is located inside a thin-walled silica discharge tube, which is inside a water-cooled microwave re-entrant cavity. The plasma is generated in an atmospheric pressure flow of helium, comprising a column flow and an additional make-up gas flow. Power is supplied by a magnetron and coupled to the plasma through a waveguide, reducing the need for frequent tuning. The power level is typically ca. 50 W and it is operated at 2.45 GHz. A diagram of the features of the cavity can be found in Fig. 5.3.

The modifications to the Beenakker cavity addressed some limitations of the original design. The water-cooling system reduced problems with high plasma temperatures, such as limited lifetime of the discharge tubes, interaction between analytes and active sites on the tube, and high oxygen and silicon background emission. The system has sealed chromatographic and optical path interfaces that prevent air leakage into the plasma but allow the addition of reagent gases to improve detection of certain elements. A purge flow can also be used at the spectrometer end of the exit chamber of the cavity, to prevent deposits on the viewing window of the spectrometer. The system is also equipped to run the make-up helium flow in reverse mode. In this mode, the helium flows from the cavity toward the column and it is used to keep the

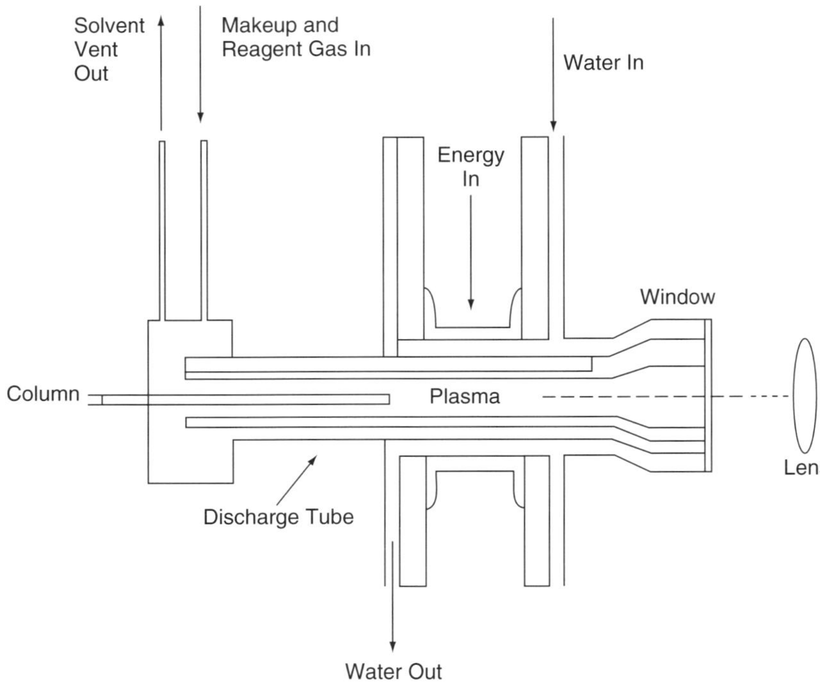

Fig. 5.3 Schematic of GC-AED microwave cavity (Agilent HP 5921A).

column flow out of the discharge tube and the plasma. This option is usually used to avoid large quantities of solvent reaching the plasma thereby causing its extinction; for that reason it is referred to as the solvent vent.

A movable photodiode array detector in the 5921A responds to emission from the plasma grating and the transducers being positioned, so that each diode corresponds to a portion of the spectrum, making it possible to monitor different wavelengths simultaneously. A computer that allows not only chromatographic conditions to be set, but also improvement or generation of selective detection recipes, and evaluation and integration of results controls the whole system. The later version of the GC-AED, the HP G2350A has a spectrometer with an optical resolution of 0.25 nm in comparison with 0.4 nm of the earlier version [48, 49]. The wavelength per diode in the new instrument is 0.08 nm/pixel compared with 0.22 nm/pixel in the former instrument. These two improvements result in increased selectivity and improved detection limits. Another improvement was the introduction of a new hydrogen flow restrictor that is five times less restrictive than the old one. This allows use of larger hydrogen flows that help to improve detection of nitrogen through measurement of the cyanogen molecular emission band at 388 nm.

Another variation of the GC-MIP combination is the use of on-column AED. As a way to reduce instrumentation costs, Pedersen-Bjergaad *et al.* reported the use of a 350 kHz helium plasma sustained inside the end of the 0.32 mm ID capillary column [52]. A feature of the small volume of the detector cell was that the energy within the miniature plasma was very high. Improved detection limits and reduction of gas consumption were among the advantages reported.

In theory, the MIP detector has enough energy to atomize and excite all the elements of eluates. Thus, response should depend on the element concentration in the plasma and not on the eluate's molecular structure, thereby enabling CIC methodology to be applied. Eluates that reach the detector have high purity so constituent elements can be detected with high selectivity and often simultaneously, thereby reducing the amount of sample required for analysis. Thus, GC-AED has been utilized for the determination of empirical formulae (EF). Although accurate and reliable microanalytical methods are available for EF measurement, they require 1–10 mg of pure material for each element to be determined. Classical microanalysis cannot be carried out for materials that are components of a mixture.

However, CIC and EF methods for GC-AED have sometimes lacked robustness due to some elemental response dependence on molecular structure. Dingjan *et al.* [53] reported good EF determination for halogen and sulfur compounds, but precision and accuracy decreased with increasing number of atoms per molecule. The consistent results improved sensitivity and selectivity obtainable from commercial GC-AED systems and increased interest in EF and CIC determinations. Sullivan and Quimby reported C, H, N and O response factors accurate at 2–7%, but mandated caution for EF or CIC because they did not apply for all compounds tested [54]. Significant errors were reported for EF determinations of hydrogen in chlorinated hydrocarbons, reflecting concentration response dependence. Huang *et al.* reported that the structure of the reference compound used for EF determination has a significant effect on the results for hydrocarbons, polynuclear aromatic, chlorinated hydrocarbons and brominated hydrocarbons [55]. A software algorithm was developed by Wylie *et al.* to calibrate and measure elemental ratios for all eluate peaks, and was applied to phenol mixtures, pesticides, pollutants and petroleum products; non-linearity for hydrogen response was again observed [56]. Kovacic and Ramus reported compound dependence of response factors for some chlorine compounds and most fluorine compounds [57]. Pedersen-Bjergaard *et al.* determined C/H and C/N ratios showing them to be affected by both the molecular structure and the concentration of the reference standard [58]. However, similarity in structure did not always improve the C/N ratio determination. CIC was used to determine the concentration of a reference compound, but errors ranging from 10 to 30% were observed [59]. Albro *et al.* concluded that CIC for determination of sulfur compounds is accurate and precise enough for the analysis of petroleum samples [60]. Janák *et al.* reported AED response to be affected by molecular structure, analyte concentration and elution temperature [61], while Webster and Cooke showed elemental response for chlorophenols as independent of molecular structure for chlorine, but not for carbon and oxygen [62]. Becker *et al.* evaluated elemental response ratios for C, H, Cl, P and O for alkylated and arylated phosphates, accurate EFs being obtained when the calibrating compound was closely related to the analyte by structure, elemental composition, molecular weight and amount [63]. Herbicides were determined by Olson *et al.*, CIC behavior being reported for Cl, N and S [64]. Independence of response to structure was shown adequate for CIC of organochlorine compounds [65]. The latest marketed version of the GC-AED system, the Agilent G2350A, was evaluated for EF determinations by Szelewnski [66]. For compounds of similar structure, elemental response factors were within $\pm 10\%$. A version of the earlier Agilent (Hewlett Packard) instrument, optimized for nitrogen determination, was used by Gonzalez and Uden for determinations of CIC and EF of nitrogen-containing compounds with N detection at 388 nm [67].

5.3.1 Analytical information from GC-AED

Elemental detection limit and sensitivity in the AED depend on its spectral intensity at the analytical emission wavelength and the signal to background observed. Each element has many emission wavelengths and the optimal must be chosen for analytical purposes. Sensitivity, the slope of the response curve, is less frequently measured than *detection limits* in C-AED, given as an absolute value of element mass (in a resolved peak), mass/time interval or in mass flow rate units.

Interelement selectivity depends on emission properties of the element and of possible interferences, and on the resolution of the spectroscopic measurement system. A useful definition is the peak area response per mole of analyte element divided by the peak area response of the *background* element, per mole of that element. Dynamic Measurement Range of response in capillary GC-AED typically extends from the upper linear analyte capacity, ca. 100 ng, to the detection limit of the target element that may be in the pg range.

Carbon selective detection may be considered as a *universal* mode of detection for organic compounds. The most often used emission wavelength is from the carbon ion line at 247.9 nm, but other spectral emission features corresponding to CN, CH, etc. may be used. This AED mode response is analogous to flame ionization but is more completely independent of carbon atom environment and exhibits as great or greater sensitivity.

Table 5.2 summarizes elemental detection limits, selectivities and linear dynamic ranges for atmospheric pressure microwave-induced helium plasma capillary GC detectors. It should be emphasized that detection limits reported

Table 5.2 Selected analytical figures of merit for GC-MIP

Element	λ (nm)	Selectivity vs carbon	LOD pg/ml	LOD pg/s	LOD pg	Linear Dynamic Range	Ref.
Al	396.2	>10 000		5.0		>1000	49
As	189.0	50 000		3.0	3	1000	143
B	250	9300		3.6		500	126
Be	313		0.3				144
Br	478	14 000		18.3			49
C	193	1		6			49
^{13}C	177	1400 (vs^{12}C)		20.1			49
Cl	479	13 000		14.3			49
Co	241	1 80 000		6.2		500	126
Cr	268	1 08 000		7.5		1000	126
Cu	325	>10 000			0.8	1000	118
D	656	112 (vsH)		0.6		500	49
F	686	30 000		40		2000	49
Fe	302	4 600 000		0.1		1000	49
Ga	294	>10 000		200		500	115
Ge	265	7600		1.3		1000	126
H	468			2.2		500	49
Hg	254	2 58 000		0.6			49
I	206	5000		21		500	126
In	326	600		8000		500	115
Mn	258	1 10 000		1.6		1000	126
Mo	282	24 000		5.5		500	126
N	174	7000		4.6			49
N	388	9 15 000		18.1			49
Nb	288	32 000		69		100	126
Ni	301	>10 000		1.0		1000	118
P	186	14 000		0.5			49
O	777	25 000		75		4000	49
O	171	18 000		43.4			49
Os	226	50 000		6.3		1000	126
Pb	261	3 14 000		0.8			49
Pd	340	>10 000		5.0		1000	118
Pt	300	1 00 000		65		1000	116
Ru	240	1 30 000		7.8		1000	126
S	181	37 000		0.6		20 000	49
Sb	218	19 000		5.0		1000	49
Se	196	50 000		4.0		1000	49
Si	252	90 000		7.0		40 000	49
Sn	326	3 300 000		0.4		30 000	49
Ti	338	50 000		1.0		1000	107
V	292	36 000		4.0		36 000	107
W	256	5400		51		100	126

are usually *best claimed* and working limits are best assumed at a higher order of magnitude. Also many of the metal detection limits were obtained with pre-commercial MIP systems and have not been confirmed with current systems; the data should provide useful guidelines in any event. In cases when poorer limits are seen (e.g. niobium, gallium and indium), chromatographic degradation and/or thermal instability may be partially at fault.

Analytical figures of merit and general developmental features of the GC-MIP methodology have been covered in a number of reviews during the past decade and readers are referred to such for more detailed discussions [44, 68–73].

5.4 GC-AED detection of non-metallic elements

Although many of the more striking applications of GC-AED are in the areas of inorganic and organometallic application, as noted subsequently, it is essential that a system provides good analytical performance for non-metals and particularly organic heteroatom elements such as the halogens, sulfur, nitrogen and oxygen. The helium MIP has been most widely used for the detection of non-metals, since for most of these elements argon metastable energy carriers show insufficient collisional energy transfer for adequate excitation. Initial studies with the commercial instrumental system mentioned earlier focused upon C, H, N and O detection including sensitivity, precision, peak shapes induced by the plasma and response variation between compounds [46, 47]. The precision ranged from 2 (C, H, N) to 7% (O).

For halogen detection, the AED has advantages over other selective GC detectors: the electron capture detector (ECD) and the Hall electrolytic conductivity detector (HECD). It does not have as great sensitivity as the ECD for some polyhalogenated compounds, but has the advantage of almost uniform elemental response for a specific halogen, irrespective of analyte molecular structure. Selective GC-ECD detection may also be precluded if electron-capturing interferents are present. A major advantage over the HECD is element specificity for different halogens, which is not shown by the halogen-specific mode of that detector. Although MIP response for chlorine is lower than for most metals, selectivity and sensitivity are adequate for many tasks (Table 5.2). Thus polychlorinated biphenyl cogeners were quantified by CIC and chlorine response varied by only 9% from tetra- to octachloro compounds [74]. A number of studies reported similar behavior for Cl and Br in aliphatic and aromatic compounds [75–77]. An example of halogenic derivatization was described by Hardas *et al.*, who brominated olefins quantitatively to determine their content in fuel oils and gasoline [78]. Polybrominated diphenyl ethers (PBDEs), as employed in flame-retardants, were quantitated by Johnson and Olson in fish tissues. Levels differing by 10-fold were noted in remote and urban samples [79].

As discussed earlier, CIC has been studied for several of the non-metal elements but rather conflicting results are recorded in the literature, one reason being that there is no general agreement on how large a discrepancy in elemental response factor (ERF) between the compounds analyzed can be tolerated, if a constant response factor can still be presumed. The investigator must decide whether CIC is demonstrated or not. A detailed description of CIC for many pesticides is available from the US Environmental Protection Agency (EPA). The elements Br, Cl, F, N, P, S and I that occur in pesticides are discussed in EPA Method 8085. For the RSD with a typical ERF, an upper limit of 10% is defined (20% for P) if CIC is to be considered valid.

As noted earlier, Wylie *et al.* developed EF calculation algorithms using both internal and external standards [56]. Figure 5.4 shows a hydrogen-selective chromatogram for a homologous series of polychlorinated biphenyls, each compound containing 12 carbon atoms and a total of 10 hydrogen and chlorine atoms. Good correspondence was seen for response factors ratioed against hydrogen peak height; values for chlorine content within 0.2 of an integer up to 5 chlorine atoms in the molecule were obtained. Such determinations do not achieve the accuracy of conventional microanalysis, but measurements are made directly on GC eluent peaks for analyte levels up to six orders of magnitude smaller.

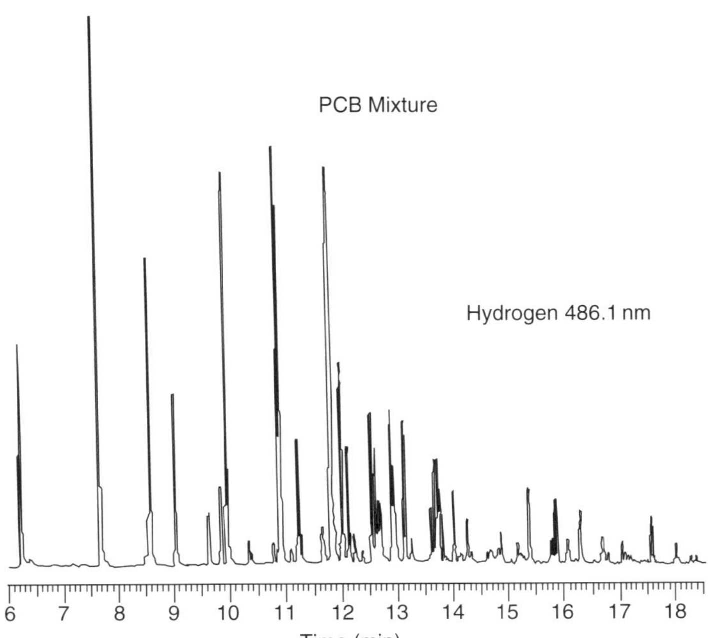

Fig. 5.4 GC-AED hydrogen response to a polychlorinated biphenyl (PCB) mixture. Reproduced by permission from Wylie *et al.* [56] Copyright 1990, Dr Alfred Huethig Publishers.

Kovacic and Ramus [57] evaluated GC-AED ERFs for compound independence; C, Cl, F, N and O showing 3, 6, 5, 6 and 5% RSDs respectively for a wide range of aliphatic, aromatic and heterocyclic compounds. Oxygen-selective detection with oxygen to carbon selectivity of 10^2 and a thousand-fold linearity was found by Bradley and Carnahan [80] with a TM_{010} cavity and a polychromator system. Background oxygen spectral emission from plasma gas impurities, leaks or back-diffusion into the plasma was minimized to give oxygen detection limits from 2 to 500 ppm in complex petroleum distillates.

Pederson-Bjergaard and Greibrokk [81] evaluated an RF plasma detector such as described by Skelton *et al.* [82] for oxygen-selective detection. Purification of plasma and purge gases was critical, and interferent oxygen signals deriving from release of the element from the discharge tube by eluting hydrocarbons could be minimized by using methane doping in a 15 W plasma. A 1000 to 1 oxygen to carbon selectivity and a 100 pg/s detection limit were obtained. A study of sulfur response found a tendency to decrease with increasing compound volatility, and the closer the chemical structure of the reference compound was to that of the analyte, the better the quantification [58]. In contrast to other findings that reported an identical response for sulfur in different functional groups [75, 83] and in 15 compounds of different molecular weights with the same functional group, the average deviation of the response factor from the real value was 4% (for 0.1 ng sulfur injected) [84]. Further examples of petroleum analysis using GC-AED are cited by Kosman [85] who notes the value of CIC wherein a simple compound such as methane can be used to calibrate the detector for carbon and hydrogen simultaneously for application in complex matrices. The importance of element-selective detection profiles in simulated distillation applications, and pattern recognition using element ratioing to type crudes and oil spills is also emphasized. Sandra has applied GC-AED for a range of environmental air pollution applications, the highly toxic compound carbonyl sulfide being determined in the presence of high levels of hydrocarbons [86]. One hundred milliliters of vapor samples was injected onto porous polymer coated capillary columns for sulfur determination at the ppb level. Sulfur species in city air were determined after concentrating on a thick silicone film trap, thermal desorption and cryofocusing. GC-AED has been interfaced with analytical pyrolysis of geochemical materials. Rowland *et al.* [87] applied sulfur-selective pyrolysis-GC-AED to diatomaceous Pleistocene sediments and applied the data together with mass spectrometry (MS) information to identify both higher and lower volatility pyrolyzates. Figure 5.5 shows representative GC-AED and GC-MS chromatograms. Samples of the order of 200 mg were pyrolyzed for 20 seconds at 610°C in a quartz probe within a platinum coil.

GC-AED has been compared critically with other element-selective detectors, the ECD, ELCD (Hall detector), NPD and FPD, for the determination of pesticide residues in agricultural products [88]. Twelve agricultural samples

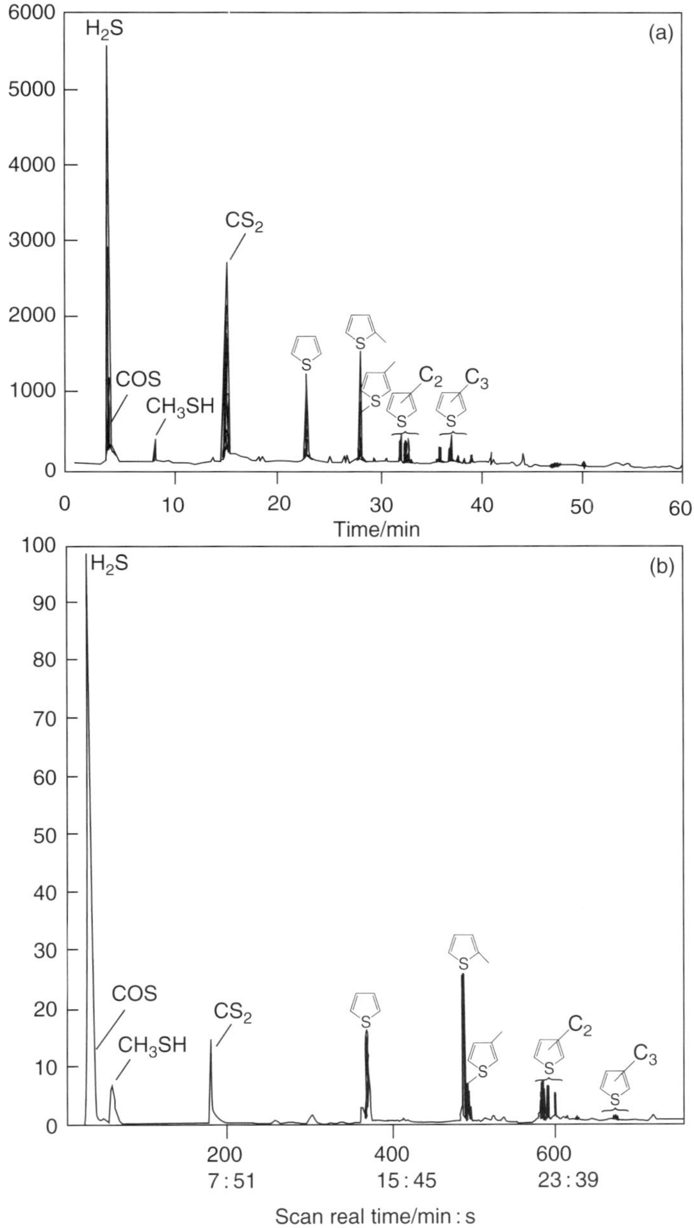

Fig. 5.5 (a) Pyrolysis-GC-AED (S emission, 181 nm) of a diatomaceous Pleistocene sediment from Peru upwelling zone. (b) Pyrolysis-GC-MS (summed mass fragmentogram m/z 34, 48, 76, 84, 98, 112). Reproduced by permission from Rowland *et al.* [87] Copyright 1993, The Royal Society of Chemistry.

were fortified with ten commonly applied pesticides that were then extracted without the use of clean-up steps. The AED was used for C, P, Cl, F, N and S analyses and its high selectivity allowed analysis in all 12 extracts. By contrast, interferences restricted the ECD use to five samples, the ELCD to eight, and the FPD and NPD to nine only. For the samples that produced the largest levels of interferences for other detectors, namely onions, strawberries and alfalfa, AED provided the greatest degree of selectivity.

5.5 GC-AED detection of metallic elements

GC-AED detection is valuable in confirming elution of metallic compounds and in obtaining valuable analytical data, since metal detection is usually more sensitive and selective than that of non-metals. Many volatile binary metal compounds, organometallic compounds and metal chelates can be quantitatively gas chromatographed, although in some instances the absence of confirmatory elution data based on elemental content of eluates has cast doubt on GC authenticity [89].

5.5.1 Main group metallic compounds

As shown in Table 5.2, GC-AED data have been obtained for many main and transition group metals. Among the former, tin, lead and mercury have been studied most frequently, as they are determinable by GC-MIP with TM_{010} cavities with sub-pg/second detection limits. In the analysis of sigma bonded main group elemental compounds, a limited range of known compounds is often the target, for example different methyl- and ethyl lead compounds. Established analytical schemes include alkylation of ionic species to give a series of methyl- and ethylalkyl lead compounds as solutes for chromatographic determination. Lead and carbon specific detection were compared for trialkyl lead chlorides extracted from an industrial plant effluent, and reacted with butyl Grignard reagent to form trialkylbutyl lead derivatives [90]. The degree of interference from carbon-containing compounds prevented any qualitative or quantitative determination of the trialkyl lead compounds by GC-FID, GC-ECD or GC-MS without extensive clean-up and loss of analyte.

Lobinski and Adams speciated organolead compounds in water at the sub-ng dm^{-3} level [91]. Sample preparation involved extraction of ionic organolead species as dithiocarbamate complexes, followed by Grignard propylation. Sample injection to the capillary column followed concentration on a Tenax trap. Figure 5.6 shows a mixture of propylated lead species present in a rain-water sample at the ng/l level.

Olsen *et al.* compared reduced and atmospheric pressure MIP systems for mercury detection and found a 1 pg detection limit for the latter, with

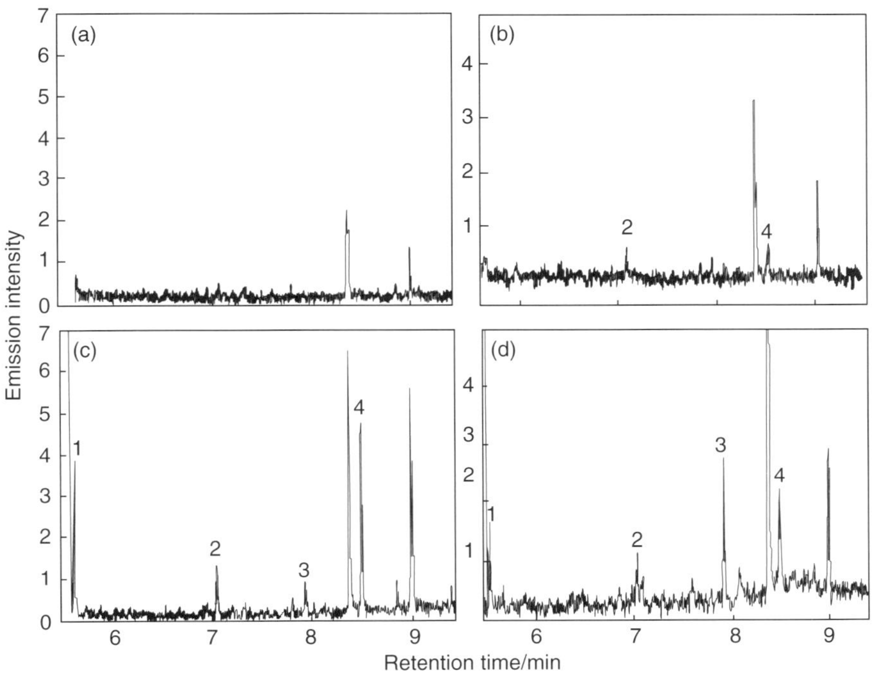

Fig. 5.6 GC-AED of propylated organolead species: (a) blank during derivatization of pure hexane with propylmagnesium chloride; (b) blank from complete analytical procedure; (c) chromatogram from tap water; (d) chromatogram of a polar snow sample. Peaks: 1 Me_3Pb^+; 2 Me_2Pb^{2+}1;3 Et_3Pb^+ and 4 Et_2Pb^{2+}. Reproduced by permission from Lobinski and Adams [91] Copyright 1992, The Royal Society of Chemistry.

selectivity of 10 000 over carbon [92]. Emteborg *et al.* described simultaneous determination of methyl- and ethyl mercury and inorganic mercury by dithiocarbamate resin concentration in a flow injection system, followed by butylation [93]. Using 500 ml water samples, a detection limit for methylmercury of 0.05 ng/l was obtained. Figure 5.7 shows response from a 20 ml water sample from the Gulf of Bothnia, Sweden, showing 0.35 ng/l of MeHg. Snell *et al.* studied mercury speciation in gas condensate fractions but encountered problems with plasma quenching and background correction due to the continually changing amount of hydrocarbons leaving the column [94]. Bouyssiere *et al.* investigated this problem further and made practical comparisons with GC-ICP-MS detection [95].

Lobinski *et al.* also reported parallel derivatization procedures for organotin compounds, allowing detection at about 0.05 pg [96]. Tributyltin species have also been determined by GC-AED utilizing solid phase extractive (SPE) concentration from sea water, followed by conversion to the corresponding hydride using solid sodium borohydride reagent, in line to the capillary column [97]. It

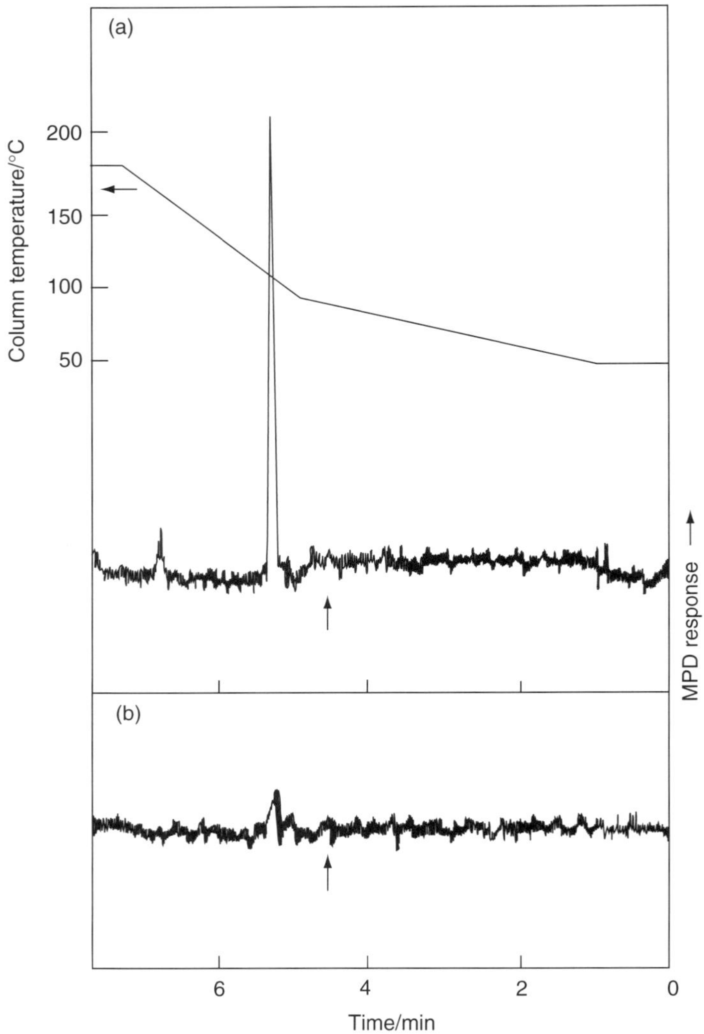

Fig. 5.7 GC-AED (mercury emission, 253.8 nm) of water from Gulf of Bothnia, Sweden. (a) 20 ml butylated water sample from 100 m level. (b) Blank. Reproduced by permission from Emteborg *et al.* [93] Copyright 1992, The Royal Society of Chemistry.

was found that interferences in the GC-AED chromatogram were due to volatile organotin residues in the SPE cartridge; these could be removed by solvent treatment of the cartridge.

As Andersson notes, the calibration plots for organometallic compounds often have different slopes, making individual external calibration necessary

[98–100]. However, CIC has been reported for tin [101]. In organotin analysis, tripentyltin chloride was used as a standard to check recoveries and tetrabutyltin as a quantification standard [102]. For many organometallic compounds a lower response is obtained for less volatile derivatives, e.g. for lead [102], mercury [103] and tin [104]. This may not be a detector phenomenon but may result from differential thermal decomposition at the end of the transfer line close to the plasma. Increasing the distance between the end of the transfer line and the plasma reduced the effect, as did increasing the temperature of the transfer line [102]. A thicker film in the transfer line can also give reduced degradation [105].

Optimization of flow rates of reactant and make-up gases is necessary to reduce the extent of deposition of metal or metal oxides on the discharge tube wall which has been found to occur for some metals [102, 106, 107]. Oxygen is often used as reactant gas to prevent carbon deposits from building up in the detector system and has been reported to have a beneficial effect on metal deposits [104]. Since the response to metals usually decreases as oxygen flow is increased, it was omitted in one study on methylmercury; carbon build-up was not observed but carbon background increased [105]. It is always important to optimize reagent gas flow rates carefully for the elements to be analyzed because each detector will need somewhat different conditions to perform optimally [98, 103, 108]. If several metals are to be analyzed, each will require individually optimized flow rates for the make-up and reactant gases for the most sensitive detection [98]. The chromatograms in Fig. 5.8 depict the simultaneous detection of carbon (at 248 nm), tin (271 nm), lead (261 nm) and mercury (254 nm) from a soil sample spiked with 23 alkyl- and phenyl-metallic compounds [103].

GC-AED has been used to detect and determine volatile organoselenium compounds present in or produced by plants. Headspace-GC-AED was used to detect selenium compounds in members of the *Allium* family such as garlic, elephant garlic, onion and broccoli (*Brassica*) [109]. A common structure seen is $R\text{-}S_x(Se_y)\text{-}R'$, where the R and R′ are methyl or allyl groups, and x or y may be 1–3. Figure 5.9 shows a typical chromatogram. Natural abundance of organoselenium compounds in human breath after ingestion of garlic was identified using Tenax trap/cryogenic-GC-AED [110]. Calle-Guntinas *et al.* reported absolute detection limits of 10 pg for dimethylselenide with the same instrument [111]. Free seleno-amino acids such as selenocysteine, Se-methyl selenocysteine, and selenomethionine in normal and selenium-enriched plants have been determined using GC-AED of ethylated derivatives [112].

5.5.2 GC-AED of transition metal compounds

The practical utility of transition metal chelate GC is determined by an optimal balance of volatility and thermal stability during injection and separation.

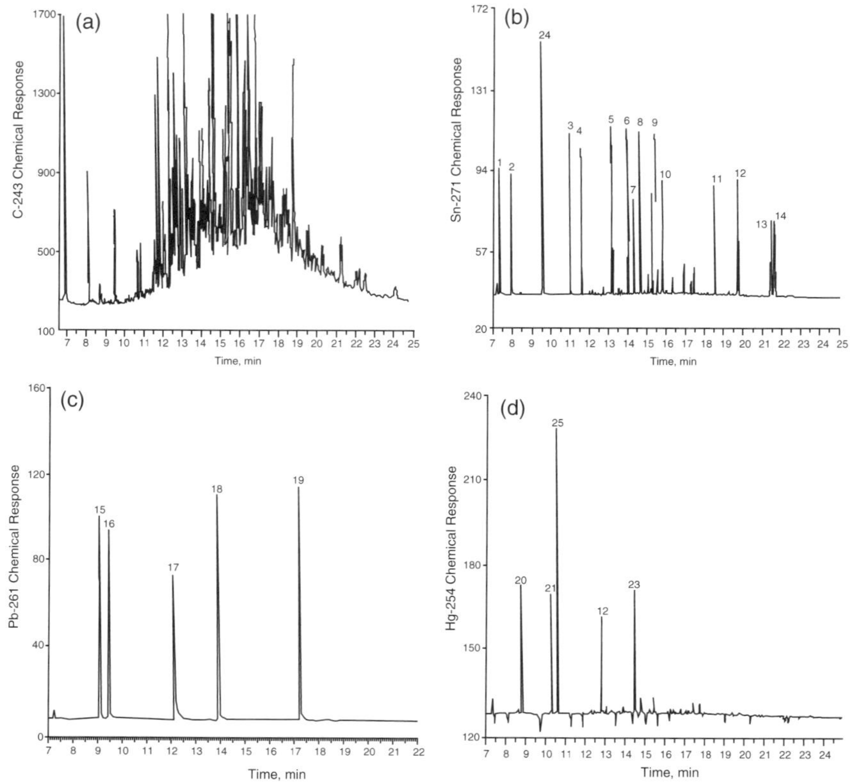

Fig. 5.8 Multi-element GC-AED of soil extract spiked with alkyl and aryl metal organometallics. (a) carbon emission, 248 nm, (b) tin emission, 271 nm, (c) lead emission, 261 nm, (d) mercury emission 254 nm. Reproduced by permission from Liu *et al.* [103] Copyright 1994, Wiley-VCH.

β-diketone, β-thioketone β-ketoamine, dialkyldithiocarbamate and dialkyldithiophosphate chelating ligand systems have all received broad study [113]. Black and Sievers used microwave plasma detection to determine chromium as its trifluoroacetylacetonate in blood plasma with excellent quantitation and precision [114]. An example of the utility of GC-AED detection was in a study of ligand redistribution and reaction kinetics of gallium, indium and aluminium chelates [115]. Figure 5.10 shows a capillary GC study of redistribution of trifluoroacetylacetone (TFA) (1,1,1-trifluoropentane-2, 4-dione) and TIB (1,1,1-trifluoro-6-methylheptane-2, 4-dione) ligands on gallium and aluminium atom centers. Al $(TFA)_3$ and Ga $(TIB)_3$ were reacted in chloroform solution for 24 hours at 25°C and the products monitored with carbon, aluminium and gallium specific GC detection. Individual peaks were identified as Al $(TFA)_2(TIB)$ – 1; Ga $(TFA)_2(TIB)$ – 2; Al $(TFA)(TIB)_2$ – 3; Ga $(TFA)(TIB)_2$ – 4; Al $(TIB)_3$ – 5 and Ga $(TIB)_3$ – 6.

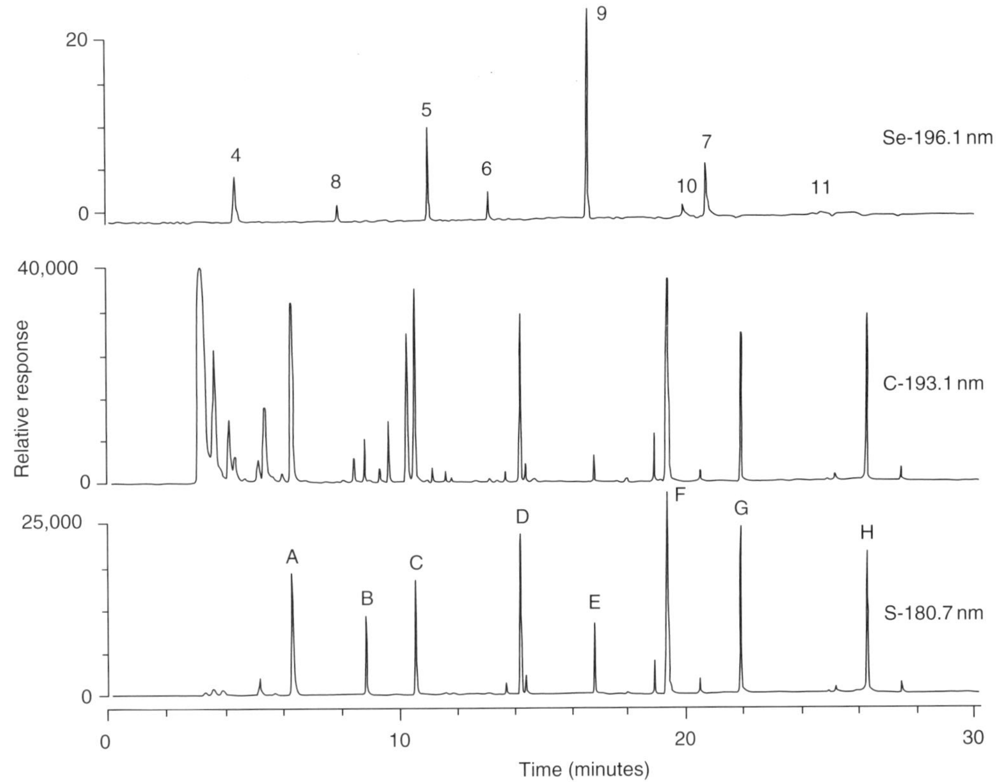

Fig. 5.9 Headspace- multi-element GC-AED (carbon emission, 193.1 nm, sulfur emission, 180.7 nm, selenium emission 196.1 nm) of volatile compounds from elephant garlic. Reproduced by permission from Cai *et al*. [109] Copyright 1994, American Chemical Society.

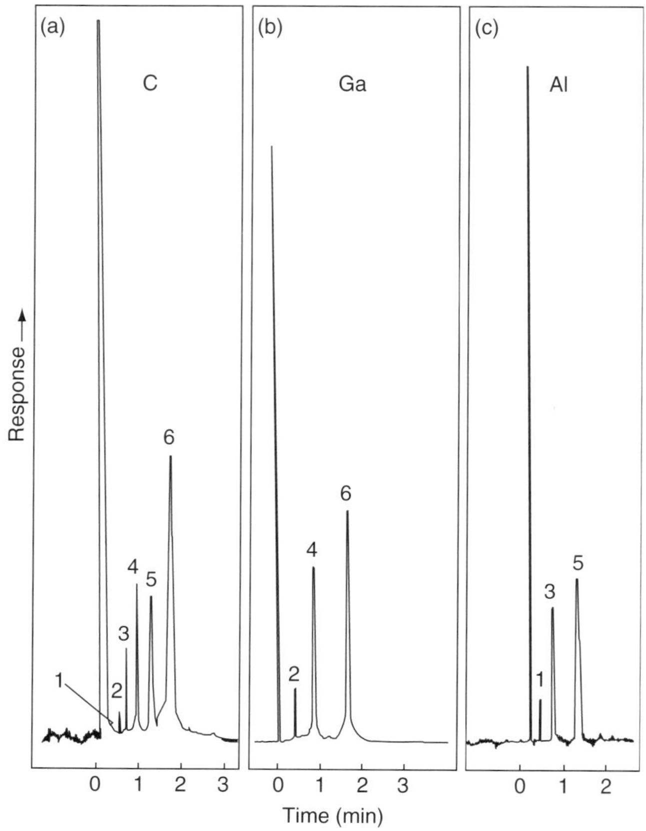

Fig. 5.10 GC-AED of redistribution of β-diketone ligands on gallium and aluminium (carbon emission 248 nm, gallium emission 294.3 nm, aluminium emission 396.1 nm). Reproduced by permission from Uden and Wang [115] Copyright 1988, The Royal Society of Chemistry.

Tetradentate β-ketoamine ligands are valuable analytically because of their selective quantitative reaction with various divalent metal ions; in many cases their chelates exhibiting better GC behavior than those of other ligands. Reagent ligands of the family of N, N′-ethylenebis (4-oxopentane-2-imine) [bis (acetylacetone)-ethylenediimine – $H_2(enAA_2)$] incorporating different substitutions on the alkylene bridge and/or the iminoketone functionality, have been utilized for packed and capillary column GC. CGC separations and determinations of copper and nickel, vanadium, palladium and platinum complexes have been reported using flame ionization or atomic emission spectral detection (AED) [107, 116–118]. GC-AED selective chromatograms for copper, nickel and palladium and vanadium complexes of N,N′-propylene-bis (trifluoroacetylacetoimine) are shown in Fig. 5.11 [107]. The distorted profile of the vanadium-specific chro-

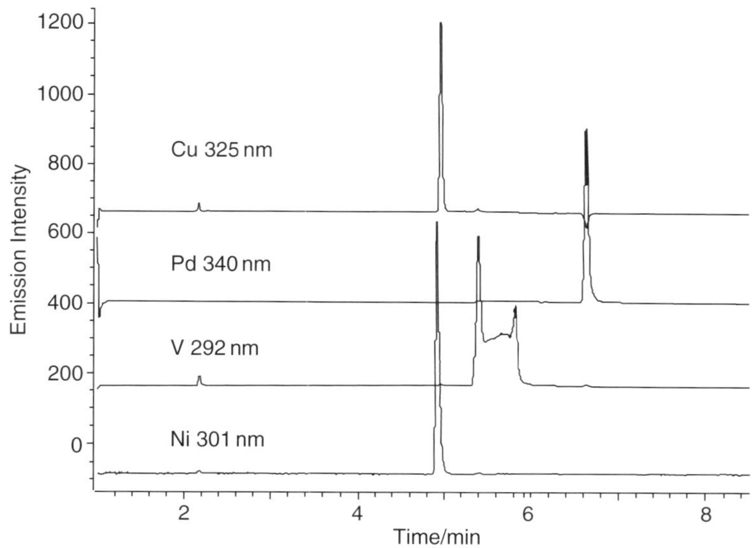

Fig. 5.11 GC-AED of copper, palladium, vanadium and nickel chelates of N,N′-propylenebis (trifluoroacetylacetoneimine). Reproduced by permission from Uden and Zeng [107] Copyright 1992, Freidr. Vieweg and Sohn, Pergamon Press.

matogram is due to the presence of stereoisomers of the oxovanadium chelate for which inter-conversion occurs on the GC column. Multi-element GC-AED for N, C, F, O and V was used to investigate this behavior.

Vanadium-specific detection has also been used in fingerprinting of metallo-porphyrins in crude oils [107]. Figure 5.12 illustrates the selectivity of vanadium detection against a carbon matrix, for a fraction extracted from a heavy crude oil. Detection using two independent wavelength recipes at 269 and 292 nm confirmed the identity of the vanadium profiles in the two spectral regions.

Many pi-bonded organometallics, such as metallocenes, are well behaved in capillary GC and provide good model compounds for optimization of GC-AED detection for their constituent elements. Sensitive and selective detection of iron, cobalt, nickel, chromium, manganese and vanadium in a series of cyclopentadienyl carbonyl/nitrosyl compounds was obtained, verifying elution of some previously unchromatographed compounds [119]. Manganese organometallics have been determined in gasoline additives by GC-AED using a DCP system [51] and GC-MIP as shown in Fig. 5.13 for a gasoline sample containing the additive MMT (methylcyclopentadienyl manganese tricarbonyl) [107]. The desirable analytical characteristics of metal specific detection are very clear and, in general, are limited far more by any shortcomings of GC column behavior than by spectral detection limitations.

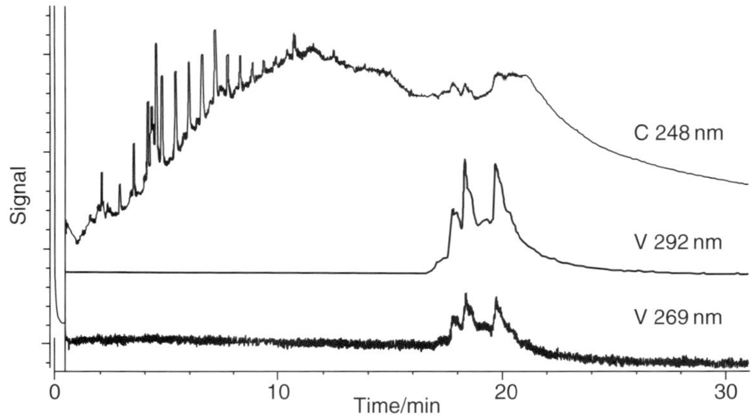

Fig. 5.12 Vanadium selective GC-AED of an extract of Boscan crude oil. Reproduced by permission from Zeng *et al*. [107] Copyright 1992, Dr Alfred Huethig Publishers.

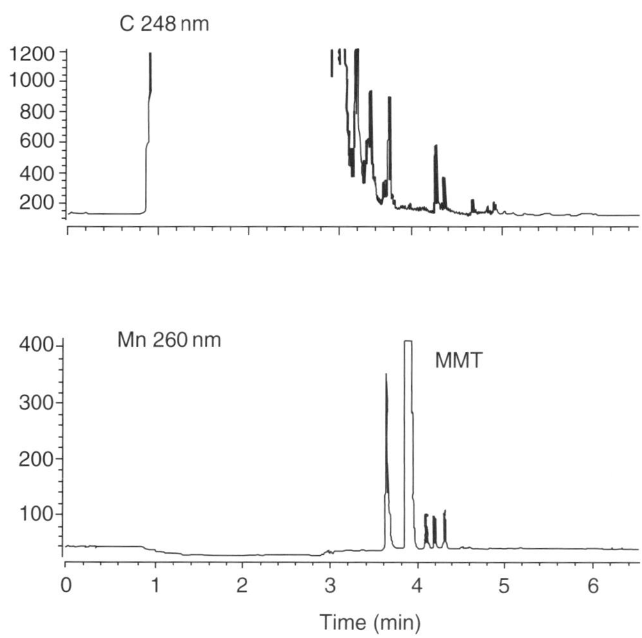

Fig. 5.13 Manganese selective GC-AED of a gasoline. Reproduced by permission from Zeng *et al.* [107] Copyright 1992, Dr Alfred Huethig Publishers.

5.6 Isotope detection by GC-AED

The ability of GC-AED to differentiate and quantify isotopes of elements of low atomic number such as hydrogen, carbon, nitrogen and oxygen has considerable potential in the area of stable isotope labeling as an analytical alternative to radioactive isotopes [120]. Quimby *et al.* [121] and Sullivan [122] reported the ability of GC-AED to determine ^{13}C selectively over ^{12}C, utilizing the molecular emission from intense C bands in the vacuum ultraviolet region. Deruaz *et al.* applied hydrogen, deuterium, sulfur and iodine detection in a study of the purity of deuteroiodomethane used for derivatization of immunoenhancing drugs, H, D and I being conveniently determined simultaneously [123]. Figure 5.14 shows ^{13}C and ^{12}C selective chromatograms in which [^{13}C] nitrobenzene was spiked into urine at the 3 ng/ml level. 30 pg of the ^{13}C compound is readily detected. The instrumental recipe is set to obtain selectivity of >2000 over ^{12}C taking into account the natural abundance (ca. 1%) of ^{13}C. ^{12}C and ^{13}C can be detected and well separated on the second order lines at 342.57

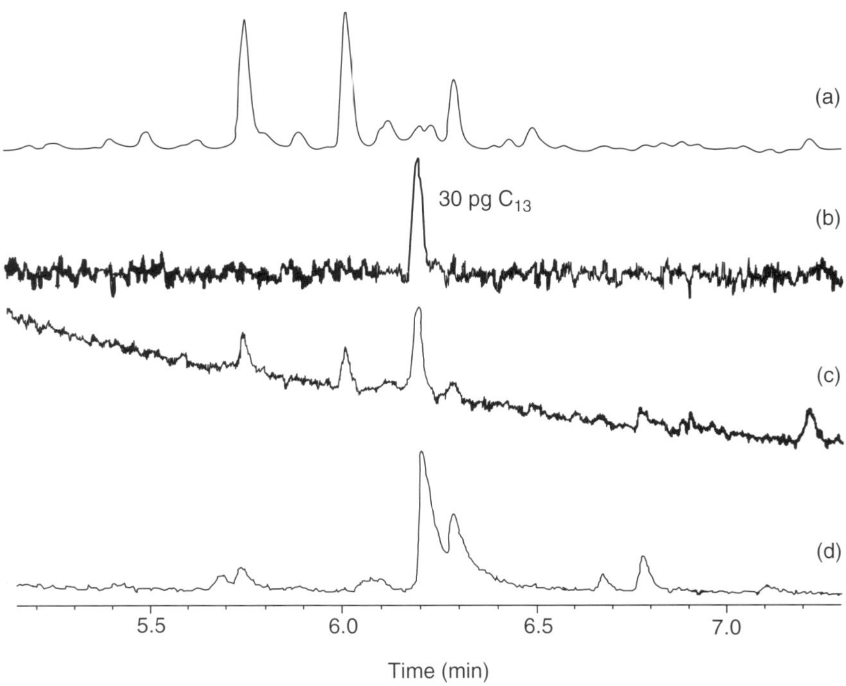

Fig. 5.14 GC-AED of carbon-13 and carbon-12. Determination by molecular emission of 30 pg of ^{13}C nitrobenzene spiked into urine and extracted. (a) carbon 193.1 nm atomic line, (b) enriched ^{13}C 171.0 nm (second order), (c) nitrogen 174.2 nm atomic line (second order), (d) sulfur 180.7 nm atomic line. Reproduced by permission from Quimby *et al.* [121] Copyright 1990, American Chemical Society.

and 341.71 nm, respectively, for CO emission [124] bands rather than atomic emission lines are preferred for nitrogen isotope detection at 388 nm. Stevens and Borgerding carried out an exhaustive H_2, O_2 and CH_4 reagent gas optimization to obtain high selectivity of ^{15}N over ^{14}N; the natural abundance of the former was compensated by software modification [125].

For deuterium, either 308 or 656.1 nm emission lines can be used, the latter requiring adequate spectral resolution from the adjacent hydrogen line at 656.3 nm [126–128]. The latter wavelength was used in a study on the fate of deuterated nicotine during smoking [129]. Although the sensitivity was poorer and the detection limit higher than at 308 nm, a linear CIC plot was obtained, unaffected by the presence of unlabeled nicotine. The hydrogen line did not interfere even when the hydrogen-containing isotopomer was present in 180-fold excess [120]. Studies with ^{13}C have included chlorinated dibenzodioxin [75], ^{13}C-labeled amino acids from whisky brewing [130], ^{13}C-labeled progesterone [131] and urinary metabolites from labeled caffeine [124, 132].

5.7 Analytical derivatization in GC-AED

Chromatographic derivatization is typically performed to enhance separation or to increase sensitivity or selectivity of detection. Both the latter characteristics may be enhanced by derivatization reactions that introduce such an element into the analyte but not into other compounds present in the sample. The benefits of tagging analytes by selective introduction of an AED-active element can be large and worthy of wider adoption. Hagen *et al.* used dual element derivative tagging by chlorofluoroacetic anhydride to enable Cl and F selective detection of acylated amines [41]. Tagging with *light* atoms is not the sole approach.

The introduction of an *AED-active* metal atom was demonstrated for alkylphenols in petroleum products [133] that were coupled with ferrocenecarboxylic acid chloride to give the esters in a rapid and simple reaction. Iron was introduced selectively into the phenols and as the AED has extremely high selectivity for iron over carbon (4 600 000, see Table 5.1), the large amounts of co-eluting hydrocarbons from the petroleum did not interfere with phenol determination. This is clearly shown by Fig. 5.15, which was recorded on the iron emission line and shows no response to the large amounts of co-injected petroleum hydrocarbons; $\mu g\ g^{-1}$ amounts of phenols were readily detected even when sample sizes below 1 mg were used. Alkenes in gasolines could be determined in the presence of alkanes and aromatic compounds after being converted to the dibromo derivatives with elemental bromine [78]. As CIC is possible for bromine, the total olefinic content of gasoline can easily be determined by GC analysis of the brominated mixture by use of the bromine emission line (Fig. 5.16). Non-ionic surfactants were determined by high-temperature GC of the trimethylsilyl derivatives at the silicon line [134]. Compounds with

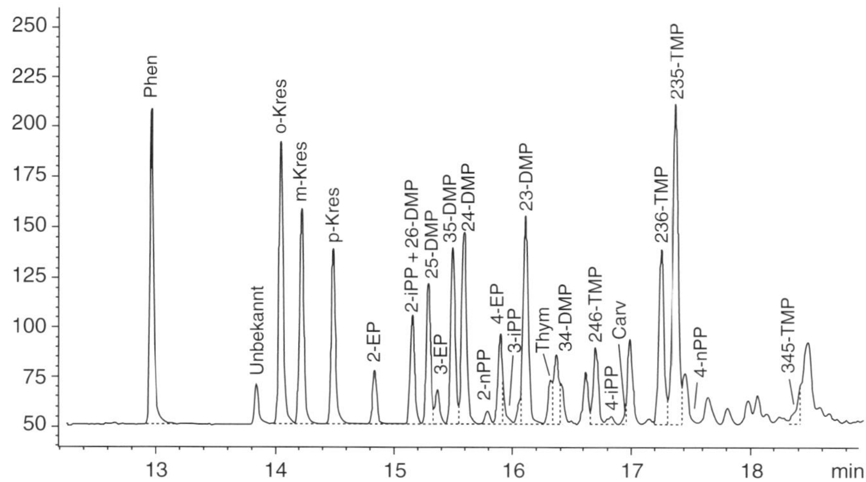

Fig. 5.15 Iron selective GC-AED of ferrocenecarboxylic acid chloride derivatized phenols in SRM 1589 shale oil, iron emission at 302 nm. Reproduced by permission from Rolfes and Andersson [133] Copyright 2001, American Chemical Society.

up to C20 alkyl chains and 16 ethylene oxide units could be determined by the use of CIC on the Si, O or C emission lines. Derivatization with alkyl chloroformates provides an effective route to volatile species for amino acid GC. The simultaneous detection of selenium at 196 nm and sulfur at 181 nm proves very valuable in speciating S- and Se-aminoacids. These have been derivatized for GC with isopropylchloroformate [135], and ethyl chloroformate, selenocysteine, Se-methyl selenocysteine, and selenomethionine being determined in normal and selenium-enriched plants using GC-AED [136].

5.8 Liquid and supercritical fluid chromatographic applications of AED detection

HPLC-plasma interfacing has been developed with the high power DCP and ICP argon plasmas that are able to tolerate mobile phase solvents at the flow rates typically used. The low-powered helium plasma is quenched by continuously introduced ml/min liquid flow streams. Some approaches to this problem have been explored. The only direct introduction of liquid into an MIP was by flowing the HPLC effluent over a heated wire and vaporizing it by a cross-stream of helium into the discharge. This system shows some potential for reversed phase separations [137]. A mixed gas oxygen–argon MIP sustained in a modified discharge tube consisting of two concentric quartz tubes was applied successfully for HPLC of mercury compounds. Methanol/water mixtures with

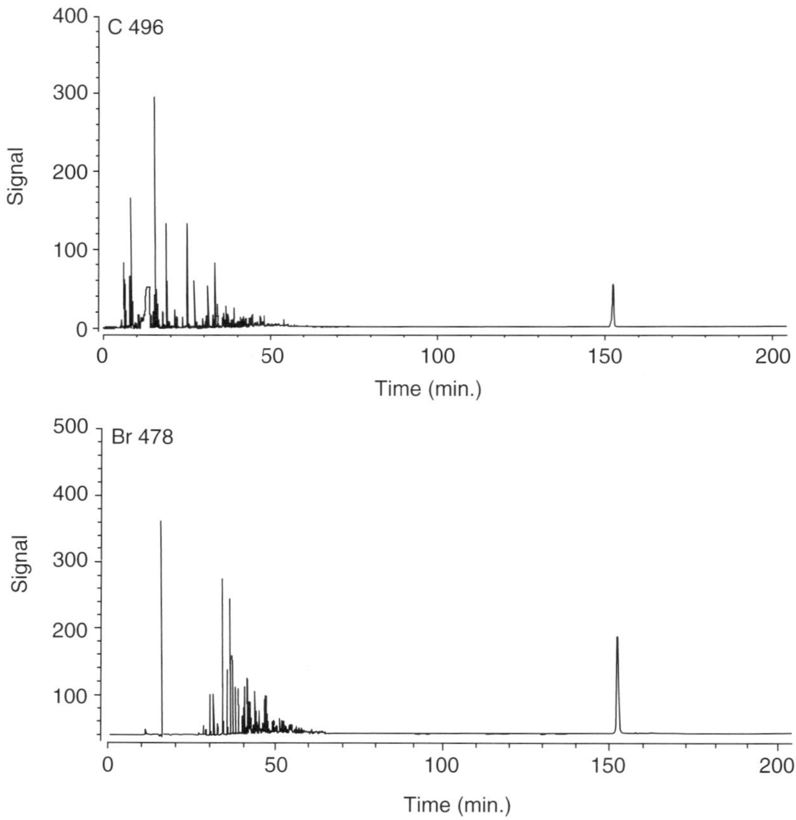

Fig. 5.16 Bromine selective GC-AED of gasoline olefin bromination products, bromine emission at 478 nm. Brominated pentadecene internal reference. Reproduced by permission from Hardas and Uden [78] Copyright 1999, Elsevier Science B.V.

up to 90% of the former solvent were tolerated; detection limits for organically bound Hg were in the ng range, but response was found to be dependent upon molecular structure [138]. A high-power (kW) discharge, operating in the RF or microwave range, accommodated continuous solvent-flows provided that nebulization was adequate [139]. The incentive for developing a viable HPLC-helium MIP interface is considerable because of the potential for monitoring of non-metallic elemental effluents; this is difficult or impossible with argon DCP and ICP systems. A moving-wheel sample transport-desolvation system has been described, in which aqueous solvent is evaporated with a flow of hot nitrogen, leaving dry analyte which is transported into the plasma, where it is volatilized, atomized and excited. The plasma used was a small volume helium MIP, operated at 100 W, with helium support gas flow of 3.1 l/min. Detection limits were in the range 0.4–20 mg of halogen [140].

Although analytical supercritical fluid chromatography (SFC) was demonstrated in the early 1960s, it has only been in recent years that the availability of viable high resolution packed and capillary SFC columns and instrumentation has led to renewed interest in the technique. High resolution SFC along with supercritical fluid extraction (SFE) promises to allow separations in areas where neither GC nor HPLC may be possible. A surfatron MIP sustained in helium was employed for SFC detection, giving sulfur-specific detection at 921.3 nm with a 25 pg/s limit for thiophene [141]. Extensive spectral characterization was carried out in this system for two common SFC mobile phases, carbon dioxide and nitrous oxide [142].

5.9 The role and future development of MIP-AED chromatographic detection

Atomic emission detection has unique properties in GC that can be used very profitably in complex analytical tasks. Not only multi-elemental selectivity, but also the possibility of CIC is of great importance. In practical terms, the wider adoption of plasma spectral detection to permit inter-laboratory comparisons of data and the development of *recommended* methods of analysis, which can be widely used will depend on the introduction of more standardized commercial instrumentation with fully integrated units which circumvent the need for analysts to interface their own chromatograph, emission device and spectrometer.

Continuing efforts are needed to clarify for which elements CIC can be used and how closely the chemical structure, concentration and retention times of the reference species employed need to be related to those of the analytes. The power of the combination of two independent techniques such as AED and MS, when used in conjunction, has been impressively illustrated to deduce much about unknown compounds in a complex mixture, establishing the empirical formula and further identifying an unknown compound.

The reports on isotope analysis and derivatization techniques discussed above illustrate the versatility of the AED that may have clear advantages over MS. Many isotope analyses could be run with the AED with great benefit. Elemental derivatization reactions can enable analytes to be detected with sample treatment, at lower concentrations, and with the benefits of CIC.

References

[1] A.E. Croslyn, B.W. Smith & J.D. Winefordner, *Crit. Rev. Anal. Chem.*, 1997, **27**, 199.

[2] J.D. Winefordner, E.P. Wagner II & B.W. Smith, *J. Anal. Atomic. Spec.*, 1996, **11**, 689.

[3] E. Hywel Evans, J.B. Dawson, A. Fisher, S.J. Hill, W.J. Pierce, C.M.M. Smith, K.L. Sutton & J.F. Tyson, Analytical spectroscopy update, *J. Anal. Atomic. Spec.*, 2001, **16**, 672, also earlier updates in this series.

[4] M. Huang, *Microchem. J.*, 1996, **53**, 79.
[5] R.D. Deutsch, J.P. Keilsohn & G.M. Hieftje, *Applied Spec.*, 1985, **39**, 531.
[6] J.J. Urh & J.W. Carnahan, *Applied Spec.*, 1986, **40**, 877.
[7] C.I.M. Beenakker, *Spectrochim. Acta*, 1976, **31B**, 483.
[8] J.F.C. Aguilar, R.P. Garcia, J.E.S. Urua & A.S. Medel, *Spectrochim. Acta*, 1994, **49B**, 545.
[9] S.R. Goode, B. Chambers & N.P. Buddin, *Spectrochim. Acta*, 1985, **40B**, 329.
[10] K.A. McCleary, G.R. Ducatte, D.H. Renfro & G.L. Long, *Applied Spec.*, 1993, **47**, 994.
[11] M.L. Bruce, J.M. Workman, J.A. Caruso & D.J. Lahti, *Applied Spec.*, 1985, **39**, 935.
[12] R.M. Bolainez, M.P. Dziewatkoski & C.B. Boss, *Anal. Chem.*, 1992, **64**, 541.
[13] H. Matusiewicz & R.E. Sturgeon, *Spectrochim. Acta*, 1993, **48B**, 515.
[14] D.M. Chambers, J.W. Carnahan, Q. Jin & G.M. Hieftje, *Spectrochim. Acta*, 1991, **46B**, 1745.
[15] N.W. Barnett, *Spectrochim. Acta*, 1987, **42B**, 859.
[16] F. Camuna, J.E.S. Uria & A.S. Medel, *Spectrochim. Acta*, 1993, **48B**, 1115.
[17] H. Matusiewicz, *Spectrochem. Acta Revs.*, 1990, **13**, 47.
[18] K. Dittrich, H. Fuchs, H. Berndt, J.A.C. Broekaert & G. Schaldach, *Fresenius J Anal. Chem.*, 1990, **336**, 303.
[19] G. Heltae, J.A. Broekaert, P. Burba, P. Tschopel & G. Tolg, *Spectrochim. Acta*, 1990, **45B**, 301.
[20] J. Yang, J. Zhang, C. Schickling & J.A.C. Broekaert, *Spectrochim. Acta*, 1996, **51B**, 551.
[21] E. Bulska, P. Tschopel, J.A. Broekaert & G. Tolg, *Anal. Chim. Acta*, 1993, **271**, 171.
[22] Z. Gong, W.F. Chan, X. Wang & F.S.C. Lee, *Anal. Chim. Acta*, 2001, **450**, 207.
[23] T. Nakahara, K. Nakanishi & T. Wasa, *Chem. Expr.*, 1986, **1**, 149.
[24] J.P. Snell, W. Frech & Y. Thomassen, *Analyst*, 1996, **121**, 1055.
[25] E.H. Evans, J.A. Caruso & R.D. Satzger, *Applied Spec.*, 1991, **45**, 1478.
[26] M.M. Abdillahi & R.D. Snook, *Analyst*, 1986, **111**, 265.
[27] M. Wu & J.W. Carnahan, *Applied Spec.*, 1990, **44**, 673.
[28] Y. Nojiri, A. Otsuki & K. Fuwa, *Anal. Chem.*, 1986, **58**, 544.
[29] M.M. Abdillahi, *Applied Spec.*, 1993, **47**, 366.
[30] E.I. Brooks & K.J. Timmons, *Analyst*, 1985, **110**, 557.
[31] T. Nakahara, T. Mori, S. Morimoto & H. Ishikawa, *Spectrochim. Acta*, 1995, **50B**, 393.
[32] K. Ng, X. Xu & M. Brechmann, *Spectrosc. Lett.*, 1989, **22**, 1251.
[33] G.L. Long & L.D. Perkins, *Applied Spec.*, 1987, **41**, 980.
[34] J.J. Urh & J.W. Carnahan, *Anal. Chem.*, 1985, **57**, 1253.
[35] J. Uebbing, A. Ciokan & K. Niemax, *Spectrochim. Acta*, 1992, **47B**, 601.
[36] M.M. Mohamed, T. Uchida & S. Minami, *Applied Spec.*, 1989, **4**, 794.
[37] J.M. Gehlhausen & J.W. Carnahan, *Anal. Chem.*, 1991, **63**, 2430.
[38] A. McCormack, S. Tong & W.D. Cooke, *Anal. Chem.*, 1965, **37**, 1470.
[39] C.A. Bache & D.J. Lisk, *Anal. Chem.*, 1965, **37**, 1477.
[40] K.S. Brenner, *J. Chromatogr.*, 1987, **167**, 365.
[41] D.F. Hagen, J.S. Marhevka & L.C. Haddad, *Spectrochim. Acta*, 1985, **40B**, 335.
[42] K. Zeng, Q. Gu, G. Wang & W. Yu, *Spectrochim. Acta*, 1985, **40B**, 349.
[43] L. Ebdon, S. Hill & R.W. Ward, *Analyst*, 1986, **111**, 1113.
[44] P.C. Uden, Atomic spectral chromatographic detection, in *Element Specific Chromatographic Detection by Atomic Emission Spectroscopy*, American Chemical Society Symposium Series # 479. Washington DC, 1992, p. 1.
[45] B.D. Quimby, P.C. Uden & R.M. Barnes, *Anal. Chem.*, 1978, **50**, 2112.
[46] B.D. Quimby & J.J. Sullivan, *J. Anal. Chem.*, 1990, **62**, 1027.
[47] J.J. Sullivan & B.D. Quimby, *Anal. Chem.*, 1990, **62**, 1034.
[48] B.D. Quimby, D. Grudoski & V. Giarrocco, *J. Chrom. Sci.*, 1998, **36**, 435.
[49] B.D. Quimby, P. Larson & P. Dryden, in *Application Note 228–363*; Hewlett Packard: Wilmington, DE, 1996, p. 4.
[50] P.C. Uden, B.D. Quimby, R.M. Barnes & W.G. Elliot, *Anal. Chim. Acta*, 1978, **101**, 99.
[51] P.C. Uden, R.M. Barnes & F.P. DiSanzo, *Anal. Chem.*, 1978, **50**, 852.

[52] S. Pedersen-Bjergaard & T. Greibrokk, *J. Microcol. Sep.*, 1994, **6**, 11.
[53] H. Dingjan & H. deJong, *Spectrochim. Acta*, 1983, **38B**, 282.
[54] J.J. Sullivan & B.D. Quimby, *J. High. Res. Chrom.*, 1989, **12**, 282.
[55] Y.-R. Huang, Q.-Y. Ou & W.-L. Yu, *J. Anal. Atomic. Spec.*, 1990, **5**, 115.
[56] P.L. Wylie, J.J. Sullivan & B.D. Quimby, *J. High Resolut. Chromatogr.*, 1990, **13**, 499.
[57] N.R. Kovacic & T. Ramus, *J. Anal. Atomic Spec.*, 1992, **7**, 999.
[58] S. Pedersen-Bjergaard, T. Asp & T. Greibrokk, *J. High Resolut. Chromatogr.*, 1992, **15**, 89.
[59] S. Pedersen-Bjergaard, T. Asp & T. Greibrokk, *Anal. Chim. Acta*, 1992, **265**, 87.
[60] T. Albro & P. Dreifuss, *J. High Resolut. Chromatogr.*, 1993, **16**, 13.
[61] K. Janak, A. Colmsjo & C. Ostman, *J. Chrom. Sci.*, 1995, **33**, 611.
[62] C. Webster & M. Cooke, *J. High Resolut. Chromatogr.*, 1995, **18**, 319.
[63] G. Becker, A. Colmsjo & C. Ostman, *Anal. Chim. Acta*, 1997, **340**, 181.
[64] N. Olsen, R. Carrel, R. Cummings, R. Rieck & S. Reimer, *J. AOAC Int.*, 1995, **78**, 1464.
[65] F. Augusto & A.P. Valente, *J. Brazilian Chem. Soc.*, 1998, **9**, 43.
[66] M. Szelewinski, in Application Note 228–382: Hewlett Packard, 1997, vol. 8.
[67] A.M. Gonzalez & P.C. Uden, *J. Chromatogr.*, 2000, **898**, 201.
[68] P.C. Uden, *J. Chromatogr.*, 1995, **703**, 393.
[69] S. Pedersen-Bjergaard & T. Greibrokk, *J. High Resolut. Chromatogr.*, 1996, **19**, 597.
[70] P.C. Uden, In Caruso, J.A., Sutton, K.L. & Ackley, K.L. (eds), *Comprehensive Analytical Chemistry*, XXXIII, Elsevier, Amsterdam, 2000, pp. 123–150.
[71] I.R. Pereiro & C. Diaz, *Anal. Bioanal. Chem.*, 2002, **372**, 74.
[72] J.T. Andersson, *Anal. Bioanal. Chem.*, 2002, **373**, 344.
[73] L.L.P. van Stee, U.A. Th. Brinkman & H. Bagheri, *Trends in Anal. Chem.*, 2002, **21**, 618.
[74] S. Pedersen-Bjergaard, S.I. Semb, J. Vedde, E.M. Bervik & T. Greibrokk, *Chromatographia*, 1996, **43**, 44.
[75] J.T. Andersson & B. Schmid, *Fresenius J. Anal. Chem.*, 1993, **346**, 403.
[76] S. Pedersen-Bjergaard & T. Greibrokk, *Anal. Chem.*, 1993, **65**, 1998.
[77] S.I. Semb, E.M. Brevik & S. Pedersen-Bjergaard, *Chemosphere*, 1998, **36**, 213.
[78] N.R. Hardas, R. Adam & P.C. Uden, *J. Chromatogr. A.*, 1999, **844**, 249.
[79] A. Johnson & N. Olson, *Arch. Environ. Contam. Toxicol.*, 2001, **41**, 339.
[80] C. Bradley & J.W. Carnahan, *Anal. Chem.*, 1988, **60**, 858.
[81] S. Pedersen-Bjergaard & T. Greibrokk, *J. High Resolut. Chromatogr.*, 1992, **15**, 677.
[82] R.J. Skelton, P.B. Farnsworth, K.E. Markides & M.L. Lee, *J. High Resolut. Chromatogr.*, 1988, **11**, 75.
[83] T.G. Albro, P.A. Dreifuss & R.F. Wormsbecher, *J. High Resolut. Chromatogr.*, 1993, **16**, 13.
[84] G. Becker & A. Colmsjo, *Anal. Chim. Acta*, 1998, **376**, 265.
[85] J.J. Kosman & R.G. Lukco, *J. Chromatogr. Sci.*, 1993, **31**, 88.
[86] P. Sandra, *Analysis*, 1992, **20**, M32.
[87] S. Rowland, R. Evens, L. Ebdon & A. Rees, *Anal. Proc.* (London), 1993, **30**, 87.
[88] S.M. Lee & P.L. Wylie, *J. Agric. Food. Chem.*, 1991, **39**, 2192.
[89] P.C. Uden, *J. Chromatogr.*, 1984, **313**, 3.
[90] S.A. Estes, P.C. Uden & R.M. Barnes, *Anal. Chem.*, 1982, **54**, 2402.
[91] R. Lobinski & F.C. Adams, *J. Anal. Atomic. Spec.*, 1992, **7**, 987.
[92] K.B. Olsen, D.S. Sklarew & J.C. Evans, *Spectrochim. Acta*, 1985, **40B**, 357.
[93] H. Emteborg, D.C. Baxter & W. Frech, *Analyst*, 1993, **118**, 1007.
[94] J.P. Snell, W. Frech & Y. Thomassen, *Analyst*, 1996, **121**, 1055.
[95] B. Bouyssiere, F. Baco, L. Savary & R. Lobinski, *J. Chromatogr. A.*, 2002, **976**, 431–439.
[96] R. Lobinski, W.M.R. Dirks, M. Ceulemans & F.C. Adams, *Anal. Chem.*, 1992, **64**, 159.
[97] T.M. Dowling & P.C. Uden, *J. Chromatogr.*, 1993, **644**, 153.
[98] R. Reuther, L. Jaeger & B. Allard, *Anal. Chem. Acta*, 1999, **394**, 259.
[99] H. Emteborg, D.C. Baxter, M. Sharp & W. Frech, *Analyst*, 1995, **120**, 69.
[100] M. Heisterkamp & F.C. Adams, *Fresenius J. Anal. Chem.*, 1998, **362**, 489.

[101] I. Rodriguez, M. Santamarina, M.H. Ballain, M.C. Mejuto & R. Cela, *J. Chromatogr. A*, 1997, **774**, 379.
[102] A. Sadiki & D.T. Williams, *Chemosphere*, 1996, **32**, 1983.
[103] Y. Liu, V. Lpoez-Avila, M. Alcaraz & W. Beckert, *J. High Resol. Chrom.*, 1994, **17**, 527.
[104] S. Tutschku, S. Mothes & K. Dittrich, *J. Chromatogr. A*, 1994, **683**, 269.
[105] M.K. Donais, P.C. Uden, M.M. Schantz & P.C. Uden, *Anal. Chem.*, 1996, **68**, 3859.
[106] Y. Zeng & P.C. Uden, *J. High Resol. Chrom.*, 1994, **17**, 223.
[107] Y. Zeng, J.A. Seeley, T.M. Dowling, P.C. Uden & M. Khuhawar, *J. High Resol. Chrom.*, 1992, **15**, 669.
[108] Y. Cai, S. Monsalud, R. Jaffe & R.D. Jones, *J. Chromatogr.*, 2000, **876**, 147.
[109] X.J. Cai, P.C. Uden, E. Block, X. Zhang, J.J. Sullivan & B.D. Quimby, *J. Ag. Food Chem.*, 1994, **42**, 2085.
[110] X.J. Cai, E. Block, P.C. Uden, J.J. Sullivan & B.D. Quimby, *J. Ag. Food Chem.*, 1995, **43**, 1751.
[111] M.B. de la Calle-Guntinas, R. Lobinski & F. Adams, *J. Anal. At. Spec.*, 1995, **10**, 111.
[112] X.J. Cai, E. Block, P.C. Uden, X. Zhang, J.J. Sullivan & B.D. Quimby, *J. Ag. Food Chem.*, 1995, **43**, 1754.
[113] P.C. Uden, *J. Chromatogr.*, 1984, **313**, 3.
[114] M.S. Black & R.E. Sievers, *Anal. Chem.*, 1976, **48**, 1872.
[115] P.C. Uden & T. Wang, *J. Anal. At. Spectrom.*, 1988, **3**, 919.
[116] M.Y. Khuhawar, A. Sarafraz Yazdi, J.A. Seeley & P.C. Uden, *J. Chromatogr.*, 1998, **824**, 223.
[117] P.C. Uden & Y. Zeng, *Chromatographia*, 1992, **34**, 269.
[118] M.Y. Khuhawar, A. Sarafraz Yazdi, Y. Zeng & P.C. Uden, *Chemia Analityczna*, 1995, **40**, 271.
[119] S.A. Estes, P.C. Uden, M.D. Rausch & R.M. Barnes, *J. High Resolut. Chromatogr.*, 1980, **3**, 471.
[120] L.C. Thomas & T.L. Ramus, *J. Chromatogr.*, 1991, **586**, 309.
[121] B.D. Quimby, P.C. Dryden & J.J. Sullivan, *Anal. Chem.*, 1990, **62**, 2509.
[122] J.J. Sullivan, *Trends Anal. Chem.*, 1991, **10**, 23.
[123] D. Deruaz, A. Bannier, M. DeSage & J.L. Brazier, *Anal. Lett.*, 1991, **24**, 1531.
[124] M.S. Boukraa, D. Deruaz, A. Bannier, M. DeSage & J.L. Brazier, *J. Pharm. Biomed. Anal.*, 1994, **12**, 185.
[125] N.A. Stevens & M.F. Borgerding, *Anal. Chem.*, 1999, **71**, 1083.
[126] S.A. Estes, P.C. Uden & R.M. Barnes, *Anal. Chem.*, 1981, **53**, 1829.
[127] D.F. Hagen, L.C. Haddad & J.S. Marhevka, *Spectrochim. Acta*, 1987, **42B**, 253.
[128] W.K. Gavlick, J.R. Stuff & P.C. Uden, *J. Chromatogr.*, 1990, **508**, 187.
[129] N.A. Stevens & M.F. Borgerding, *Anal. Chem.*, 1999, **71**, 2179.
[130] K. Oshita, M. Kubota, M. Uchida & M. Ono, *Proc. 25th Eur. Brew. Conv.*, 1995, 387–394.
[131] B. Deruaz, D. Deruaz, A. Bannier, M. DeSage & J.L. Brazier, *Analusis*, 1994, **22**, 241.
[132] W. Elbast, M.S. Cuabet, D. Deruaz & J.L. Brazier, *Anal. Lett.*, 1999, **32**, 3111–3130.
[133] J. Rolfes & J.T. Andersson, *Anal. Chem.*, 2001, **73**, 3073.
[134] C. Asmussen & H.J Stan, *J. High Resolut. Chromatogr.*, 1998, **21**, 597.
[135] H. Kataoka, Y. Miyanaga & M. Makita, *J. Chromatogr.*, 1994, **659**, 481.
[136] X.J. Cai, E. Block, P.C. Uden, X. Zhang, B.D. Quimby & J.J. Sullivan, *J. Agric. Food Chem.*, 1995, **43**, 1757.
[137] H.A.H. Billiet, J.P.K. van Dalen, P.J. Schoemakers & L. DeGalen, *Anal. Chem.*, 1983, **55**, 847.
[138] D. Kollotzek, D. Oeschsle, G. Kaiser, P. Tschopel & G. Tolg, *Fresenius Z. Anal. Chem.*, 1984, **31B**, 485.
[139] D.L. Haas, J.W. Carnahan & J.A. Caruso, *J. Appl. Spectrosc.*, 1983, **37**, 82.
[140] L. Zhang, J.W. Carnahan, R.E. Winans & P.H. Neill, *Anal. Chem.*, 1989, **61**, 895.
[141] D.R. Luffer, L.J. Galante, P.A. David, M. Novotny & G.M. Heiftje, *Anal. Chem.*, 1988, **60**, 1365.
[142] L.J. Galante, M. Selby, D.R. Luffer, G.M. Heiftje & M. Novotny, *Anal. Chem.*, 1988, **60**, 1370.
[143] J.A. Seeley, Y. Zeng, P.C. Uden, T.I. Eglinton & I. Ericsson, *J. Anal. At. Spec.*, 1992, **7**, 979.
[144] H. Tao, A. Miyazaki & K. Bansho, *Anal. Sciences*, 1988, **4**, 299.

6 X-ray fluorescence analysis (XRF)

Bruno Vrebos

6.1 Introduction

Wavelength dispersive X-ray spectrometers became commercially available around 1940. These early instruments were based on the goniometer from X-ray diffractometers. They evolved into specifically designed instruments in the late 1940s. In wavelength dispersive X-ray spectrometers, the selection of a desired analytical wavelength and the determination of the intensity of each wavelength of interest are separated into two different subsystems. This separation of wavelength (or energy) determination and photon counting is the main difference between wavelength and energy dispersive X-ray fluorescence spectrometry. Wavelength dispersive spectrometers (WDS) nowadays are highly reliable instruments performing qualitative and quantitative elemental analysis in a wide range of industries. They are used, for example, in cement factories to analyze the incoming materials and the final product; in metal-producing industries such as the iron and steel industries for a rapid determination of the concentration of alloying elements and in oil industries for quality control and process monitoring. Also, they found a widespread application in the geoanalytical sciences and in materials research, where the high specificity and the accuracy of the method are well appreciated. The technique has been the subject of several textbooks such as Jenkins and de Vries [1], Bertin [2] and more recently van Grieken and Markowicz [3]. Also, many textbooks on instrumental analytical chemistry deal with the method to varying degrees.

In the 1960s, progress in semiconductor technology has made it possible to use solid-state devices as detectors. The field of X-ray analysis using X-ray spectrometers fitted with such energy dispersive detectors is called energy dispersive spectrometry (EDS). These systems are very different in a number of their basic components, their analytical capabilities (though the gap between them is narrowing) and overall complexity. EDS systems find wide application in research and in industrial quality control.

6.2 Physics of X-ray absorption and excitation

6.2.1 Introduction

Only the physical processes directly relevant to the analytical X-ray fluorescence methods will be discussed. For a more complete treatment of physical processes and interactions between X-rays and matter, please consult textbooks such as, for example, references [2–5]. We will limit ourselves here to a short treatment of absorption, the photoelectric effect, scatter and some aspects of X-ray tubes.

6.2.2 *Absorption*

In the energy range of X-rays (rather arbitrarily *defined* as between 0.1 and 100 keV), the interactions between electromagnetic radiation and matter can be divided into three different processes: coherent scattering, incoherent scattering and the photoelectric effect. The probability that a photon and an atom interact is given by the atomic cross-section of the interaction considered. The total cross-section or attenuation coefficient μ is thus composed of the cross-sections for the photoelectric effect (τ), the coherent scattering (σ_{coh}) and the incoherent scattering (σ_{incoh}):

$$\mu = \tau + \sigma_{coh} + \sigma_{incoh} \tag{6.1}$$

Atomic cross-sections are expressed in barns/atom, which is equal to 10^{-24} cm^2. The conversion from barns/atom to the (in X-ray analysis) more common units of cm^2/g can be done by multiplying with Avogadro's number and dividing by the atomic weight of the element concerned.

The number of X-ray photons (of a given energy) absorbed by matter (of a specific nature) is proportional to the number of incident photons (I) and the mass per unit area. The mass per unit area can be calculated from the density ρ and the thickness. For a layer of infinitesimal thickness dt, the mass per unit area is given by ρdt and the variation in intensity dI can thus be described by

$$dI = -\mu I \rho dt \tag{6.2}$$

where the proportionality constant μ is called the mass attenuation coefficient. The minus sign is to indicate that the intensity I decreases as dt increases. After integration, Beer's law of absorption is obtained:

$$I(t) = I_0 \exp(-\mu \rho t) \tag{6.3}$$

where I_0 is the intensity of the incident beam, and $I(t)$ is the remaining intensity after travelling a distance t through the material.

The mass attenuation coefficients μ are specific for each element and exhibit a strong energy dependence as shown in Fig. 6.1. Also, discontinuities can be seen clearly. These are due to the photoelectric effect. Tables and algorithms to calculate mass attenuation coefficients can be found in the literature

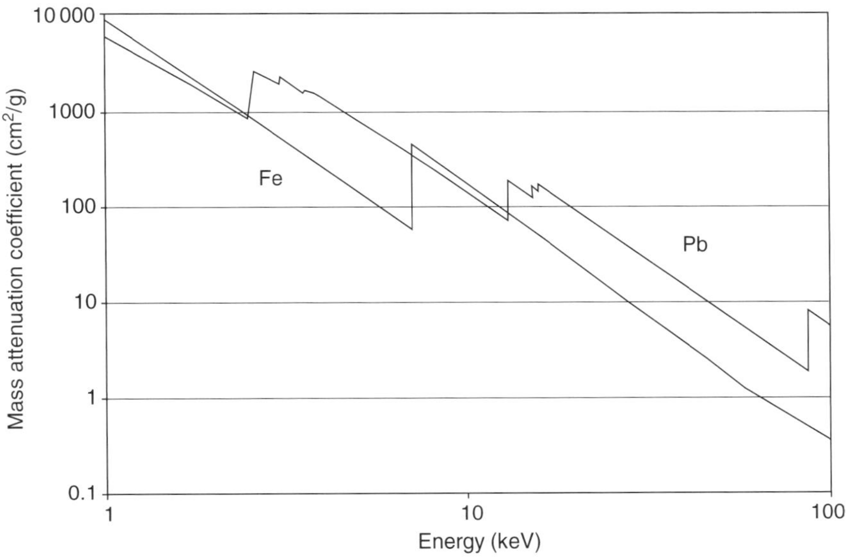

Fig. 6.1 Mass attenuation coefficient (cm^2/g) versus energy in keV (log-log scale) for iron and lead. The discontinuities are due to the photoelectric effect.

[6–10]. An overview has been published by Hubbell [11]. The attenuation coefficient for a compound material $\mu(E)$ at a given energy E can be calculated as the weighted sum of the elemental attenuation coefficients $\mu_j(E)$:

$$\mu(E) = \sum_{j=1}^{n} W_j \mu_j(E) \qquad (6.4)$$

where W_j is the mass fraction of element j.

6.2.3 *The photoelectric effect*

The largest contribution to the total X-ray attenuation is due to the photoelectric effect. In this process, the total energy of the incident photon is absorbed by an electron in an inner shell of an atom and the photon is annihilated. The electron is ejected from its shell, with a kinetic energy E_{kin} given by

$$E_{kin} = E_{photon} - E_{shell} \qquad (6.5)$$

where E_{photon} is the energy of the incident photon and E_{shell} is the binding energy of the electron. The ejected electron is called photoelectron. Quite clearly, the photon energy must be larger than the binding energy of the electron, for the photoelectric effect to be feasible. The relaxation of the ionized atom is done by filling the vacancy with an electron from a shell with a lower binding energy. This creates another vacancy, which can be relaxed by an electron from another shell (or the conduction band) etc. The energy difference between the two shells involved in the relaxation can be given off as a characteristic X-ray photon. The vacancy that was created in the first instance is called the primary vacancy. The vacancy shifts to shells with lower binding energy as the electron moves to fill the vacancy.

The process described whereby an electron from one shell fills the vacancy in another is governed by a number of selection rules; these have their origin in the quantum-mechanics. In general, when the electron transition involves a change in the quantum numbers j and l as given by

$$\begin{cases} -1 \leq \Delta j \leq 1 \\ \Delta l = \pm 1 \end{cases} \tag{6.6}$$

the radiative transition has a reasonable probability to occur. Such transitions are electric dipole transitions. X-ray emission following a transition with $\Delta l = +2$ or -2 is also observed; such transitions are electric quadrupoles and the resulting X-ray intensity is very low [12].

Not all vacancies, however, involve the emission of an X-ray photon. The probability that the electron transition to a primary vacancy in a given shell involves X-ray emission is given by the fluorescence yield and denoted by ω. The fluorescence yield depends on the shell involved and on the element; tabulated values can, for example, be found in the work by Bambynek *et al.* [13].

The energy of the emitted X-ray photon E_{photon} can be calculated from the energy difference between the atomic energy levels concerned:

$$E_{\text{photon}} = E_{\text{initial}} - E_{\text{final}} \tag{6.7}$$

The atomic energy levels are characteristic to the element, thus the energy difference is also a characteristic. Hence, the resulting X-ray photon is called a characteristic photon. The energy of the characteristic photons is varying in a systematic way with the atomic number Z of the elements; a fact that can be represented using Moseley's law [14]:

$$E_{\text{photon}} = k(Z - s)^2 \tag{6.8}$$

where k is a constant for a given series (as defined by the shell with the initial vacancy) and s is the screening constant. A complete tabulation of the energies of characteristic radiation can be found in the work by Bearden [15].

The relationship between the energy of a photon E and the wavelength of the electromagnetic radiation λ is given by

$$\lambda = \frac{hc}{E} \tag{6.9}$$

where h is Planck's constant and c is the velocity of light in vacuum. Equation 6.9 reduces to

$$\lambda = \frac{1.2398}{E} \quad [\text{nm, keV}] \tag{6.10}$$

when the energy is expressed in keV and the wavelength in nm.

At a given photon energy, the photoelectric effect is the largest contribution to the total attenuation. The requirement that the photon energy must be larger than the binding energy of an electron in a particular shell causes sharp increase in the values of the attenuation coefficients: at energies lower than, for example, the binding energy of the K-shell, the photoelectric effect with that shell is (quasi-) non-existing, while at energies in excess of the binding energy the photoelectric effect is a major contributor to the total absorption. The value of the attenuation coefficient thus increases abruptly when the photon energy exceeds the binding energy. The discontinuities in the attenuation coefficients are referred to as absorption edges. This is evidenced by the appearance of sharp discontinuities in the graphs in Fig. 6.1, where the absorption edge for Fe at 7.11 keV can clearly be seen. Similarly, for lead, the K-absorption edge at 88 keV, the L1, L2 and L3 edges (15.86 keV, 15.20 keV and 13.04 keV respectively) and finally the M edges (around 2.5 keV) can be distinguished. There is only one edge associated with the K-shell, there are three edges for the L-shell and five for the M-shell, provided the element has enough electrons to warrant the existence of the relevant shells. High atomic number elements have also N and O shells, with corresponding absorption edges, but these are of virtually no relevance in the analytical X-ray region.

6.2.4 *Scattering of X-rays*

The photoelectric effect describes interactions between a photon and an electron in the inner shell, which lead to the emission of a photoelectron. Interactions between photons and electrons in outer shells without subsequent emission of a photoelectron are also possible. These phenomena are grouped under the name of scatter. Two different processes can be distinguished: coherent and incoherent scatter. Coherent scatter is the process in which the incident photon interacts with an electron and is deviated without change in energy. This process is called Rayleigh scattering. When the interaction involves an energy (and momentum) transfer from the photon to the electron, the scattering process is incoherent (Compton scatter). In the latter case, the laws of conservation of energy and conservation of momentum, when applied to the system consisting of the electron and the photon, yield an expression relating the angle ψ

over which the photon is scattered to its energy decrease (or increase in wavelength $\Delta\lambda$):

$$\Delta\lambda = 0.00243(1 - \cos\psi) \quad [\text{nm}] \tag{6.11}$$

In Equation 6.11, the numerical value is for wavelength differences in nanometers. The derivation of Equation 6.11 assumes that the electron is initially at rest (i.e. it has no kinetic energy). Compton scattered radiation can be observed in X-ray spectra. Its intensity is dependent upon the composition of the material and the wavelength (or energy) of the incident photon. It is more intense for elements with low atomic number than for elements with high atomic number, and it is more intense with increasing energy of the incident photon.

6.2.5 *Nomenclature of characteristic X-rays*

The traditional system of nomenclature used in X-ray fluorescence analysis is the system developed by K. Siegbahn [16]. This system is based on a combination of apparent X-ray intensities and the initial shell involved. This system lacks in clarity and some of the designations are somewhat illogical. The International Union of Pure and Applied Chemistry (IUPAC) has proposed a nomenclature system based on the designation of both edges involved: the initial and the final state. A complete description can be found in the work by Jenkins *et al.* [17]. This system, however, is lacking in general acceptance.

6.2.6 *The fluorescence yield*

After excitation of an electron from an inner shell, the atom (ion) exists in an excited state. It can revert to a more stable state with lower energy by filling the vacancy with one of its electrons from a less-energetic shell. The energy released in this relaxation can be emitted as a characteristic photon. This is, however, not the only process. An alternative mechanism involves the ejection of an electron from a less energetic shell. In this process the energy to be emitted is spent on the ejection of the electron, providing it with kinetic energy. This process is known as the Auger process and the emitted electron is the Auger electron. The kinetic energy of the Auger electron forms the basis for Auger electron spectroscopy. The ratio of the number of emitted X-rays to the number of initially created vacancies in a given shell is the fluorescence yield, ω. For the K-shell:

$$\omega_K = \frac{I_K}{n_K} \tag{6.12}$$

where I_K is the total number of characteristic K X-rays and n_K is the number of vacancies originally created in the K-shell. The definition of the fluorescence yield is straightforward for the K-shell, but is complicated for the other shells by the presence of the sub-shells and Coster-Kronig transitions. The latter are radiationless transitions involving electrons with the same principal quantum number n.

The fluorescence yield for the K-shell is very small for the low atomic number elements and increases rapidly with atomic number between $Z = 20$ and $Z = 50$. The fluorescence yield for the different L-shells is lower than the one for the K-shell, and does not increase rapidly. Bambynek *et al.* [13] have published extensive tables of values of the fluorescence yield and the Coster-Kronig transitions. Coster-Kronig transitions result in a redistribution of initial vacancies, whereby part of the vacancies from, for example, the L1- and L2-shells are shifted toward the L3-shell. Additional vacancies in the L3-shell are created by the emission of a $K\alpha_1$ (KL3) characteristic photon: the KL3 transition fills the vacancy in the K-shell, but leaves the atom with a vacancy in the L3-shell. After the emission of an Auger electron there are two vacancies in shells with lower binding energy. This reshuffling of vacancies must be taken into account when calculating the distribution of vacancies. After a K-shell vacancy is relaxed by emission of a characteristic X-ray, not all X-rays are $K\alpha_1$ (KL3). For each element, the ratio of the emitted lines in a given series is constant.

6.2.7 Excitation of X-rays

Excitation of X-rays in X-ray fluorescence involves a beam of incident X-rays. Three common modes are the use of X-ray tubes, radio-isotopes and synchrotron radiation. Note that electron-induced X-ray emission – the excitation of characteristic X-rays by electrons as in electron microscopes – is, *stricto sensu*, not X-ray *fluorescence*. The physical processes involved in the interaction between electrons and matter are different from the ones with X-ray interactions. Once the atom is excited, the processes are identical again. The electron probe micro-analyzer (EPMA) has been pioneered by Castaing [18] who also described its application to quantitative analysis. More information on EPMA can be found in the textbooks by van Grieken and Markowicz [3] and Goldstein *et al.* [19].

6.2.7.1 Radio-isotopes

Radio-isotope sources emit X-rays or γ-rays following radioactive decay. γ-rays are emitted directly from the nucleus, while X-rays are generated following the ejection of an atomic electron. The rate of intensity loss is determined by the half-life of the isotope considered. The half-life and the energies emitted are given in Table 6.1 for some commonly used isotopes. The half-life

Table 6.1 Half-life and energy of emitted radiation for some commonly used radio-isotopes

Radio-isotope	Half-life (years)	Emitted radiation
^{55}Fe	2.7	Mn K X-rays
^{109}Cd	1.3	Ag K X-rays (a small fraction of the radiation consists of γ-rays of 88 keV).
^{241}Am	241	γ-rays of 59.5 keV
^{57}Co	0.74	γ-rays at 122 keV and 136 keV

is defined as the period after which the emission rate is by 50%. Radio-isotope sources are usually replaced after a time period equal to one or two half-lifes.

Radio-isotope-based system can be produced at low cost because there is no need to provide a high-voltage generator and an X-ray tube. The incident radiation is stable over time, the only drift is due to the progressive decrease of the activity with time. The size of the radio-isotope is generally very small, typically measured in millimetres. The intensity, however, is generally very low. This is compensated to some extent by the extreme close coupling that is possible to be realized between the source and the specimen. In some geometries, the source is annular shaped with the detector in the middle. A collimator is provided between source and detector to avoid direct radiation from the source reaching the detector.

The background in X-ray measurements originates to a large degree from incident radiation that has been scattered by the specimen. From the data in Table 6.1, it can be seen that the emitted radiation consists of photons with a few discrete energies. The background in the measurements is thus quite low. This is another advantage of radio-isotope excited systems over systems using photons with a continuous range of energies. Some radio-isotope systems have more than one radio-isotope fitted, which can be selected (i.e. brought in front of the specimen) by microprocessor control. Such systems allow to use a given isotope for a given range of analytes and another isotope for other elements. Such tuning of the source to the analytes can yield significant benefits compared to systems with a single isotope source. An introduction to radiation protection can be found in the work by Martin and Harbison [20]. In most countries the use and the disposal of radio-isotopes are regulated by a government agency.

6.2.7.2 X-ray tubes

Vacuum X-ray tubes are the method of choice when it comes to providing a high photon flux at a reasonable price. X-ray tubes come in a range of power, varying from a few watts to several kilowatts. The flux is much higher than that achievable with radio-isotopes. The first X-ray tube with a hot cathode has been made by Coolidge [21] and the basic principles of operation are still the same. The X-ray tube consists of a filament, an anode and a window. The filament serves as the source of electrons (cathode) which are accelerated towards the anode due to an electric field between anode and cathode. When the electrons impinge on the anode they are decelerated by the atoms. This deceleration produces mostly heat. A fraction of the energy is converted into photons with a continuous energy distribution: the Bremsstrahlung [22]. Also, the incident electrons excite electrons from the atoms which, in turn, can give rise to the emission of characteristic radiation. These two components – the continuum and the characteristic radiation – make up the tube spectral output. A description of the shape of the continuum has been given by Kramers [23]:

$$dI_\lambda = KiZ\left(\frac{\lambda}{\lambda_{\min}} - 1\right)\frac{1}{\lambda_2}\,d\lambda \tag{6.13}$$

where dI_λ is the intensity in the wavelength interval $d\lambda$, K is a proportionality constant, i is the current through the tube, Z is the atomic number and $\lambda_{\min}$ is the minimum wavelength, which can be calculated from Equation 6.9 with the energy E equal to the potential over the tube. This description is still the basis of many more recent algorithms, such as the methods developed by Pella *et al.* [24] or Ebel [25]. From Equation 6.13 it follows that the spectral output of a tube increases proportional to the current and to the atomic number. Furthermore, it increases with the voltage applied over the tube. Note that Kramers' law does not include the characteristic radiation from the tube's anode, nor does it take the attenuation by the tube's exit window into account.

The requirements imposed on X-ray tubes for X-ray fluorescence (XRF) analysis are a good stability of the photon flux and a high spectral purity (i.e. absence of characteristic lines from construction materials). Most of the energy of the electron beam is converted into heat with only about 0.1% *converted* to electromagnetic radiation. This causes extreme heat loading on the anode, especially with high power tubes. This is the main limiting factor in tube design and, for high power tubes, it forces the anodes to be cooled. To allow even higher power ratings, the anode can be made to rotate quickly. In such rotating anode tubes, each section of the anode is illuminated by the electrons only a fraction of the time, allowing such tubes to run at a power exceeding 10 kW.

6.2.7.3 Synchrotron radiation

The acceleration of charged particles results in the emission of electromagnetic radiation. In synchrotron storage rings, charged particles, such as electrons, are radially accelerated which causes radiation to be emitted. To compensate for the energy losses due to radiation, an additional field is applied. As a result, the beam can be sustained for many hours. Typical beam energies are in the GeV region, with currents up to 1 A. The resulting electromagnetic radiation is highly polarized. The spectrum is very intense from very low energies up to around 10 keV; at higher energies, the intensity drops rapidly. Because of the small cross-section of the particle beam, the brilliance is very high and the beam is thus well suited to experiments requiring small source sizes, such as X-ray microscopy. Furthermore, the high intensity over a fairly large energy range enables the use of monochromators to select a certain energy. More information on synchrotron radiation can be found, for example, in the textbooks by Kunz [26] or Margaritondo [27].

6.3 Principal components of wavelength dispersive XRF spectrometers

6.3.1 Introduction

All wavelength dispersive XRF spectrometers have one component in common: a highly orientated structure placed in the secondary beam, between the specimen and the detector (system). This device is usually a single crystal of a naturally occurring compound (this is why this type of spectrometers is sometimes referred to as a *crystal* spectrometer); for the analysis of longer wavelengths, artificial structures are used. These devices are fairly large, their dimensions are typically measured in centimetres (2–3 cm wide and 5–8 cm long). The purpose of this crystal (or artificial structure) is to diffract a selected wavelength into the detector, while suppressing other wavelengths. The phenomenon of diffraction is most easily explained in terms of the wavelength of the X-ray photons, rather than their energy. This is the reason why this type of XRF spectrometer is called *wavelength dispersive*, although exactly the same electromagnetic radiation is analyzed as on their *energy dispersive* counterparts. The analytical range of XRF spectrometers starts from about 100 eV to over 40 keV. This covers the characteristic radiation of the elements from berillium (atomic number 4) to the elements beyond uranium. The elements from beryllium to tin are usually measured using their K characteristic lines, while for the elements beyond tin the L lines are used. Routinely wavelength dispersive X-ray spectrometers are used to determine the elements from sodium (atomic number 11) upwards, although there are some applications involving routine determination of elements such as boron, carbon and fluorine.

The most common type of wavelength dispersive X-ray spectrometer is the flat crystal spectrometer; its main components are identified in Fig. 6.2. Most of the wavelength dispersive X-ray spectrometers use a sealed X-ray tube for excitation. The primary beam from the X-ray tube is directed to the specimen. The distance between the X-ray tube and the specimen is kept as small as possible, since the intensity of the primary beam at the surface of the specimen varies inversely with the square of the distance (if the source is a point source). Putting the X-ray tube close to the specimen thus increases the sensitivity. The only components that are sometimes put between the tube and the specimen are primary beam filters. These are thin foils of a metal, such as aluminum,

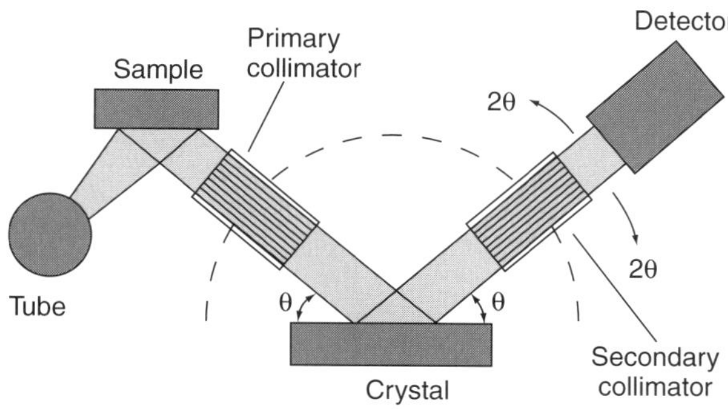

Fig. 6.2 The basic components of the planar wavelength dispersive X-ray spectrometer.

brass or rhodium; the thickness typically varies between 100 and 500 μm. Their purpose is to modify the primary beam to improve on the lower limit of detection [28], or to allow measurement of the analytes whose characteristic radiation is interfered with by the characteristic radiation of the tube's anode element(s). The sealed X-ray tubes are operated at high voltages between 20 and 100 kV, at power settings varying from a few watt to several thousand watt.

The flat crystal is the analyzing crystal in Fig. 6.2. Its purpose is to diffract a selected wavelength while suppressing others. The operating principle is based on Bragg's law for diffraction (see Section 6.3.2). The detector is coupled to the rotation of the crystal: if the angle between the plane of the crystal and the incident beam is θ, then the detector is rotated to an angle of 2θ.

6.3.2 Bragg's law for diffraction

Wavelength dispersive spectrometers (WDS) are based on Bragg's law for diffraction (see Fig. 6.3) on a single crystal, which can be represented by

$$n\lambda = 2d \sin\theta \qquad (6.14)$$

where n is the order of diffraction, λ is the wavelength of the characteristic radiation considered, d is the spacing of the lattice planes of the crystal and θ is the angle of diffraction. In nearly all practical cases $n = 1$; the first order is used. This is because the intensity of the second and higher orders is much lower than that of the first order. Some crystals do not exhibit diffraction at all orders. The crystal (in this context sometimes referred to as an analyzer crystal) is placed in the secondary beam; that is the beam of X-rays leaving the specimen. The analyzer crystals are invariably used with the lattice planes of interest parallel to the surface (symmetrical reflection). The diffracted beam is directed towards the detector. For a given crystal (thus for a given $2d$ spacing),

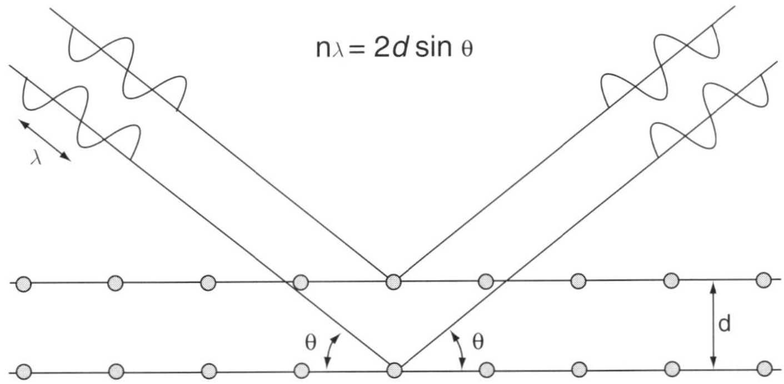

Fig. 6.3 Illustrating Bragg's law. Diffraction occurs when the difference in pathlength (calculated from the spacing d and the angle θ) is equal to an integer multiple of the wavelength λ.

the angle θ defines the wavelength of the spectrum incident on the crystal which will be diffracted toward the detector. In order to cover the range of interest, more than one analyzer crystal (each with different spacing) is required. The maximum wavelength that can be analyzed depends on the value for the $2d$ spacing and on the maximum angle allowed by the mechanical construction. Note that θ cannot be equal to 90 degrees as this implies that the detector and the specimen occupy the same space. In practice, θ is limited to a maximum of 75–80 degrees.

The expression for *angular dispersion* can be obtained by differentiating Equation 6.14 with respect to λ:

$$\frac{d\theta}{d\lambda} = \frac{n}{2d\cos\theta} \qquad (6.15)$$

from which it can be seen that decreasing the $2d$ spacing will cause a larger difference in angle θ (i.e. the angular difference between two emission lines of slightly different energies will be larger), thus reducing the overlap between the two lines. Increasing the order n has a similar effect, but at the expense of sensitivity.

There are two possible implementations: one involves planar crystals and requires the incident beam to be parallel. This is achieved by a parallel plate collimator between the specimen and the crystal. Alternatively, the crystals can be curved. If this latter configuration is used, the crystal is positioned on the focusing circle of the spectrometer and slits are positioned between the specimen and the crystal, and between the crystal and the detector. Several variations of curved crystal implementations exist, employing crystals with a constant curvature [29], crystals that are curved and then ground [30] and logarithmically curved crystals where the radius of curvature varies along the crystal's surface. The planar crystal geometry is most often used for sequential spectrometers, while curved crystal geometries are popular for simultaneous spectrometers.

6.3.3 Analyzer crystal

The analyzer crystals most often used in WDS are naturally occurring compounds. Since these crystals are positioned in the beam, it is important that the constituent elements do not fluoresce or, if they do, that their characteristic fluorescent radiation does not interfere with the measurement. Very popular crystals are LiF crystals for the analysis of the elements from potassium (atomic number 19) upwards. The fluorescence of both lithium and fluorine is extremely weak, so their characteristic radiation does not interfere with the measurement of the elements of interest. LiF crystals can be obtained in several orientations, such as (100), (110) and (210) with intense reflections. LiF is a face-centered cubic crystal, belonging to the space group Fm3m, and thus reflections on (100), (110) and (210) lattice planes are extinguished while strong reflections exist on (200), (220) and (420) planes. Hence, some manufacturers use the indications LiF 200, LiF 220 and LiF 420 to indicate the actual orientation of the crystal used. The value of the $2d$ spacing used in Bragg's law (Equation 6.14) is the one corresponding to the strongest reflection with n equal to 1. The reason for the use of different orientations is the resolving power; see Equation 6.15. The $2d$ spacing is 0.403 nm for LiF 200, 0.285 nm for LiF 220 and 0.180 nm for LiF 420. The planes with higher indices have a lower reflectivity, which is a penalty to be paid for the higher angular resolution. An example can be seen in Fig. 6.4, where a scan is made over the wavelength range covering the Mn Kα (0.213 nm) and the Cr Kβ radiation (0.209 nm). The specimen is a steel alloy containing about 17% of Cr and 1.5% of Mn; all measurement conditions were the same for both scans, except for the analyzer crystal used. The better peak resolution obtained with the LiF 220 crystal compared to the LiF 200 can be clearly seen. The intensity, however, is lower by a factor of about 3.

Another commonly used crystal is penta erythritol ($C_5H_{12}O_4$, PE) for the elements aluminium (atomic number 13) and higher. Its constituent elements have no characteristic wavelengths in the region of interest. It is a tetragonal structure belonging to space group I4, with a very intense reflection on the (002) planes. Similar to LiF crystals, reflections on (001) planes are extinguished. For PE, the $2d$ spacing between the lattice planes used is 0.874 nm. Other naturally occurring crystals, such as TlAP (thallium hydrogen phosphate) and ADP (ammonium dihydrogen phosphate) can be used for elements from Mg upwards. Ge and InSb crystals are sometimes used for specific applications: Ge provides better angular resolution than pentaerythritol for P, S and Cl, and InSb has a very high reflectivity for Si Kα radiation.

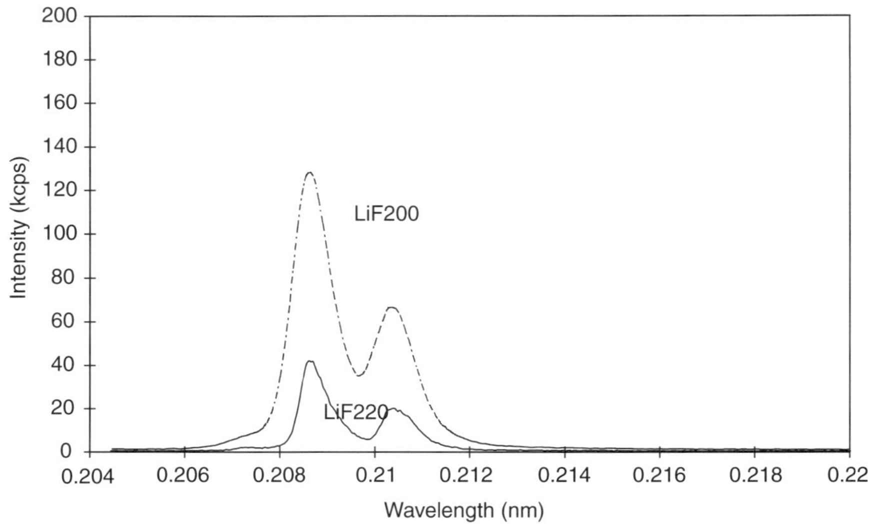

Fig. 6.4 Wavelength scan over the range covering the Mn Kα and the Cr Kβ radiation. Note the better peak separation obtained with LiF220 compared to LiF200, at the expense of some sensitivity.

The crystals used are not perfect single crystals: they consist of areas with high degree of monocrystallinity with slightly different orientations. The domains are typically 10 nm across. This mosaicity increases the integrated reflectivity compared to a crystal with a lower degree of mosaicity. It also widens the peak width of the diffracted beam; the width of the rocking curve is increased. For maximum sensitivity, the divergence of the incident beam is matched to the rocking curve. The integral reflectivity is mainly determined by the space group of the crystal, the atomic scattering factors of its constituents elements, the number of atoms and their arrangement in the crystal lattice and the degree of mosaicity. The peak width, on the other hand, is determined by the divergence of the incident beam and the rocking curve.

Crystals such as gypsum and mica have been used in the past for the longer wavelengths, but their reflectivities are quite low. These have now been superceded by the use of synthetic structures [e.g. 31, 32]. These synthetic structures are now mostly made by depositing a layer of a material with a high electron density on top of a layer with a low electron density. This process is then repeated 20 to 100 times. Their operation is based on the fact that X-rays can also be reflected in ordered materials consisting of layers with different electron densities. Common multi-layered structures consist of a stacking of Mo and Si layers, or B_4C layers alternated with Mo. The choice of the elements is mainly based on parameters such as the difference in electron density, the compatibility of the elements, the deposition behavior of the elements (the layers must be smooth) and diffusion between layers should not occur. Furthermore, the resulting structures must be able to withstand a high X-ray flux without sustaining radiation damage and they must be thermally and chemically stable. By carefully selecting the composition of the layers and their interplanar distances, the reflectivity can be optimized. Due to the fact that the long wavelengths involved are easily absorbed, the useful wavelength range of these multilayers is often rather limited. Some multilayers are element-specific, in the sense that they exhibit high reflectivity only in a narrow band around the characteristic radiation of a specific element [33, 34]. Beryllium, the element with the lowest atomic number that still emits characteristic radiation, has been measured first in 1987 [35] using a multilayer.

6.3.4 Goniometer

Most wavelength dispersive instruments use a goniometer to adjust the angle θ at will. The position of the specimen is fixed, while the analyzer crystal is at the center of rotation, and the diffracted beam is kept directed toward the detector, which is fitted to the goniometer arm. When the crystal makes an angle θ with the secondary beam, the detector arm is actually at an angle of 2θ with respect to the secondary beam. The angular range of such goniometers is typically from a few degrees to 150 degrees (2θ). Higher angles are excluded because then the specimen and the detector occupy the same area in space. The angular accuracy of modern goniometers is often better than 0.001 degree, while the reproducibility is 10 to 100 times better. Each of the wavelengths of interest requires different settings of the angle and possibly a different crystal as well. All wavelengths are thus measured sequentially in time. Qualitative analysis can be performed by slowly rotating the crystal, effectively changing the angle θ and thus the wavelength as seen from Equation 6.14. Quantitative analysis can be done by moving the crystal to a given angle θ, corresponding to the wavelength of interest and determining the intensity. Most goniometers are equipped with a crystal mill that can be fitted with several crystals (typically 6–10), one of which can be moved into the plane of diffraction depending upon the requirements.

For speed-critical applications, a separate class of instruments has been developed where the positions of the crystal and the detector are fixed. These assemblies do not have moving parts. One such assembly is then dedicated to the detection of a single wavelength (in essence, a small wavelength band). For each of the characteristic lines to be measured such an assembly has to be present. These instruments allow the measurement on all assemblies simultaneously; hence they are called *simultaneous instruments*. Since they do not require a goniometer, they are generally considered more robust and are dedicated to a single application. They can, however, be equipped with one or more goniometers, making them more versatile.

6.3.5 Collimators

Flat crystal spectrometers, as schematically illustrated in Fig. 6.2, use collimators to render the beam (somewhat) parallel. A collimator in WDS is a stack of thin plates (typically 100 μm in thickness) arranged parallel to each other at a fixed distance. The interplanar distance and the length of the plates determine the divergence of the collimator. The degree of divergence allowed in the plane of Fig. 6.2 is determined by the width of the rocking curve of the crystal used. If the divergence of the collimator is wider than the rocking curve, then the peak to background ratio is adversely affected. If, however, the divergence of the collimator is smaller than the rocking width, then a sizeable reduction of sensitivity is the consequence. In practice, the divergence of collimators varies from about 0.05 to 4 degrees. With a length of about 10–12 cm, this corresponds to interplanar spacings of 100 μm to a few mm. It is clear that the construction of a collimator with a spacing of 100 μm between the plates is not self-evident, as a high degree of parallelism is required. The finer collimators are used in combination with crystal such as LiF 220 and LiF 420, which are specifically treated for high resolving power, while the coarser collimators are used in combination with the synthetic multi-layered structures.

6.3.6 Detectors

The detectors used in WDS have a fairly simple function compared with their siblings in EDS. Since the analyzer crystal monochromatizes the X-ray beam, ideally the detector should only count the photons. In practice, there are some phenomena that are complicating the picture. Among these are higher order lines, escape peak phenomena and pulse shrinkage.

Higher order peaks can occur when photons with a wavelength λ of half the desired wavelength, λ_0, impinge on the crystal. According to Bragg's law (refer to Equation 6.14) these photons can also be diffracted in second order, with $n = 2$. Indeed, 2λ equals $1\lambda_0$, where '1' is used to explicitly denote that the first order is used. These photons then enter the detector and must not be counted, as they have another wavelength than the desired one. Hence, the detector must have some means to distinguish these energies: some form of energy determination is required. The diffraction in second order is usually very weak (or close to non-existent); it is most commonly a problem when analyzing an element at a low concentration level, when a major element has an interfering line in second order.

Escape peak phenomena are caused by the fact that in order for a photon to be counted, it must be absorbed by the active part of the detector. The absorption process can cause fluorescence of the constituent atoms. This fluorescence radiation can escape the detector body, and thus the energy determined is less than the energy of the incident photon. This can lead to significant counting losses under certain circumstances.

Pulse shrinkage can be caused by the fact that with increasing count rates, the individual electric pulses are no longer completely resolved from one another. Most of the detector systems used in wavelength dispersive XRF are based on the principle that the photon is absorbed within the active body of the detector. Depending upon the type of detector, different phenomena are involved in counting. The most commonly used detectors in WDS XRF are the gas-filled detectors and the scintillation detectors.

6.3.6.1 Gas-filled detectors

Gas-filled detectors consist of a cylindrical volume (typically between 1 and 4 cm in diameter and 3–7 cm in length) filled with a gas mixture, the main component of which is a noble gas, such as Ar or Kr; and a wire running through the center. A photon can remove an outer electron of such inert gas atoms. For an Ar-filled detector, the following equation holds:

$$Ar \xrightarrow{h\nu} Ar^{+} + e^{-} \tag{6.16}$$

The energy required for the ionization, e_{i}, depends on the element; for Ar about 26 eV is required to create an ion–electron pair. The total number of such primary ionizations, n_{p}, caused per incident photon with energy E is then given by

$$n_{\mathrm{p}} = \frac{E}{e_{\mathrm{i}}} \tag{6.17}$$

This phenomenon is subject to a Poison statistic. This spread in the number of primary ionizations is the limiting factor determining the resolution of this type of detectors. In order to avoid that the Ar ion and the electron quickly recombine, a constant electric field is applied between the wire and the detector housing. The wire is kept at a voltage of about 1–2 kV, while the cylindrical housing is kept at earth potential. Under the influence of the electric field, the electrons are accelerated toward the anode wire while the positive ions migrate toward the housing. The primary electrons will collide with other gas atoms. The field strength is such that the electrons get sufficient kinetic energy between collisions to cause additional ionizations. The result of this avalanche effect is that for each individual electron originally created by the incident photon, many more are finally collected at the anode wire. The ratio between the number of electrons collected and the number of initial ion–electron pairs is called the gas amplification factor and is typically around 10^4 to 10^5. The positive ion cloud, which is formed after one photon has been absorbed, is moving to the cathode at a slower speed than the free electrons. This cloud actually decreases the electric field strength experienced by the electrons and thus a subsequent photon will be registered as having a lower energy. This is one factor contributing to pulse shrinkage: under the influence of the diminished electric field, electrons will create less additional ionizations. When the ion cloud has disappeared, the photons are counted at the correct energy again. At low count rates, the electric field is fully recovered before the next photon is registered and pulse shrinkage is thus not a problem. At higher count rates, close loop compensation circuits are used to compensate for the diminished field strength.

In order for the photons to be able to enter the detector, gas-filled detectors are provided with a thin entrance window, the material of which is specially selected for low absorption of X-rays. Beryllium can be used if the range of interest is above, say, 2 keV. Beryllium is air tight and strong enough (at thickness of about 25 μm; this does depend on the actual size of the window) to withstand the pressure difference that exists between the vacuum of the spectrometer chamber and the detector. For the detection of lower energy radiation, beryllium is no longer suited (because of its absorption) and a thin (a few microns thick) polymer foil, such as polypropylene or polycarbonate is used. These foils (as well as a beryllium window) must be coated at the interior with a conductive coating for the electric field within the detector to be homogeneous. Gas-filled detectors with Be windows can be sealed, in the sense that they are filled with the counting gas by the manufacturer and then hermetically sealed. The useful lifetime of such sealed detectors can be several years. Polymer windows, however, often have pinholes, so the counting gas will gradually leak away. In order to compensate, a constant flow of counting gas is used. The pressure and the temperature of the counting gas should be kept constant. Changes in gas temperature or pressure would alter the absorption properties of the detector and also infer changes on the gas amplification. The gas leaking from the detector is vented into the spectrometer chamber while the excess gas is leaving the detector through a tube and is vented outside the spectrometer. Such detectors are therefore called gas flow detectors and are especially used for the detection of X-rays with very long wavelengths. The efficiency of the gas-filled detectors is determined by the attenuation in the entrance window and the absorption efficiency of the counting gas for the wavelengths of interest. This absorption efficiency depends on the path length available within the detector, the composition of the counting gas and its pressure. The efficiency can thus be adapted according to the wavelength range to cover. They are rarely used, however, for the detection of photons with wavelengths shorter than 0.03 nm, where the efficiency is near zero.

In some instruments, a sealed detector is placed closely behind the gas flow detector. The gas flow detector is then equipped with a thin Be window at the back. This permits the photons that are not absorbed by the gas detector to enter the sealed detector. If this is the case, the gas flow detector will be used for the longer wavelengths, while the efficiency of the sealed detector (which is then usually filled with an inert gas such as xenon) will gradually increase with decreasing wavelength, as the fraction of photons absorbed in the gas flow detector diminishes. In some instruments, this *second* detector is a scintillation detector rather than a sealed gas detector.

6.3.6.2 Scintillation detectors

The scintillation detector is based on a two stage process. In the first step, the incoming photon is converted to photons of much longer wavelength by a phosphor, most commonly an NaI crystal, doped with Tl. For such phosphors, the excitation potential required, e_i, is about 3 eV. The photons emitted have a wavelength of about 410 nm. The NaI crystal is closely coupled to a photo multiplier, which is sensitive to radiation in that wavelength region. The efficiency of the scintillation detector is determined by these two components. The NaI crystal should be thick enough to absorb the radiation of interest with high efficiency, while it must be as transparent as possible to the re-emitted photons, so that they can reach the photo cathode of the photo multiplier. The photo multiplier consists of a photo cathode and a series of dynodes. At the photo cathode, the 410 nm photons are generating free electrons, which are then accelerated to the first of the dynodes. Upon impact, more free electrons are created, which are then all accelerated toward the next dynode. In general, 10 dynodes are used, each successive one placed at a higher voltage than the preceding one. The total amplification thus obtained is about 10^6.

The NaI crystal is hygroscopic, so it must be mounted in a sealed container. The entrance window is then usually made of beryllium. With a thickness of 2 mm for the NaI crystal, the efficiency of the detector is quite high for photons with wavelengths down to 0.05 nm. The long wavelength limit is imposed by the thickness and the material of the entrance window. The scintillation detector and the gas-filled detectors are complimentary and in many systems, both are provided whereby the gas-filled detector is used for the long wavelength region and the scintillation detector covers the short wavelength region. In some spectrometer systems, they are actually used in tandem in an arrangement, where the gas-filled detector is provided with an exit window and coupled to a scintillation detector. The scintillation detector is then used to detect those photons that are not absorbed by the gas flow detector. Depending upon the dimensions of the gas flow detector and the thickness of the windows in the X-ray path, the wavelength at which the efficiency of the scintillation detector is higher than that of the gas flow detector varies between 0.2 and 0.15 nm. For longer wavelengths the gas flow detector is used, for the intermediate range both detectors are used and for the shorter wavelengths only the scintillation detector is used.

6.3.7 Other arrangements

Other arrangements have been used. Some involved two analyzing crystals in sequence, to obtain high resolution spectra, at the expense of sensitivity. The main application of this instrument was research on the shift of the emission lines due to changes in valency of the atoms and speciation [36]. One quite different instrument has been developed by De Bokx and Urbach [37]. In their instrument, the fluorescent beam is taken at a glancing angle from the sample surface. This reduces the intensity from the incident beam after being scattered from the specimen dramatically, thus leading to very low minimal detection limits, usually associated with instruments based on total reflection (TXRF), while maintaining the resolution of a WDS. A comparison of this method compared to TXRF and PIXE has been given by Claes *et al.* [38]. The instrument has been used for applications involving the analysis of thin coatings. More information regarding TXRF can be found in the work by Klockenkämper [39].

6.4 Principal components of EDS

6.4.1 Introduction

The main difference between WDS and EDS is in the method used to discriminate between energies. In WDS, the two functions of energy discrimination and photon counting are basically separated: the energy discrimination is mostly done by the analyzing crystal (see Sections 6.3.2 and 6.3.3); the detectors used in those systems are selected for counting capability and efficiency at the wavelengths concerned; their energy resolution is of minor importance. In EDS, the energy resolution of the detector is a prime selection criterion, as there is no other component to assist in energy resolution.

The basic EDS system is thus quite simple; it consists of three components: the X-ray tube with its associated high voltage generator, the sample stage and the detector with the detector electronics, including a multi-channel analyzer. There are still EDS systems available, where the X-ray tube is replaced by a radio-isotope. Such systems are quite often portable. The main problem with such spectrometers is the relatively low sensitivity (due to the limited activity of the source) and the regulations involved when applying/handling radioactive material.

The maximum count rate that energy dispersive detectors can handle is limited. It is possible to adjust the shaping time in the pre-amplifier to achieve high count rates but this is at the expense of resolution. High resolution work is usually done at fairly low count rates; typically not exceeding a few thousand counts per second. Digital pulse processors have been instrumental to allow for higher count rates without compromising resolution to the same extent as the analog processors. Combining a silicon drift chamber detector (see Section 6.4.2.4) with a digital pulse processor yields the best compromise between count rate and resolution. In the following sections, only some common detector systems will be highlighted.

6.4.2 Detectors

6.4.2.1 Introduction

Most detectors used in EDS nowadays are solid-state detectors. This is due to their superior energy resolution capability compared to their gas-filled brethren. The gas-filled detectors have been used first. They exhibited low energy resolution, but their application was made fairly easy because of the gas amplification that made for reasonable signal-to-noise ratios. They are still applied successfully in portable instruments and instrumentation for quality control [40]. Sodium iodide scintillators, coupled with photo multipliers (see Section 6.3.6.2) have been used extensively in the field of γ-rays and very energetic X-rays.

6.4.2.2 Si(Li) detectors

In the mid-1960s, the Si(Li) detector was introduced to the field of EDS. Such a detector consists of a bulk of high quality silicon, with rectifying *p*- or *n*-contacts on either side. The bulk of the material should have as few free charge carriers as possible. This can be achieved either by selecting material of high purity (as is currently done for the high purity germanium detectors – HPGe) or by compensating the charge of *p*-type silicon with lithium donor atoms. By reverse-biasing the diode, the free charge carriers move away from the bulk under the influence of the electric field and an active depletion region is created. Only the thermally generated charge carriers (these consist mainly of electrons excited above the band gap of the semi-conductor) contribute to the current running through the device under these conditions. By cooling the detector, the noise generated by these free charge carriers can be reduced. In general, these detectors are cooled by liquid nitrogen. In optimal cases, the resolution (expressed as FWHM) can be better than 140 eV at 5.9 keV (Mn Kα) at low to moderate count rates (1000 cps).

When an X-ray photon is absorbed within the active volume of the detector, it interacts primarily by the photoelectric effect. This creates a photoelectron and a inner-shell vacancy. The photoelectron produces several low-energy ionizations by interacting with the atoms in the crystal lattice. The inner-shell vacancy decays by emitting Auger electrons or low-energy X-rays. Most of these are re-absorbed and also create multiple ionization events. All the free charge carriers are subjected to the electric field. The holes migrate toward the front contact while the electrons migrate toward the back contact, where their charge are converted by means of a field effect transistor (FET) into a voltage pulse. The amplitude of this voltage pulse is directly proportional to the number of free charge carriers and thus to the energy of the incident photon.

6.4.2.3 Si-PIN detectors

Si-PIN detectors are based on the use of intrinsic silicon, requiring no compensation by lithium atoms. There is no need for cooling down the devices of temperatures, such achievable with liquid nitrogen, and often thermo-electric cooling is used. In those cases, the resolution is about 190–220 eV FWHM at 5.9 keV, depending upon thickness and size of the detector.

6.4.2.4 Si-drift chamber detectors

The main advantage of Si-drift chamber detectors [41] compared to other silicon-based detectors is that the capacitance is very small, in the order of 100 fF. This is due to the unique layout of the electrodes. This small capacitance leads to very good signal-to-noise ratios, even for fairly large dimensions and at temperatures well within the range of thermo-electric cooling. Also, the achievable count rate is much higher than with more conventional detectors [42].

6.4.2.5 High-resolution detectors

Considerable effort has been spent on the development of high-resolution detectors for EDS. Two main approaches seem promising; in fact they have led to commercial products. The *micro-calorimeter* [43] is based on the conversion of the photon's energy to heat. The temperature increase is then measured. Because of the low energies involved, the heat capacity of the detector must be extremely small. This leads to detectors with very small dimensions. The detectors are cooled to very low temperatures of about 100 mK. Because of these dimensions, the efficiency at higher energies is often very small. Another approach is based on the *superconducting tunnel junction* [44]. This detector is also cooled to below a tenth of the superconducting transition temperature. To reach those temperatures, a multi-stage cooling is used e.g. involving liquid nitrogen, liquid helium and adiabatic demagnetization. The energy resolution obtained is generally around 10 eV or better. The main disadvantages are the need for the extremely low temperatures and the limited count rates that can be processed. To increase the count rate, the detectors can be used in an array.

6.5 Applications

6.5.1 Qualitative analysis

The resolving power (often referred to as resolution) of the WDS is better than the resolution of the energy dispersive spectrometer in the energy range up to about 20 to 25 keV. (This depends on the configuration of the wavelength dispersive system used, most notably the analyzing crystal and the collimator used; and the detector system used in the EDS. In the latter system, the detectors based on the calorimeter [43] are promising exceptional resolution, outperforming the classic wavelength dispersive systems over the complete energy range.) The superior resolution of the wavelength dispersive system renders qualitative analysis easier. In practice, however, it has only been recently that wavelength dispersive X-ray spectrometers have come to be used routinely for qualitative analysis. This was partly caused by the fact that the cost of WDS is substantially higher than the cost of EDS and therefore they are usually procured for routine quantitative analysis, based on peak measurements. Nowadays, the computing power available is such that it is possible to perform qualitative analysis in a highly automated fashion. This, and the advent of ever more demanding legislation regarding environmental issues has created a large demand for qualitative and semi-quantitative analysis on WDS as well as on EDS. Software packages have recently been developed that – together with some calibration standards or *setting-up samples* – will perform automated qualitative and semi-quantitative analysis.

6.5.2 *Quantitative analysis*

Quantitative analysis involves the conversion from measured intensities to the concentration of the analyte. When analyzing dilute specimens (such as metals in oils), this conversion can be done by employing linear calibration graphs. This also applies when very thin specimens are analyzed; for example, aerosols deposited on a filter. In many other cases the specimens used for analysis are thicker, from several millimeters to a few centimeters. Note that in solid samples the thickness of the layer from which a significant contribution to the measured signal can emerge is limited from a few μm to a few mm. All the other elements (matrix) present in the specimen interfere with the emitted intensity of the analyte element, and some form of matrix correction is thus required. Two types of matrix effects can be distinguished: absorption and enhancement.

6.5.2.1 *Absorption*

The relationship between a monochromatic incident beam of wavelength λ_0, the composition of the specimen and the emitted intensity λ_i (in the absence of enhancement) of the analyte i can be written as

$$P_i(\lambda_0) = K_i I_0(\lambda_0)\csc\psi' \frac{\mu_i(\lambda_0)}{\frac{\mu_s(\lambda_0)}{\sin\psi'} + \frac{\mu_s(\lambda_i)}{\sin\psi''}} W_i \tag{6.18}$$

where $P_i(\lambda_0)$ is the intensity of the characteristic radiation of the analyte element i caused by the primary fluorescence effect; K_i is a proportionality constant, $I_0(\lambda_0)$ is the intensity of the incident beam with wavelength λ_0, μ_s and μ_i are the mass attenuation coefficients of the specimen and the analyte respectively for the wavelengths indicated, W_i is the weight fraction of the analyte and ψ' and ψ'' are the angle of the incident beam and of the exiting beam with the surface respectively. When a polychromatic spectrum is used, such as from an X-ray tube commonly used with wavelength dispersive XRF spectrometers, Equation 6.18 must be integrated over the incident spectrum; this is done by numerical integration. The denominator in Equation 6.18 indicates the main source of matrix effects: as the mass attenuation coefficients of the specimen can vary drastically depending upon the composition, the intensity of the characteristic radiation is not a simple function of the analyte's concentration W_i but is also affected by the concentration of other elements present. Hence, when the mass attenuation coefficients between the matrix and the analyte vary significantly, the emitted characteristic intensity is no longer a simple function of the concentration of the analyte alone. This effect can be quite large and needs to be accounted for in most practical analyses of bulk material. This can be readily seen from the graph in Fig. 6.5, where the measured intensity of Ag Kα is plotted versus the Ag concentration. The specimens are tin and lead–tin alloys, with varying concentrations of lead.

The values of the mass attenuation coefficients for Ag Kα in lead is 69.9 cm^2/g, versus 16.8 cm^2/g in tin. As a result, the mass attenuation coefficient for Ag Kα varies between 122 cm^2/g for specimens without little or no lead to 367 cm^2/g for a specimen with 37% lead. After correction for matrix effect, an adequate calibration line is obtained, as shown in Fig. 6.6.

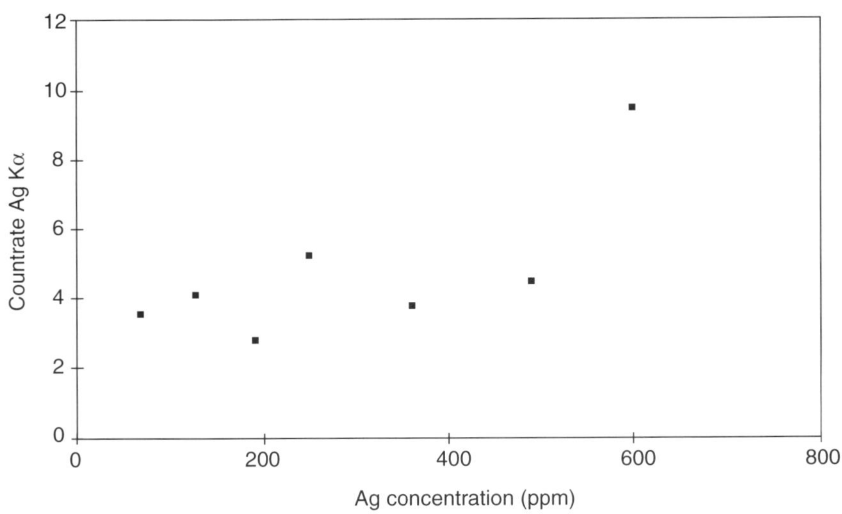

Fig. 6.5 Plot of Ag Kα versus concentration in lead–tin alloys. Note the large spread due to difference in mass attenuation coefficient of lead (69.9 cm^2/g) and tin (16.8 cm^2/g) for Ag Kα.

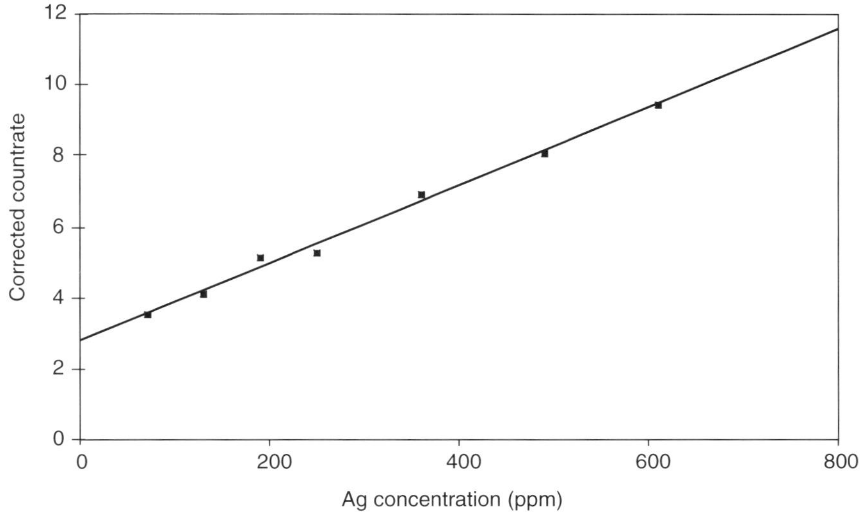

Fig. 6.6 Calibration line for Ag Kα in tin and lead–tin alloys, after matrix correction. Compared to Fig. 6.5 the spread is reduced markedly.

6.5.2.2 Enhancement

Enhancement is the phenomenon whereby part of the incident beam is absorbed by an element j, which subsequently emits characteristic radiation. This characteristic radiation can be absorbed by atoms of the analyte i, causing additional fluorescence of the analyte element. This effect is also referred to as secondary fluorescence. Elements that are subject to enhancement usually have a higher sensitivity compared to the situation where secondary fluorescence is absent. A necessary condition for this effect to occur is that the characteristic radiation of element j is sufficiently energetic to cause primary fluorescence of the analyte's atoms. For infinitely thick specimens, the contribution to the intensity of characteristic radiation of the analyte i caused by secondary fluorescence $S_{ij}(\lambda_0, \lambda_j)$ is given by

$$S_{ij}(\lambda_0,\lambda_j) = K_{ij}\frac{I_0(\lambda_0)\csc\psi'\,\mu_j(\lambda_0)W_j\mu_i(\lambda_j)W_i}{2(\mu_s(\lambda_0)\csc\psi' + \mu_s(\lambda_i)\csc\psi'')} \times$$
$$\left(\frac{\sin\psi'}{\mu_s(\lambda_0)}\ln\left[1+\frac{\mu_s(\lambda_0)}{\mu_s(\lambda_j)\sin\psi'}\right] + \frac{\sin\psi''}{\mu_s(\lambda_j)}\ln\left[1+\frac{\mu_s(\lambda_j)}{\mu_s(\lambda_j)\sin\psi''}\right]\right) \quad (6.19)$$

where K_{ij} is a proportionality constant, involving fundamental parameters of both element i and j. The other symbols have been explained earlier. A similar equation exists for each characteristic line of the enhancing element, and all these contributions need to be summed. When polychromatic radiation is used, Equation 6.19 needs to be numerically integrated over the incident spectrum.

This enhancement effect is particularly strong for the transition elements (atomic numbers Z between 20 and 32) where elements with atomic number Z + 2 are very efficient in exciting atoms of elements with atomic number Z. On the other hand, the characteristic radiation of the element with atomic number Z + 2 is severely absorbed by the element with atomic number Z. This can be seen in Fig. 6.7, where the relative intensity of Fe Kα is plotted as a function of its concentration in mixtures of Fe (Z = 26), Ni (Z = 28) and Cr (Z = 24). The solid line is for Fe in FeNi and shows a clear upward curvature, due to the enhancement of Fe Kα by nickel radiation. The lower (dashed) curve is for Fe Kα in FeCr and shows a severe depression of the Fe Kα intensity, due to the absorption in Cr. The characteristic radiation of Fe is enhancing Cr but is severely absorbed in the process. The straight line in the middle represents the linear relationship between relative intensity and concentration in the absence of matrix effects. The dots are representing ternary mixtures of Cr, Fe and Ni. For any given concentration of Fe, which can be read from the axis, the remainder is taken by Cr and Ni, in steps of 0.1 weight fraction. The data in Fig. 6.7 clearly indicates the magnitude of matrix effects and its effect on intensity. For a specimen with 0.1 Fe, the relative intensity varies between 0.04

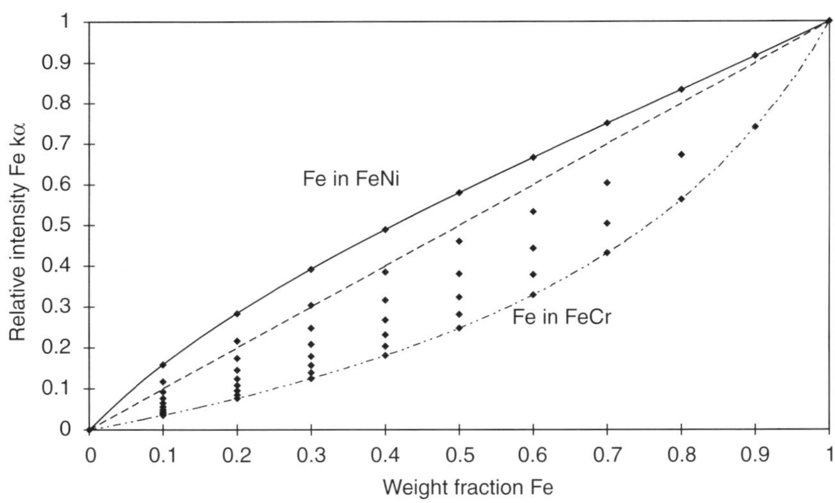

Fig. 6.7 Relative intensity of Fe Kα as a function of its concentration in CrFeNi alloys. The solid line is for Fe Kα in FeNi binary alloys (enhancement), while the dashed line is for Fe Kα in FeCr alloys (absorption). The straight line in the middle would be the intensity of Fe Kα in the absence of matrix effects.

and 0.16, a factor of 4, depending upon the matrix elements. It is clear that these matrix effects must be corrected for if accurate quantitative analysis is envisaged.

6.5.3 Correction for matrix effects

Many different methods have been developed over the years to correct matrix effects. A generally applicable classification of these methods is difficult to give. Only the main methods will be described in the following sections.

6.5.3.1 Fundamental parameter method

The net intensity of a given analyte's characteristic radiation in a specimen of a known composition can be calculated by combining Equations 6.18 and 6.19 and integrating them numerically over the wavelength distribution of the incident tube spectrum. The resulting equation is referred to as the fundamental parameter equation. The value of the proportionality constant in Equations 6.18 and 6.19 is determined from a measurement on one or more standard specimens. By applying those equations in an iterative way, the composition of unknown specimens can be obtained. The method was pioneered in the 1960s [45, 46] and although excellent results were obtained, it was not in general use due to the large amount of computation required. Its use has become more general with the advent of powerful mini- and micro-computers.

6.5.3.2 Compton correction

The intensity of the radiation scattered by the specimen can be used to obtain an estimate of the mass attenuation coefficient of the specimen at a specific wavelength λ_s, provided absorption is the main cause for matrix effect. Indeed, it can be shown that the intensity of the scattered radiation is inversely proportional to the mass attenuation coefficient μ_s. From Equation 6.18 it follows that the intensity of the characteristic radiation is also inversely proportional to a linear combination of mass attenuation coefficients and proportional to the concentration of the analyte element. The method [47] is further based on the fact that – between absorption edges – mass attenuation coefficients at different wavelengths are also proportional to each other [48, 49]. Hence, the ratio of the characteristic intensity of the analyte and the scattered radiation is proportional to the analyte's concentration. This method finds general application for the analysis of trace elements in specimens of geological origin, such as soils, ores, etc. Most of the trace elements of interest in those situations are elements with atomic number above 26 (iron), whose characteristic radiation is subject *mainly* to absorption (there is some enhancement from trace elements with higher atomic number on those with lower atomic numbers, but this contribution can be neglected due to the low levels of concentration). More elaborate methods allow for accurate analysis of trace elements beyond absorption edges of major elements [50]. Such methods tend to work on a calculation of the mass attenuation coefficients.

6.5.3.3 Influence coefficient algorithms

A very popular class of methods to correct for matrix effects consists of the influence coefficient algorithms. Over the years, many different approaches have been proposed, but nowadays, the most common ones all have the same basic equation; they differ only in the way that the values of the coefficients are calculated.

The basic equation for influence coefficient algorithms is

$$W_i = R_i \left[1 + \sum_{\substack{j=1 \\ j \neq e}}^{n} \alpha_{ij} W_j \right] \tag{6.20}$$

where W_i and W_j are the weight fractions of the analyte i and the interfering element j respectively and α_{ij} is the influence coefficient, expressing the influence of interfering element j on the analyte i. There is such an equation for each of the analytes, leading to a set of simultaneous equations. The summation in Equation 6.20 covers $n - 1$ terms, one of the constituent elements, e, has been eliminated. This equation has been pioneered by Lachance and Traill [51] in 1966. Lachance and Traill eliminate the analyte from each of the Equation 6.20 of the set, based on the fact that

$$W_i = 1 - \sum_{\substack{j=1 \\ j \neq 1}}^{n} W_j \tag{6.21}$$

A full description of an influence coefficient algorithm, including the calculation of the influence coefficients in all cases (polychromatic exciting beam, primary and secondary fluorescence) has been given by de Jongh [52], who developed his algorithm starting from a Taylor expansion of the theoretical relationship between specimen composition and the incident beam (the fundamental parameter equation). De Jongh eliminates the same element for all equations describing the specimen; he recommends eliminating the major element, such as iron in steels or SiO_2 in soils, or non-measurable elements, such as the loss on ignition when analyzing fused beads. The influence coefficients, calculated with the method of de Jongh are calculated around a reference composition; which is sometimes also called the *average* composition in the sense that the reference composition is more or less in the center of the range of the unknown specimens. The influence coefficients can then be considered constants for analysis in a reasonable range around the reference composition.

Lachance [53] has proposed a method to calculate the value of the influence coefficients as a function of the specimens composition. The coefficient α_{ij} in Equation 6.20 is now calculated from

$$\alpha_{ij} = \alpha_{ij1} + \frac{\alpha_{ij2} W_m}{1 + \alpha_{ij3}(1 - W_m)} \tag{6.22}$$

W_m is the total concentration of the matrix elements; i.e.

$$W_m = 1 - W_i \tag{6.23}$$

The values for the factors α_{ij1}, α_{ij2} and α_{ij3} are calculated from fundamental parameters. This approximation is very adequate for binary specimens. For use in multi-component specimens, cross-product coefficients are included, which leads to

$$W_i = R_i \left[1 + \sum_{j \neq 1}^{n} \left(\alpha_{ij1} + \frac{\alpha_{ij2} W_m}{1 + \alpha_{ij3}(1 - W_m)} \right) W_j + \sum_{j \neq i}^{n} \sum_{k \neq i, k>j}^{n} \alpha_{ijk} W_j W_k \right] \tag{6.24}$$

The value of α_{ijk} is calculated from fundamental parameters. The advantage of this method is that the coefficients can be calculated for given instrumental conditions before the calibration is performed. Equation 6.24 is valid over the complete concentration range from 0 to 100% by weight, although it is not recommended to correct for matrix effect when it is not significant, such as for the determination of trace concentrations in alloys. The calculation of the composition of the unknown specimens involves solving a set of equations such as Equation 6.24 for each analyte. This is a very fast and relatively simple calculation, which is why methods based on influence coefficients are more popular than the fundamental parameter method. A comparison of the accuracy of several common influence coefficients methods can be found in [54].

6.6 Notes on specimen preparation

Due to the wavelengths of the analytical radiation considered and the physical properties, XRF is essentially a surface sensitive method. Although it is commonly said that analysis of bulk specimens is possible, the actual layer from which the measured signal is originating in solid samples varies from less than a μm to less than 100 μm (when fused beads are used, the layer thickness is somewhat larger). It is thus essential that the surface of the specimens is prepared in a reproducible manner.

Liquids can often be analyzed as is, by pouring them in a specially designed cup, with a thin foil to separate the liquid from the atmosphere within the spectrometer chamber. In some instruments, this is done by a thin foil on the cup itself, while others have a thin beryllium window between the cup and the spectrometer chamber. The best sensitivity is obtained by flushing the spectrometer chamber with helium. The use of vacuum is possible, but the window (or foil) need then to be of such a thickness that the sensitivity is adversely affected.

Metallic specimens usually need only a surface treatment with e.g. abrasive paper. In *powder samples*, the grain size is important, especially if the grains have different chemical composition; a comprehensive study has been made by Feret [55]. Grinding is almost always required. The powder can then be pressed into a pellet, a binder might be added for rendering the pellets stronger. Excellent results can be obtained on pressed powder specimens, especially on the trace elements in specimens of a geological origin [28]. Higher accuracies, especially on major elements can be obtained by grinding the powder sample and then fusing it with a flux such as $Li_2B_4O_7$. Mixtures of lithium tetra borate and lithium meta borate can be used to lower the fusion temperature. This is a rather time-consuming method, but it enables one to obtain homogeneous specimens from loose powders. If accurate determination of low atomic number elements is mandatory, it is often the only practical way to proceed. Another drawback is that the dilution affects the sensitivity somewhat. Fortunately, the fluxes used consist of the elements Li, B and O, which are fairly poor absorbers in the X-ray region, so the sensitivity is not affected in a detrimental way. More information regarding specimen preparation can be found in [56].

6.7 Conclusion

X-ray fluorescence is a robust analytical technique for elemental analysis. It is characterized by a high element specificity, a high reproducibility and good detection limits. XRF spectrometers are used in a wide variety of circumstances, varying from industry for routine process control application to materials research. Many of the instruments are equipped with a programmable specimen loader, which allows them to be operated unattended for extended periods of time. Others are integrated in a comprehensive automated laboratory, where everything, from the taking of the sample, transporting it to the laboratory, specimen preparation and analysis is done automatically.

References

[1] R. Jenkins & J.L. de Vries, *Practical X-Ray Spectrometry*, 2nd edn, Springer-Verlag, New York, 1967.

[2] E.P. Bertin, *Principles and Practice of X-Ray Spectrometric Analysis*, 2nd edn, Plenum Press, New York, 1975.

[3] *Handbook of X-Ray Spectrometry – Methods and Techniques*, R.E. van Grieken & A.A. Markowicz (eds), Marcel Dekker, NY, 2002.
[4] A.H. Compton & S.K. Allison, *X-Rays in Theory and Experiment*, D. van Nostrand, Princeton, NJ, 1935.
[5] B.E. Warren, *X-ray Diffraction*, Addison Wesley, Reading, Mass, 1969.
[6] W.H. McMaster, N.K. Delgrande, J.H. Mallet & J.H. Hubbell, *Compilation of X-Ray Cross Sections*, UCRL 50174, Sec II, Rev 1, 1969.
[7] K.F.J. Heinrich, *The Electron Microprobe*, T.D. McKinley, K.F.J. Heinrich & D.B. Wittry (eds), Wiley, New York, 1966, p. 296.
[8] J. Leroux & T.P. Thinh, *Revised Tables of Mass Attenuation Coefficients*, Corporation Scientifique Claisse, Quebec, Canada, 1977.
[9] W.J. Veigele, *Handbook of Spectroscopy*, J.W. Robinson (ed.), 1, CRC Press, Cleveland, USA, 1974, p. 28.
[10] D.K.G. de Boer, *Spectrochim. Acta*, 1989, **44B**, 1171.
[11] J.H. Hubbell, *X-Ray Spectrom.*, 1999, **28**, 215.
[12] N.A. Dyson, *X-rays in Atomic and Nuclear Physics*, Longman, London, UK, 1973.
[13] W. Bambynek, B. Crasemann, R.W. Fink, H.U. Freund, H. Mark, C.D. Swift, R.E. Price & P. Venugopala Rao, *Rev. Mod. Phys.*, 1972, **44**, 716.
[14] H.G.J. Moseley, *Philos. Mag.*, 1913, **26**, 1024.
[15] J.A. Bearden, *Rev. Mod. Phys.*, 1967, **39**, 78.
[16] R. Jenkins, *An Introduction to X-Ray Spectrometry*, Heyden, London, 1974.
[17] R. Jenkins, R. Manne, R. Robin & C. Semenaud, *Pure Appl. Chem.*, 1991, **63**, 735.
[18] R. Castaing, PhD Thesis, University of Paris, 1951.
[19] J.I. Goldstein, D.E. Newbury, P. Echlin, D.C. Joy, C.E. Fiori & E. Lifshin, *Scanning Electron Microscopy and X-Ray Microanalysis*, Plenum Press, New York, NY 1981.
[20] A. Martin & S.A. Harbison, *An Introduction to Radiation Protection*, Chapman and Hall, London, 1986.
[21] W.D. Coolidge, *Phys. Rev.* 1913, **2**, 409.
[22] S.M. Seltzer & M.J. Berger, *Nucl. Instr. Methods* B, 1985, **12**, 95.
[23] H.A. Kramers, *Philos. Mag.*, 1923, **46**, 836.
[24] P.A. Pella, L. Feng & J.A. Small, *X-Ray Spectrom.*, 1986, **15**, 251.
[25] H. Ebel, *X-Ray Spectrom.*, 1999, **28**, 255.
[26] C. Kunz, *Synchrotron Radiation: Techniques and Applications*, Springer-Verlag, Berlin, 1979.
[27] G. Margaritondo, *Introduction to Synchrotron Radiation*, Oxford University Press, New York, NY, 1988.
[28] M.N. Ingham & B.A.R. Vrebos, *Adv. X-Ray Anal.*, 1994, **37**, 717.
[29] H.H. Johann, *Z. Physik*, 1931, **69**, 185.
[30] T. Johansson, *Z. Physik*, 1933, **82**, 507.
[31] J. Nicolosi, J.P. Groven & D. Merlo, *Adv. X-Ray Anal.*, 1987, **30**, 183–192.
[32] T. W. Barbee Jr., *MRS Bulletin.*, 1990, 37–44.
[33] A. van Eenbergen & B. Volbert, *Adv. X-Ray Anal.*, 1987, **30**, 201.
[34] H.A. van Sprang & M.H.J. Bekkers, *X-Ray Spectrom.*, 1998, **27**, 31.
[35] J.A. Nicolosi, J.P. Groven & D. Merlo, *Adv. X-Ray Anal.*, 1987, **30**, 138.
[36] Y. Gohshi, Y. Hukao & K. Hori, *Spectrochim. Acta*, 1982, **27B**, 135.
[37] P.K. de Bokx & H.P. Urbach, *Rev. Sci. Instrum.*, 1995, **66**, 15–19.
[38] M. Claes, R. van Grieken & P. de Bokx, *X-Ray Spectrom.*, 1997, **26**, 153–158.
[39] R. Klockenkämper, *Total-reflection X-Ray Fluorescence Analysis*, John Wiley & Sons, New York, NY, 1997.
[40] P. Lemberge, P. van Espen & B. Vrebos, *X-Ray Spectrom.*, 2000, **29**, 297.
[41] E. Gatti & P. Rehak, *Nucl. Instrum. Methods*, 1984, **A225**, 608.
[42] A. Castoldi, C. Fiorini, C. Guazzoni, A. Longini & L. Strüder, *X-Ray Spectrom.*, 1999, **28**, 312.

[43] D.A. Wollman, G.C. Hilton, K.D. Irvin & J.M. Martinis, *Proceedings of Microscopy and Microanalysis*, 488–489, San Francisco Press, San Francisco, 1996.
[44] M. Kurakado, *X-Ray Spectrom.*, 2000, **29**, 137.
[45] J.W. Criss & L.S. Birks, *Anal. Chem.*, 1968, **40**, 1080.
[46] T. Shiraiwa & N. Fujino, *Adv. X-Ray Anal.*, 1968, **11**, 63.
[47] D.L. Taylor & G. Andermann, *Anal. Chem.*, 1971, **43**, 712.
[48] R.C. Reynolds, *Amer. Mineral.*, 1963, **48**, 1133.
[49] R.C. Reynolds, *Amer. Mineral.*, 1967, **52**, 1493.
[50] R.W. Nesbitt, H. Mastings, G.W. Stolz & D.R. Bruce, *Chem. Geol.*, 1976, **18**, 203.
[51] G.R. Lachance & R.J. Traill, *Canad. Spectrosc.*, 1966, **11**, 43.
[52] W.K. de Jongh, *X-Ray Spectrom.*, 1973, **2**, 151.
[53] G.R. Lachance, paper presented at the International Conference on Industrial Inorganic Elemental Analysis, Metz, France, June 1980.
[54] B. Vrebos & J. Helsen, *X-Ray Spectrom.*, 1986, **15**, 167–171.
[55] F. Feret, *X-Ray Spectrom.*, 1994, **23**, 130.
[56] A practical guide for the preparation of specimens for X-ray fluorescence and X-ray diffraction analysis, V.E. Burhke, R. Jenkins & D.K. Smith (eds), ISBN 0–471–19458–1, Wiley, 1998.

7 Electrothermal atomic absorption spectrometry

Albert Kh. Gilmutdinov

7.1 Introduction

All techniques of analytical atomic spectrometry (atomic absorption, emission, fluorescence and mass spectrometry) can detect atoms or ions of analyte elements only in the gaseous phase and thus they require prior transformation of a sample from the condensed state into the atomic vapor state. In atomic absorption spectrometry (AAS), two types of atomizers are widely used: high temperature flames and electrothermal atomizers.

In flame AAS (described in Chapter 8) only a small fraction of the sample reaches the flame, and analyte atoms pass quickly through the probing radiation beam. This results in lack of sensitivity, which is the major drawback of the flame version of AAS. Insufficient sensitivity of the flame technique was the major driving force for the development of alternatives, among which electrothermal atomic absorption spectrometry (ETAAS) has become the most popular.

Application of electrothermal atomization in analytical chemistry began with the classical work of L'vov published in 1961 [1]. In the proposed technique, the sample was deposited on to the tip of a carbon electrode, which was introduced into a tubular heated furnace through an aperture. Since then, dozens of modifications of electrothermal atomizers have been developed. Electrically heated tubes, rods, filaments, and boats made of carbon, tungsten, tantalum or molybdenum were used for sample atomization. The dimensions and shapes of the atomizers also exhibited significant differences. Nowadays all commercially available electrothermal atomizers have a tubular shape, approximately 20–30 mm length and 5–6 mm internal diameter, and are heated resistively by mounting them between water-cooled electrodes. The tubes are made of high density polycrystalline electrographite. Polycrystalline graphite by itself, however, is subject to two major drawbacks: high porosity and a tendency for carbide formation. Therefore at elevated temperatures it is a rather reactive material, making it unsuitable for ETAAS. Numerous investigations were made in the protection of the surface of the electrographite with various protective coatings. It was found that due to the extremely low permeability, the best protective material is a pyrolytic graphite coating of about 50 μm thickness. The coating can be made, for example, by heating the tubes in a mixture of argon and methane.

7.2 Graphite atomizers

A typical graphite furnace atomizer, shown schematically in Fig. 7.1, consists of the following major components: the graphite tube, electrical graphite contacts, enclosed water-cooled housing and inert gas flows. The graphite tube, being the heating element of the furnace, is enclosed in a housing and protected from atmospheric oxygen by being surrounded with a protective gas. The housing includes two water-cooled stainless steel electrodes that are used to resistively heat the tube. The atomizer is placed in the optical path of the spectrometer, and radiation at characteristic wavelengths from a primary source (usually a hollow cathode lamp or electrodeless discharge lamp) passes coaxially through the graphite tube via quartz windows. Transmitted radiation is further isolated by a monochromator, and subsequently detected and recorded. The graphite tube serves as the spectrometer sampling cell. A sample, typically 5–50 μL of a liquid, is dosed through a hole in the tube center onto the inner tube wall or a platform. All graphite atomizers operate only in an inert gas atmosphere. There are three reasons for this:

(i) At elevated temperatures carbon reacts intensively with oxygen resulting in the formation of carbon mono-oxide and carbon di-oxide (burning of carbon). Therefore the external protective gas flow passes continuously around the graphite tube, preventing the ingress of atmospheric oxygen.
(ii) The majority of analyte atoms have a high affinity for oxygen and form oxides, resulting in a decrease in sensitivity.
(iii) During sample vaporization, large quantities (compared to the mass of analyte) of concomitants are volatilized into the atomizer which must be removed from the absorption volume prior to the atomization step.

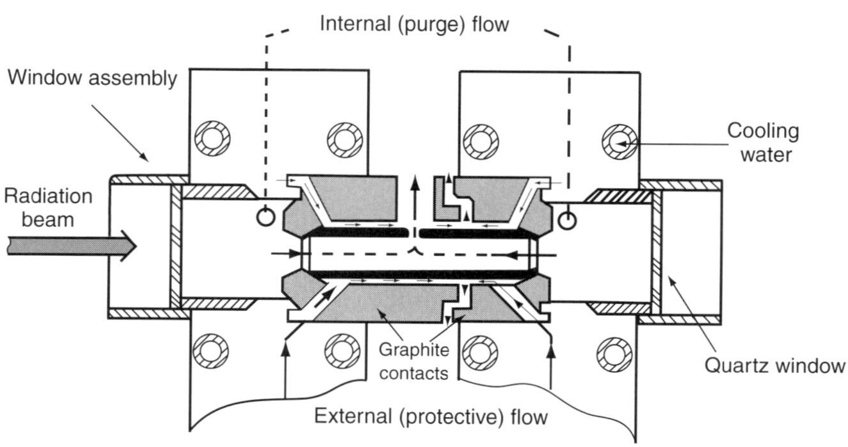

Fig. 7.1 Schematic diagram of a graphite furnace atomizer.

For this purpose, an internal purge gas flow that is independent from the external protective flow is used. The purge gas enters from both tube ends and exits through the sample dosing hole halfway along the tube. Argon of high purity (typically 99.996%) is used almost exclusively as the purge and protective gas.

Depending on the means of heating, modern electrothermal atomizers are divided into two major groups of longitudinally and transversely heated graphite furnaces. In the first case the atomizer is heated longitudinally, i.e. it is clamped at both ends by contact cylinders and electric current is passed through the length of the tube (Fig. 7.2a). The furnace is heated by applying low voltage of about 10 V and high current up to 500 A. This type of furnace is similar to the original design of Massmann [2]. Since electrical contacts at each end must be cooled, such atomizers exhibit a significant temperature gradient over their length, with the central portion being several hundred degrees hotter than the ends (Fig. 7.2b). The analyte and matrix concomitants vaporized at the central hotter part of the atomizer can then condense or recombine with other species at its cooler ends. This in turn results in the tailing of the atomization signal and memory effects for refractory and carbide-forming elements. Also, if condensation is not prevented, sample constituents can re-volatilize during subsequent atomization, leading to substantial interferences. Thus, for such a design it is especially important that the purge gas flow

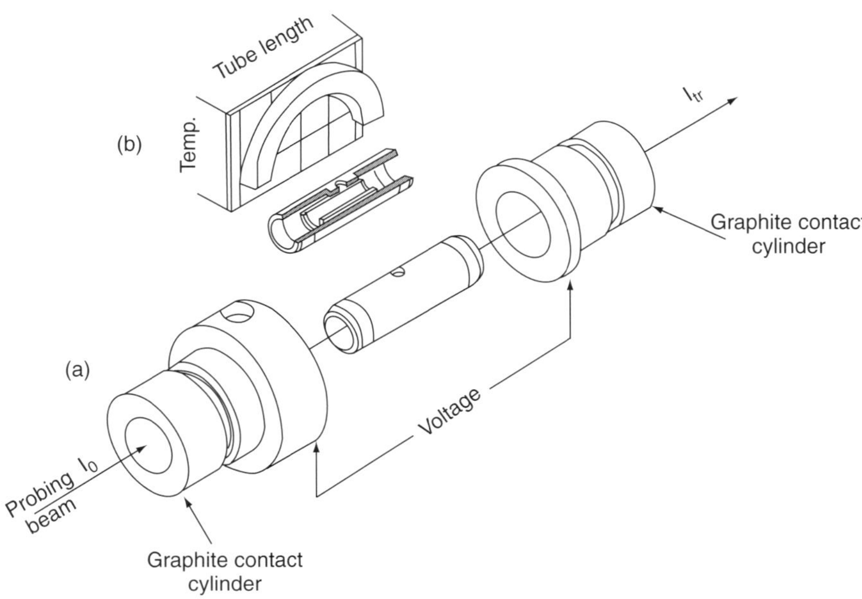

Fig. 7.2 (a) Longitudinally heated graphite atomizer. I_0 and I_{tr} stand for intensities of incident and transmitted radiation, respectively. (b) Cut out of the longitudinally heated graphite tube and temperature distribution along its length.

prevents the condensation of volatilized concomitants at cooler spots in the atomizer unit. The internal purge gas, therefore, flows from the cooler ends to the hottest center as shown in Fig. 7.1.

The transversely heated graphite furnace eliminates many of the problems associated with the longitudinally heated systems. The graphite tube of the transverse heated furnace, shown in Fig. 7.3, includes integrated tabs that protrude from each side. These tabs are inserted into the electrical contacts and, when power is applied, the tube is heated transversely across its circumference. With such a design, the tube is heated uniformly over its length, eliminating or significantly reducing the sample condensation problems seen with the end-heated furnaces.

Frech and Jonsson [3] proposed a third type of electrothermal atomizer, a two-step furnace, that allows separate control of the vaporization and atomization processes. In this system, a graphite cup is positioned below a hole in the bottom of a graphite tube. Separate power supplies permit independent control of cup heating where the sample is vaporized, and tube heating where the vapors are subsequently atomized. The tube is first heated to a predefined constant atomization temperature, and then the cup is heated to the appropriate temper-

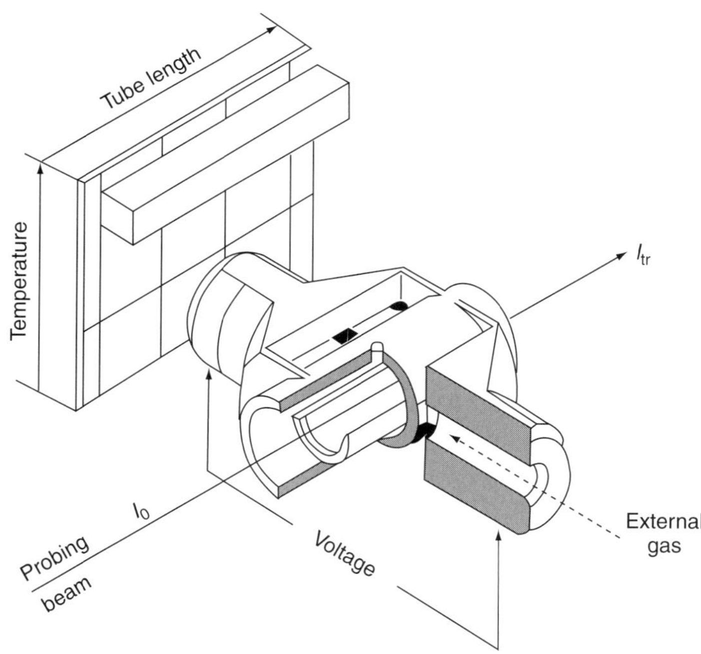

Fig. 7.3 Schematic diagram of a transversely heated graphite furnace and temperature distribution along its length.

ature for sample vaporization. In such a system, the sample vapors enter the nearly isothermal tube environment where absorbance measurements are taken. Although the two-step atomizer has in many cases demonstrated better performance than the traditional one-step systems, it was not commercialized because of technical complexity.

In the early furnace techniques, the sample was deposited directly on to the tube wall. When heating the tube, the analyte atoms form as soon as the wall reaches atomization temperature. The gas within the tube is somewhat cooler, so when the atoms leave the wall, they may recombine with matrix constituents and cannot be detected at the atomic line. It means that analytically they are lost. This results in numerous interferences that for many years caused serious problems in ETAAS. The solution to the problem is introduction of the analyte atoms into a preheated environment. To achieve this, L'vov suggested sample vaporization from the surface of a graphite platform placed loosely inside the graphite tube. As the platform has only slight contact with the inner wall of the tube, it is heated mainly by radiation from the tube walls and thus its heating is retarded. As a result of the delay in heating of the platform, the analyte is not atomized until the tube and the gas phase have attained a final temperature. In this way quasi-isothermal conditions can be obtained, even when the tube ends are still cooler. Another advantage of platform atomization is that more reproducible positioning of the sample solution is achieved and the spreading of the solution is more easily prevented than with wall atomization. Platforms of different shapes and dimensions were investigated. It was found that the major requirement for the platform technique is a stable fixing of the platform at minimal contact with the tube. An optimum solution to this problem was finally achieved by making both longitudinally and transversely heated tubes with an integrated platform from a single piece of electrographite and coating it with pyrographite (Figs 7.2b and 7.3). In these cases, the platforms are attached to the tube at only one point, and the requirement for minimal contact and the prevention of the direct flow of electric current through the platform are met.

Under the conditions of platform atomization, the analyte absorbance is much less affected by the heating rate of the graphite tube and by the sample matrix.

7.3 Graphite furnace analysis

A known volume of liquid sample in the range of 5–50 μL is placed into the furnace via dosing hole in the tube center, normally using an auto sampler. Prior to atomization, the sample is subjected to a multi-step temperature program that serves largely to separate the analyte element from the concomitants. When the temperature is increased to the point where sample atomization

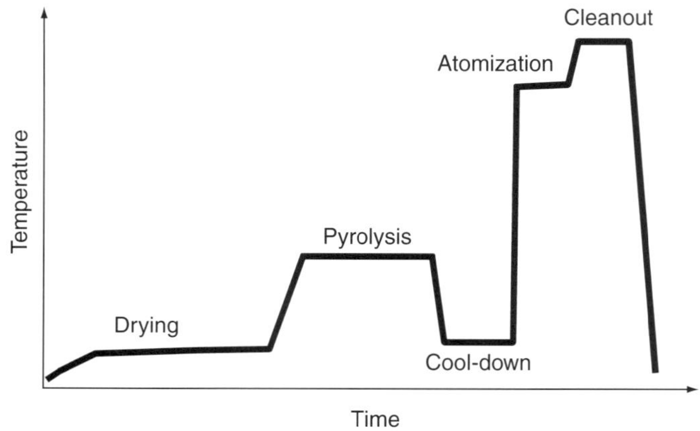

Fig. 7.4 A typical graphite furnace temperature program.

Table 7.1 An example of furnace temperature program for determination of aluminium

Step No.	Step	Temperature, °C	Time, s		Internal gas flow, mL/min
			Ramp	Hold	
1	Drying	100	10	30	300
2	Pyrolysis	900	10	20	300
3	Cool-down	100	1	9	300
4	Atomization	2300	0	5	0
5	Clean out	2650	1	5	300

occurs, the atomic absorption measurement is made. While the number of steps in a temperature program is variable, the following steps make up the typical graphite furnace program (Fig. 7.4, Table 7.1).

7.3.1 Drying

After the sample is dosed in the furnace, it is dried at temperatures around 100 °C. The use of a temperature ramp provides a variable time over which the temperature is increased. A longer ramp time provides a slower heating that is required to avoid sample sputtering. After the temperature ramp, the furnace is held at the pre-selected drying temperature until drying of the sample droplet is complete. Normally, only a few microliters of sample is used for analysis, so that the drying hold time is less than a minute. During this step, the internal gas flow is kept at its maximum value of 200–300 mL/min to purge the vaporized solvent out of the tube.

7.3.2 Pyrolysis

The purpose of this step (sometimes also referred to as the char, ashing, or pretreatment step) is to selectively vaporize matrix components from the sample, leaving the analyte element in a less complex matrix for further analysis. During this step, the temperature is increased at a preselected rate as high as possible largely to vaporize matrix constituents, but below the temperature at which analyte loss would occur. The temperature selected for this step depends on the analyte and the matrix. The internal gas flow is again left at the maximum rate of 200–300 mL/min to drive off all the volatilized matrix components. For some samples, i.e. biological samples, it may be useful to change the internal gas during this step to aid more efficient sample decomposition. Air or oxygen is normally used as an alternative purge gas for such samples.

7.3.3 Cool-down (optional)

With end-heated furnaces, it is frequently advantageous to cool the furnace prior to atomization. It was found that for high heating rates the initial uniform temperature profile was retained for about a second, after which the cooling effect of the contacts became apparent. It was therefore proposed that the tube should be allowed to cool to ambient temperature prior to atomization, since a flat temperature profile then prevails, and thereafter to heat rapidly. Also, the heating rate is a function of the temperature range to be covered. With the cool-down step the temperature range is increased, so that the rate of heating also increases. Thus, the use of a cool-down step prior to atomization maximizes the heating rate and extends the isothermal zone within the tube immediately after heating. This results in an improved sensitivity and reduction of peak tailing for a number of elements. The cool-down step is not normally required for transversely heated furnaces since the isothermal zone extends along the length of the tube.

7.3.4 Atomization

The purpose of the atomization step is to transfer the analyte elements from the condensed state on the atomizer surface into a gas-phase vapor of free atoms, thereby allowing atomic absorption to be measured. The temperature in this step is increased to the point where complete dissociation of volatilized molecular species should occur. The atomization temperature is a property of the analyte element. Excessively high atomization temperatures should be avoided, as the analyte residence time in the tube will be decreased resulting in a loss of sensitivity. Also, the use of excessively high atomization temperatures shortens the useful lifetime of the graphite tube. To obtain a well-shaped analytical signal, it is desirable to increase the atomization temperature as quickly as possible. Therefore, the ramp time of this step is set to a minimum value to provide the highest heating rate of about 2000°C/s. It is desirable to interrupt the internal gas flow during atomization, since it would otherwise interfere with the formation of a local thermal equilibrium in the atomizer. Gas-stop conditions also increase the analyte residence time, maximizing sensitivity and reducing some interference effects.

7.3.5 The clean out

After atomization, the graphite furnace is normally heated to higher temperatures to burn off any sample residue that may remain within the atomizer. During this step, the purge gas flow is turned on again to its maximal level to completely remove the analyte and any possible concomitants volatilized earlier during atomization.

The analytical signal generated in a graphite furnace depends on analyte mass, so that the measured absorbance can also be controlled by the dosed sample volume. Larger volumes of sample solution contain more analyte and thus result in greater absorbances. For very low analyte concentrations, the maximum sample volume can be used, while for higher concentrations, the sample volume can be reduced. Smaller sample volumes are also used in cases where background absorption is excessively high or sample availability is limited (for example, in the analysis of blood). Platform furnaces can accommodate up to 50 μL of sample solutions. Where larger volumes are required, i.e. for improved detection limits, multiple injections with appropriate drying and pyrolysis steps between each injection can be used to increase the effective sample size.

7.4 Electrothermal atomization

A major difference of electrothermal atomization from atomization in flames is that some matrix components are removed at the pyrolysis step and the atomization takes place within a confined geometry in an inert gas atmosphere. During atomization, the purge gas flow is normally shut off so that analyte atoms remain within the probing beam as long as possible.

In ETAAS, atomic absorbance, $A(t)$, is transient and at any instant is proportional to the number of analyte free atoms along the probing beam. In turn, the changing number of analyte atoms in the atomizer volume, $N(t)$, can be described by the convolution of supply and removal processes [4]:

$$A(t) \sim N(t) = \int_0^t S(t')R(t - t')dt' \tag{7.1}$$

Here, $S(t')$ is the function of atom supply that gives the rate (atoms s^{-1}) of primary generation of the analyte from the deposited sample. $R(t - t')$ is the removal function that characterizes subsequent transfer of the vaporized atoms in the atomizer volume. As a first approximation, the deposited sample can be considered to form a monolayer on the atomizer surface. In this case, analyte supply is determined by desorption from the atomizer surface and the supply function can be quantified as [5]

$$S(t) = N_0\, v \exp\left\{-\frac{E}{RT(t)} - v\int_0^t \exp\left[-\frac{E}{RT(t')}\right]dt'\right\} \tag{7.2}$$

where N_0 is the total number of atoms in the deposited sample, v and E are the frequency factor and the activation energy for the desorption process respectively, R is the gas constant, and $T(t)$ is the time dependent surface temperature. Figure 7.5 gives a typical example of the atom supply function in ETAAS. The removal function $R(t - t')$ is the probability for an analyte atom that is

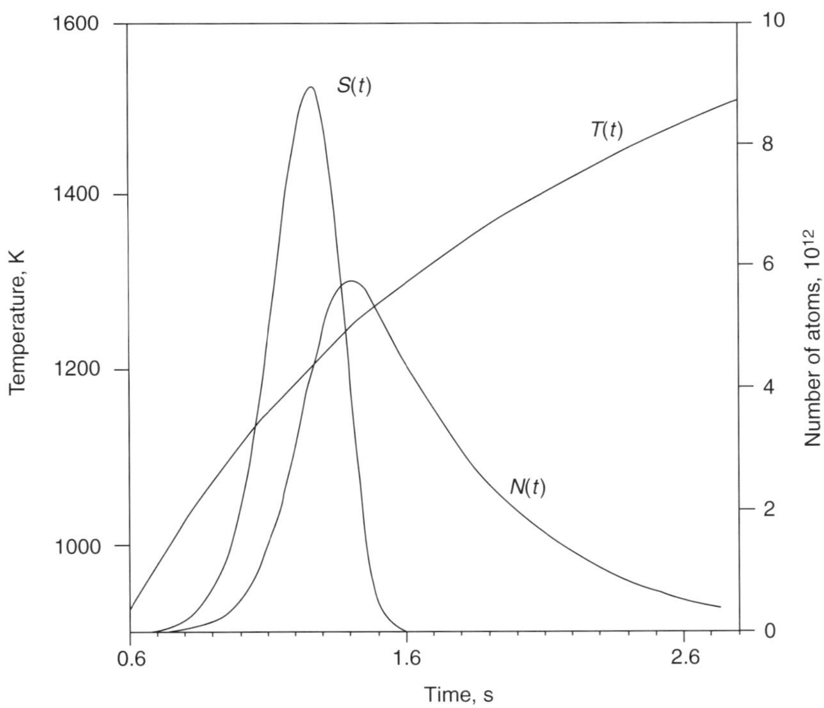

Fig. 7.5 Supply function $S(t)$ and the change in the total number $N(t)$ of analyte atoms in an electrothermal atomizer computed for the given change of the atomizer temperature $T(t)$ (on the example of silver atomization with E = 280 kJ mol^{-1}, $\nu = 10^{13}\ s^{-1}$).

vaporized at the time, t', to be present in the atomizer volume at a later instant, $t > t'$. Obviously, early in time, the probability is equal to unity and decreases with increasing time because of loss processes. The loss processes in ETAAS include concentration diffusion, thermal diffusion (analyte diffusion caused by temperature gradients) and convection. Based on the removal function, the mean residence time, τ, of analyte atoms in the atomizer volume is defined as $\tau = \int_0^\infty R(t)dt$. For the platform atomization, where the analyte loss is primarily determined by concentration diffusion through the tube ends, the residence time is given by the relationship [4]:

$$\tau = \frac{l^2}{8D}\left[1 - \frac{1}{3}\left(\frac{a}{l}\right)^2\right] \tag{7.3}$$

where l is the tube length, D is the analyte diffusion coefficient and a is the length of the area occupied by the sample on the atomizer surface. In the case of platform atomization, the length of the platform can be taken as an estimation of this parameter. It can be seen that the length of the tube plays a decisive role in the analyte residence time. To achieve the longest possible residence time and thereby highest sensitivity, the graphite tube should be as long as possible. Longer tubes, however, would require more power and time to achieve an atomization temperature. It was found empirically that the optimum tube length is within 20–30 mm. Based on the above relationship, a typical analyte residence time in a tube graphite atomizer is a few tenths of a second. This is about 100 times longer than that in a flame. A longer residence time, along with the high degree of analyte atomization caused by the reducing environment, results in a 10–1000 times increase in graphite furnace sensitivity compared with flame AAS.

A typical change in the total number of analyte atoms, $N(t)$, within a tube electrothermal atomizer obtained by the convolution of the supply function with the diffusion removal function is given in Fig. 7.5. The absorbance measured at any instant of time t by an AAS would be proportional to $N(t)$.

Two important consequences can be derived from the general Equation 7.1 [5]. The first is the equation for the integrated absorbance $A_{int} = \int_0^\infty A(t)dt$:

$$A_{int} = k(x)\int_0^\infty N(t)dt = k(\lambda)N_0\tau \tag{7.4}$$

where $k(\lambda)$ is a coefficient depending on spectral properties of the analysis line and τ is the mean residence time of the atoms. It is seen from this equation that under conditions of isothermal atomization, integrated absorbance is directly proportional to the total amount of analyte atoms in the sample and to the analyte residence time. This is the basis for the nearly exclusive use of integrated absorbance as an analytical signal in ETAAS. There are many factors (heating rate, temperature, variations in matrix, etc.) that can affect the shape of the absorbance pulse and thus the peak absorbance. However, if atomization occurs in a steady-state environment, the shape of the peak is of little importance for the integrated absorbance.

The second equation represents the measurement of $N(t)$, when sample vaporization is practically finished. In this case $S(t) = 0$ and

$$A(t) \sim N(t) \sim R(t)$$

Thus the tailing part of the absorbance signal represents removal function.

The actual analyte atom formation and dissipation processes in electrothermal atomizers are much more complex than that described above. In many cases, especially in complex matrices, the analyte is atomized as a result of a set of multi-stage processes taking place simultaneously. Basic processes involving the analyte in a graphite furnace are shown schematically in Fig. 7.6. Here, solid arrows stand for pathways of free analyte atoms while dotted arrows show the pathways of analyte in a molecular form. The list of processes include [6]:

- *Primary generation* of the analyte vapor from the site of sample deposition as an atomic (1) or a molecular (1′) species;
- Irreversible *loss of analyte* from the furnace through its ends (2) and through the sample dosing hole (2′) by diffusion and convection;
- *Physical adsorption/desorption* at the graphite surface (3);
- Gas phase *condensation* (4) at the cooler parts of the atomizer;
- *Gas phase reactions* (5) that bind free analyte atoms into stable molecules or those (5′) that increase the free atom density;
- *Heterogeneous reactions* of analyte vapor with the atomizer walls including both production (6) and loss (6′) of free atoms at the furnace wall.

The listed processes are summarized in Table 7.2. The last column of the table gives the elements for which the respective processes were documented.

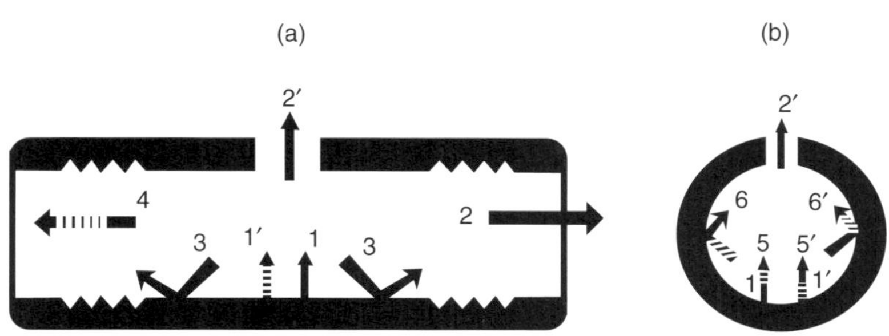

Fig. 7.6 Schematic diagram of the basic physical and chemical processes occurring in a tube graphite furnace.

Table 7.2 Selection of processes involving analyte atoms in a graphite tube atomizer

Process*	Ref. in Fig. 7.6	Elements
Me (s,l) $\rightarrow$Me (g)	1	Ag, Au, Co, Cu, Ni, Pb, Pd
MeO (s,l) $\rightarrow$Me (g)	1	Al, Cd, In, Pb
MeO (s,l) + C(s) $\rightarrow$Me (g)	1	Mn, Mg
Me C (s,l) $\rightarrow$Me (g)	1	Al, Ba, Ca, Cu, Mo, Sr, V
MeOH (s,l) $\rightarrow$Me (g)	1	Rb
MeO (s,l) $\rightarrow$MeO (g)	1′	Al, As, Ga, In, Tl
Me (s,l) $\rightarrow Me_2$ (g)	1′	Se, Co
Me (ad) $\rightarrow$Me (g)	3	As, Cu
Me (g) $\rightarrow$Me (s,l)	4	Ag, Au, Cu, Mg, Mn, Pd
Me (g) + O (g) $\rightarrow$MeO (g)	5	Al, Si, Sn
MeO (g) $\rightarrow$Me (g)	5′	Al, As, Cd, Ga, In, Mn, Pb, Zn
MeO (g) + C(s) $\rightarrow$Me (g) + CO	6	Al, Ga, In, Tl
Me (g) + C(s) $\rightarrow$MeC (g)	6′	Al

* s, l, g, ad stand for solid, liquid, gaseous and adsorbed species, respectively.

The complexity of electrothermal atomization can be illustrated by the example of aluminum atomization. It is seen from the table that atomization of this element involves practically all the listed processes.

7.5 Interferences in graphite furnace AAS

In the early days of the graphite furnace AAS (GFAAS), it was widely recognized as a highly interference-prone analytical technique. Significant progress achieved in understanding the nature of graphite furnace interferences resulted in the development of instrumentation and procedures that allow the analysis of a large variety of complex samples using simple acidified standards for quantification.

Generally, an interference effect is a change of the analytical signal by the sample matrix as compared to the reference standard, typically an acidified aqueous solution. There are two types of interferences in AAS – spectral and non-spectral. Spectral interferences originate from incomplete isolation of the radiation absorbed by the analyte atoms from other radiation detected by the spectrometer. In graphite furnace technique, the most obvious spectral interference may arise from thermal (*black body*) radiation of incandescent walls and platform when it reaches the instrument's detector. Since the maximum intensity of thermal radiation lies in the near-infrared region, the potential for observing this emission interference increases at longer wavelengths. It is especially noticeable at barium 553.6 nm analysis line, resulting in an increased signal noise that degrades analytical performance. Emission interference is largely controlled by the use of reduced monochromator slit height and proper furnace alignment in the optical path of the spectrometer. If these measures are taken, then the angle of view of the spectrometer is narrowed so that only transmitted radiation from the primary source reaches the detector.

The most severe spectral interference problem encountered with graphite furnace technique is the so-called *background absorption*. Background absorption is a non-specific attenuation of probing radiation at the analyte wavelength caused by matrix constituents (molecules, non-volatilized microparticles or by atoms other than those of the analyte element) in the sample. By its spectroscopic nature, background absorption is broad band, covering dozens of nanometers. Therefore, the chance of spectral overlap with the analyte wavelength is significant. In modern GFAAS, the problem is solved by the use of a two-step procedure. The first is in situ reduction of background-creating species as much as possible at the sample pretreatment steps and the second is optical compensation of the residual background.

As was stated in the previous section, the main purpose of the pyrolysis step in the furnace temperature program is selective volatilization of matrix constituents from the sample. If all the sample components that cause background were driven off prior to atomization, then there would be no background absorption. The highest possible pyrolysis temperature is required to effectively separate the concomitants. However, analyte atoms must not be lost during sample pretreatment. Therefore, the maximum pyrolysis temperature is limited to the value at which analyte atoms begin to evaporate. Thus the degree of matrix removal during the pyrolysis step highly depends on the relative volatilities of the analyte and matrix constituents. When analyzing volatile elements such as As, Cd, Pb, Se, Tl, etc. such a removal is insufficient in a variety of matrices. Fortunately, a procedure called *chemical modification of the matrix* (often referred to as matrix modification) was developed [7] that allows control of relative volatilities of the matrix and analyte. By adding a special reagent, *matrix modifier*, to the sample it is possible either to increase matrix volatility or decrease analyte volatility. Since the action of the modifier is analyte- and matrix-dependent, dozens of different modifiers have been proposed that extremely complicate the proper choice. The problem was significantly simplified when a *universal* $Pd/Mg(NO_3)_2$ was introduced [8]. It has been proven that this modifier can effectively stabilize 21 elements most commonly analyzed by GFAAS [9]. The maximum pyrolysis temperatures achieved with this modifier along with optimum atomization temperatures are presented in Table 7.3. The table clearly demonstrates the stabilizing effect of the modifier: usable pyrolysis temperatures are significantly increased, especially for highly volatile elements (As, Cd, P, Se and Te).

Varying the sample volume provides an additional means of background control. Larger sample volumes contain proportionally more matrix components. When the sensitivity of the analysis is not an issue, it is desirable to reduce the sample size, thereby reducing the mass of the background-producing constituents.

Table 7.3 Maximum usable pyrolysis temperatures without modifier and with the use of the Pd/Mg modifier

Element	Maximum pyrolysis temperature, °C		Atomization temperature, °C
	Without modifier	Pd/Mg modifier	
Ag	650	1000	1600
Al	1400	1700	2350
As	300	1400	2200
Au	700	1000	1800
Bi	600	1200	1900
Cd	300	900	1700
Cu	1100	1100	2600
Ga	800	1300	2200
Ge	800	1500	2550
Hg	100	250	1000
In	700	1500	2300
Mn	1100	1400	2300
P	200	1350	2600
Pb	600	1200	2000
Sb	900	1200	1900
Se	200	1000	2100
Si	1100	1200	2500
Sn	800	1200	2400
Te	500	1200	2250
Tl	600	1000	1650
Zn	600	1000	1900

Although the described techniques are effective in the reduction of matrix components, often background-producing species cannot be eliminated completely. This necessitates correction for the residual background to obtain unbiased analytical results. Continuum source background correction used in flame AAS can also be used in the graphite furnace technique. Although fully adequate for practically all flame applications, this means of background correction has the following limitations that are especially noticeable when used in GFAAS:

(i) Precise optical alignment of the continuum and primary sources is critical for accurate correction. This requirement originates from the fact that absorption layers are often highly non-uniform in the furnace radial cross section. The background component of the absorption signal is measured from continuum source, while the gross uncorrected absorbance is measured with the line source. If the two different sources do not view exactly the same portion of the non-uniform absorbing cloud, then correction will be inaccurate.

(ii) In the case of structured background absorption, continuum source technique is inaccurate. Since the two sources are spectrally different, background absorption exhibiting spectral structure will attenuate the two sources to different degrees.

(iii) Two continuum sources are required to cover the full spectral range. The wavelength range used in AAS extends from 190 nm to about 900 nm and it can only be covered by two different continuum sources. Usually deuterium arc lamp is used for wavelengths shorter than 350 nm and tungsten lamp for longer wavelengths.

(iv) Intensities of the primary and continuum sources should be similar.

For applications where the above limitations of the continuum sources technique are significant to the analysis, the background correction system based on Zeeman effect may be preferable. Zeeman effect consists in splitting of an atomic line into two or more components displaced symmetrically about the normal position when the atom is placed in a strong magnetic field. Thus by placing the atomizer between the poles of a magnet and making alternating absorption measurements with the magnet on and off, the uncorrected gross absorbance (magnet off, Fig. 7.7a) and background absorbance (magnet on, Fig. 7.7b) can be measured. Only one radiation source is used in this background correction system; thus the problems of matching source intensities and coincident alignment of optical paths are fully eliminated. Further, with Zeeman effect background correction, the emission profile of the line source is identical for both atomic and background absorbance measurements. This

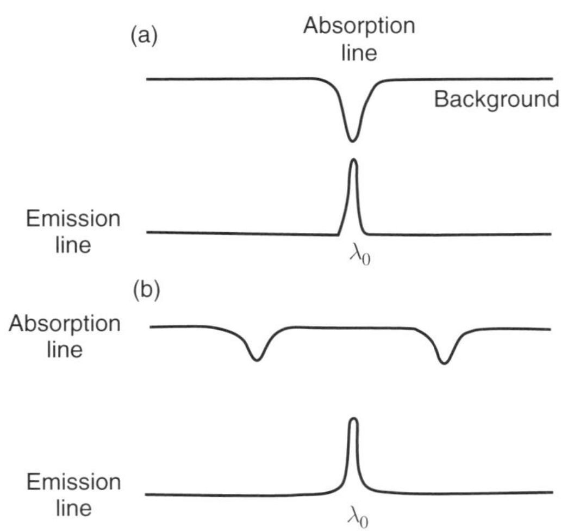

Fig. 7.7 The principle of Zeeman effect background correction. (a) Magnetic field is off: the incident radiation at the analysis wavelength λ_0 is absorbed by both analyte atoms and background species; (b) Magnetic field is on: the absorption line is split into two shifted components and the incident radiation is absorbed by only background species.

allows much more complex structured background situations to be accurately corrected.

For non-spectral interferences, the analyte signal itself is directly affected by the sample constituents. In AAS, the analytical signal is atomic absorbance that is caused by free atoms of the analyte element. Thus, non-spectral interferences will occur in all the cases when diverse components in the sample matrix inhibit the formation of free analyte atoms. They can generally be represented by the following equation:

$$A + M \rightarrow AM$$

where A stands for the free analyte atom and M is a diverse matrix component atom. Non-spectral interferences are best classified according to the stage at which the particular interference occurs, i.e. condensed-phase (often referred to as solute-volatilization interferences) and gas-phase interferences. The majority of condensed-phase interferences are caused by the formation of volatile compounds of the analyte in the condensed phase that are later vaporized and lost during the pyrolysis step. For example, beryllium and aluminum form volatile hydroxides; germanium, thallium and tin form volatile suboxides, etc. Often, preatomization losses can be detected using a spike recovery or the method of standard additions.

Gas-phase interferences originate from the effect of matrix on the atomization efficiency of the analyte and can usually be recognized by spike recovery experiments.

Graphite furnace interferences can be eliminated or significantly reduced under the conditions of the so-called *Stabilized Temperature Platform Furnace* (STPF) [10]. The STPF concept is an integrated set of measures to achieve interference-free determinations in the following three steps:

(i) Separation of the analyte from matrix concomitants as far as possible prior to atomization.
(ii) Control of reactions in the condensed phase to prevent analyte losses.
(iii) Minimization of the effect of residual concomitants on the analyte in the gas phase by as complete atomization as possible.

As proposed by Slavin and Manning [10], the STPF system includes the following instrumental factors and analytical procedures:

- *Fast spectrometer electronics* – to obtain undistorted absorbance pulses.
- *High quality hydrolytically coated graphite tubes* – to provide an impervious, non-reactive atomization surface.
- *Platform* – to delay analyte atomization until stable temperature conditions are achieved.

- *Fast furnace heating* – to quicken the establishment of atomization temperature and increase temperature lag between the tube wall and the platform.
- *Internal gas stop during atomization* – to maximize analyte residence time.
- *Peak area measurement* – to quantitate all the volatilized analyte atoms independently of the volatilization rates.
- *Matrix modification* – to separate the analyte from the matrix.
- *Zeeman effect background correction* – to correct for high level of structured sample background.

7.6 The relative merits of ETAAS

ETAAS has become a well-developed spectrochemical technique with interference-free methods in most applications, including the direct analysis of solids and slurries. The major advantages of the technique can be summarized as follows.

Detection limits provided by the graphite furnace fall in the ng/L range for most elements and are exceeded only by inductively coupled plasma-mass spectrometry (ICP-MS) technique. Atomization into a confined and reducing area of the heated graphite tube results in a greatly increased sensitivity.

Possibility of direct analysis of solid samples. It reduces total analysis time, minimizes the potential for contamination or analyte loss that are likely in trace analyses. Solids can be introduced either directly or more often in the form of slurries. With slurry sampling, the sample is reduced to a fine powder that is weighed directly into an autosampler vial and a measured amount of liquid is added. After agitation, the slurries are further analyzed as though they were true solutions.

Decreased sample size. For electrothermal atomization, a sample size as low as 1 μL is sufficient for determinations. This results in significant improvements in absolute sensitivities, and measurements of sub-picogram amounts of analyte are possible. Also, in cases when sample size is limited, GFAAS may be the only choice.

Modern graphite furnace technique is highly automated. Even though heating programs and procedures can be very sophisticated, the entire process is automated once the analysis program is initiated. Unattended, overnight operation of the system is a routine practice nowadays.

Inexpensive and safe in operation. Compared to other techniques, graphite furnace is relatively inexpensive to operate. The main reason for this is that the inert argon gas is used at a very low consumption rate.

The final and most obvious advantage of the graphite furnace is its wide applicability. About 50 elements of the periodic table can be determined at trace and ultratrace levels in a wide variety of matrices. The analytical applications of GFAAS are discussed in detail in [11].

However, none of the spectro analytical techniques is ideal. The major disadvantage of GFAAS is that it is a relatively slow technique. While flame or plasma spectrometries provide determination rates of 500–3000 analyses per hour, conventional single channel graphite furnaces can usually perform only 20–30 determinations per hour. Although fast furnace temperature programs have been developed, where drying and pyrolysis times are greatly reduced, they find only limited applications.

For a long time, GFAAS was a single element technique allowing determination of only one element at a time, whereas emission and mass spectrometry are inherently simultaneous multi-element techniques. This weakness of electrothermal atomization has been significantly reduced after the introduction of multi-element GFAAS instruments that can determine simultaneously up to six elements.

Like flame AAS, graphite furnace has a narrow working range not exceeding three orders of magnitude of analyte concentrations to be analyzed. Inductively coupled plasma atomic emission spectrometry (ICP-AES) and ICP-MS nowadays offer six and up to nine orders of magnitude of working range, respectively. This allows ultratrace and macrocomponent determinations to be run in one measurement that is impossible in atomic absorption.

References

[1] B.V. L'vov, *Spectrochim. Acta*, 1961, **17**, 761.
[2] H. Massmann, *Spectrochim. Acta*, 1968, **23B**, 215.
[3] W. Frech & S. Jonsson, *Spectrochim. Acta*, 1982, **37B**, 1021.
[4] A.Kh. Gilmutdinov & I.S. Fishman, *Spectrochim. Acta*, 1984, **39B**, 171.
[5] W.M.G.T. Van den Broek & L. de Galan, *Anal. Chem.*, 1977, **49/14**, 2176.
[6] K.W. Jackson (ed.), *Electrothermal Atomization for Analytical Atomic Spectrometry*, John Wiley & Sons Ltd, New York, NY, USA, 1999.
[7] R.D. Ediger, *At. Absorpt. Newsl.*, 1975, **14**, 127.
[8] G. Schlemmer & B. Welz, *Spectrochim. Acta*, 1986, **41B**, 1157.
[9] B. Welz, G. Schlemmer & J.R. Mudakavi, *J. Anal. Atom. Spectrom.*, 1992, **7/8**, 1257.
[10] W. Slavin & D.C. Manning, *Spectrochim. Acta*, 1980, **35B**, 701.
[11] G. Schlemmer & B. Radziuk, *Analytical Graphite Furnace Atomic Absorption Spectrometry*, Birkhauser Verlag, Basel, Switzerland, 1999.

8 Flame atomic absorption spectroscopy, including hydride generation and cold vapor techniques

Julian F. Tyson and Emily Yourd

8.1 Introduction

The characteristic that distinguishes the techniques discussed in this chapter from the others in this book is the use of flame atomizers. This chapter is concerned, therefore, almost exclusively with flame atomic absorption spectrometry (FAAS). There are at least two commercially available instruments in which a flame atomizer produces atoms for atomic fluorescence spectrometry (AFS), and so this topic will be briefly mentioned. In addition, mercury is nearly always determined by a procedure in which mercury atoms are generated at room temperature (the so-called cold vapor technique). Commercial instruments dedicated to the determination of mercury are based on either AAS or AFS. This chapter will also include a brief discussion of cold vapor mercury determination. The generation of volatile hydrides is now well established for the determination of arsenic and selenium (and one or two other elements). Although the hydrides can be introduced into a variety of atomizers, the most widely adopted is the heated quartz tube. For the two most popular elements, the atomization mechanism involves reaction with hydrogen atoms generated in a flame inside the tube and so a brief discussion of hydride generation (HG) AAS is also included.

Following the sections on the basic principles and some comments about the practical aspects of making measurements by AAS, a selection of examples of determinations and instrumental developments described in the recent primary literature is presented. A more general annotated bibiliography is also provided enabling readers to find more detailed information about any of the topics mentioned in this chapter.

8.1.1 Why flame atomic absorption spectrometry?

There are lots of reasons why FAAS is a popular instrumental technique. Instruments are rugged, reliable and capable of quantitative measurements to single digit parts per million (mg L^{-1}) concentrations and below for a large number of elements. Current instrumentation is capable of multi-analyte determinations in a wide variety of liquid sample matrices, including solutions containing high dissolved solid contents and/or significant portions of non-aqueous solvents. The instruments are relatively inexpensive in terms of capital, running and maintenance costs. They are simple to operate and have numerous safety features. They do not require any special power supplies, cooling water or thermostatically controlled laboratory facilities. The interferences are well understood and it is possible to get accurate analyses without complicated, or time-consuming, calibration procedures such as standard additions or internal standards. In approaching to devise a method for trace element determinations, FAAS should be the first choice of technique considered. Only if there is some feature of the analysis that makes one of the other techniques a better choice, should FAAS be passed over.

8.1.2 The current status of FAAS

Large numbers of procedures have already been developed and validated, and are available in compilations of standard methods, or in the form of application notes from the various manufacturers. Although there is still a significant number of papers published each year in the analytical chemistry original literature, most of the papers concern the coupling of some novel and/or clever sample pretreatment procedure directly with the spectrometer. There are still some research groups working on sample introduction procedures and a few papers appear each year concerned with new nebulization devices, such as the hydraulic, high-pressure nebulizer. There is current interest in the possibilities of diode lasers as alternatives to the hollow cathode lamps (HCLs) or electrodeless discharge lamps (EDLs) that are currently the standard light sources. Essentially FAAS is a mature technique: there are numerous publications in the original literature every year in which FAAS features as the instrumental technique in the methodology that provides the information about chemical composition that is needed for the investigation. Mostly, these papers are not in the analytical chemistry literature, but are in the literature related to further understanding of the subject to which the sample material relates. Thus, the literature dealing with relevant aspects of food science, environmental science or clinical chemistry contains publications in which data were obtained by FAAS. This literature is reviewed on a rolling annual basis by the *Atomic Spectrometry Updates of the Journal of Analytical Atomic Spectrometry*, published by the Royal Society of Chemistry. Details of these publications and other sources of further information are provided in the bibliography at the end of the chapter.

Manufacturers continue to refine their instruments in terms of ease of use, computer control, computer-based data handling, report generation software, self diagnostic, ease of repair and so on. Manufacturers also continue to work on burner and nebulizer design, the efficiency of the optical system, and accessories that can be interfaced directly, such as autosamplers, diluters and hydride generation systems. According to Strategic Directions International (www.strategic-directions.com), the global market for atomic spectroscopy will grow from $1.9 billion in 2001 to $2.4 billion by 2006. Of this AAS (both flame and furnace) accounts for 20%, while ICP-OES accounts for 16%, ICP-MS for 10% with elemental analyzers, X-ray methods and arc-spark procedures accounting for the remaining 54%. The breakdown is somewhat different if the situation for environmental laboratories is examined, where AAS accounts for 37%, ICP-OES for 23% and ICP-MS for 18%.

8.1.3 Closely related techniques: hydride generation, cold vapor mercury, AFS and AES

As interest in selenium, arsenic and other elements in groups 5A and 6A in the periodic table continues to grow, the determination of these elements by HG coupled to a heated quartz tube atomizer device is becoming a standard methodology. This mode of atomization is much more efficient than the generation of atoms in a standard combustion flame (for reasons which will be explained later in Section 8.4). Although the quartz tube is often heated in a flame, giving rise to the description *tube-in-flame* atomizer, the reality is that atoms are formed from the hydrides by decomposition in a hydrogen–oxygen flame inside the tube, and so all devices should really be termed *flame-in-tube* atomizers. These same elements can be efficiently determined by AFS in hydrogen diffusion flames from the hydride precursors and there is a niche market for AFS instruments dedicated to the determination of just a limited number of hydride-forming elements (see Section 8.6) and of mercury, another topic to be discussed later (see Section 8.5). If the analytical workload calls for the determination of just these elements on a regular basis, especially at single digit parts per billion ($\mu g\ L^{-1}$) concentrations or lower, it is worth investigating the capabilities of AFS.

Mercury is unusual in that it is one of the few elements that anyone is interested in, which can be determined by atomic spectrometry at room temperature, as the vapor pressure of mercury is relatively high and the vapor is monatomic. Other elements with this property include the elements in group 8 (rare gases) of the periodic table, for which there is little call for their determination in commonly encountered samples. It is also possible to determine cadmium by atomic spectrometry of the room temperature atoms (a thermodynamically unstable species, generated by the reaction of the metal aquo ion with borohydride) that have sufficiently long lifetime to allow useful spectroscopic measurements to be made. This so-called cold vapor (CV) technique is sufficiently well established in the case of mercury that there are several commercial instruments available dedicated solely to the determination of the element by either AAS or AFS.

Flame atomic emission spectrometry (FAES) is no longer used to any great extent, having been largely superseded by FAAS or inductively coupled plasma (ICP) source emission spectrometry. It is still possible to buy low-cost dedicated flame AES instruments, often referred to as flame photometers. These are suitable for the determination of Ba, Ca, K, Li and Na in the approximate concentration range 1–100 mg L^{-1}. Such instruments are still used for clinical applications, though there has been a trend in recent years to replace this determination with measurement by ion-sensitive electrodes, as these are more compatible with continuous-flow, automated analyzers. Many FAAS instruments can also be used in the emission mode. Although the performance may be inferior than would be obtained with an instrument specifically designed for emission spectrometry, this feature can be useful for the *one-off* determination of an element for which no HCL is immediately available.

Instrumental detection limits are given in Table 8.1 for the techniques of FAAS, HG-AAS, electrothermal atomization AAS and both ICP emission and MS. Detection limits are discussed in more detail later (see Sections 8.7.2 and 8.9.5).

8.2 Basic principles

Just as for other absorption spectrometric procedures, AAS is based on the concept that the extent to which radiation is absorbed by a derivative of the target analyte species may be directly related to the amount of the analyte species in the material under investigation. All absorption spectrometries are based on the relationship between the power of the radiation incident on the absorbing medium and the power transmitted, which is formulated in terms of some fundamental characteristics of the absorbing species and their number per unit volume (or number density). Provided certain boundary conditions are met, the logarithm of the ratio of the incident power, measured by the detector when no absorbing species attenuate the beam, to the intensity transmitted, following absorption of radiation, is directly proportional to the number of absorbers per unit volume in the beam. The spectrometer is constructed so that the boundary conditions are met. This logarithm of the intensity ratio is known as the absorbance, A.

In the case of AAS, the derivative of the analyte species that is responsible for the absorption of radiation is the free gaseous atom. This species, while not particularly easy to generate, has some extremely useful properties which make it eminently suitable as the basis for a practical technique of chemical analysis: atomic absorption is both efficient and specific. Efficient in the sense that the phenomenon is of sufficient magnitude that it forms the basis of the measurement of analyte species present at trace concentrations in the original material for analysis, and specific in the sense that the atomic absorption of one element can be detected without the interference from the

Table 8.1 Atomic absorption spectrometry detection limits (μg L^{-1})

Element	Flame AA	Hg/Hydride	Element	Flame AA	Hg/Hydride
Ag	1.5		Mo	45	
Al	45		Na	0.3	
As	150	0.03	Nb	1500	
Au	9		Nd	1500	
B	1000		Ni	6	
Ba	15		Os		
Be	1.5		P	75000	
Bi	30	0.03	Pb	15	
Br			Pd	30	
C			Pr	7500	
Ca	1.5		Pt	60	
Cd	0.8		Rb	3	
Ce			Re	750	
Cl			Rh	6	
Co	9		Ru	100	
Cr	3		S		
Cs	15		Sb	45	0.15
Cu	1.5		Sc	30	
Dy	50		Se	100	0.03
Er	60		Si	90	
Eu	30		Sm	3000	
F			Sn	150	
Fe	5		Sr	3	
Ga	75		Ta	1500	
Gd	1800		Tb	900	
Ge	300		Te	30	0.03
Hf	300		Th		
Hg	300	0.009	Ti	75	
Ho	60		Tl	15	
I			Tm	15	
In	30		U	15000	
Ir	900		V	60	
K	3		W	1500	
La	3000		Y	75	
Li	0.8		Yb	8	
Lu	1000		Zn	1.5	
Mg	0.15		Zr	450	
Mn	1.5		Mo	45	

All detection limits are given in micrograms per liter and were determined using elemental standards in dilute aqueous solution. All detection limits are based on a 98% confidence level (3 standard deviations).

All flame atomic absorption (AAnalyst™ 800) detection limits were determined using instrumental parameters optimized for the individual element, including the use of System 2 electrodeless discharge lamps where available.

Cold vapor mercury detection limits were determined with an FIAS™-100 or FIAS-400 flow injection system with amalgamation accessory. The detection limit without an amalgamation accessory is 0.2 μg L^{-1} with a hollow cathode lamp, 0.05 μg L^{-1} with a System 2 electrodeless discharge lamp. (The Hg detection limit with the dedicated FIMS™-100 or FIMS-400 mercury analyzers is <0.005 μg L^{-1} without an amalgamation accessory and <0.0002 μg L^{-1} with an amalgamation accessory.)

Hydride detection limits shown were determined using an MHS-15 Hg/Hydride system.

atomic absorption of any other element. The instrument is constructed so that the number of atoms (absorbers) per unit volume in the atom cell is directly related to the number of precursor species per unit volume in the sample solution.

For most analytical chemistry applications, it is sufficient to think of the process of absorption of radiation by atoms as one in which (a) the light beam behaves as though it consisted of stream of photons, and (b) the atoms behave as though they contained discrete electronic energy levels. The process of absorption can thus be considered as the collision between an electron and a photon, whose energy just matches the difference between the energy of the level from which the electron is removed and the energy of the level to which the electron is promoted. While, in principle, there are a great many energy levels from which an electron could be removed and a great many energy levels into which an electron could be placed, the conditions which prevail for the techniques that are commonly referred to under the umbrella of *AAS* are based on moving a valence electron in an atom in its lowest energy state (the ground state) to an easily accessible excited state. That is, AAS is based on resonance absorption. For all the elements that are determinable by AAS, the energies of the photons involved correspond to wavelengths in the ultraviolet and visible parts of the electromagnetic spectrum. The transitions are almost always referred to by the wavelength of maximum absorption in nanometers.

This basic model of the absorption of radiation of atoms will be revisited later, when the question of why every element cannot be determined to the same extent by AAS is discussed (see Section 8.7.2).

8.3 How do FAAS instruments work?

Two current generation atomic absorption spectrometers are shown in Fig. 8.1. Unlike previous generations of instruments there are very few, if any, manual controls. Most of the important operating parameters, and the data collection and processing, are done by computer software through suitable interfaces. Instruments have two basic functions: generate a population of atoms from the sample and interrogate the atoms with a beam of light that only the atoms of interest absorb. Secondary functions inherent in all spectrometers are (a) discrimination against light emitted from the flame and (b) removal of stray light. All of the possible flames that might be used in the atom production process have been extensively investigated with the result that only the air–acetylene and nitrous oxide acetylene flames are considered of practical importance for atomization from aqueous solution (though there might be some benefits in employing an air–hydrogen diffusion flame for the determination of tin). The

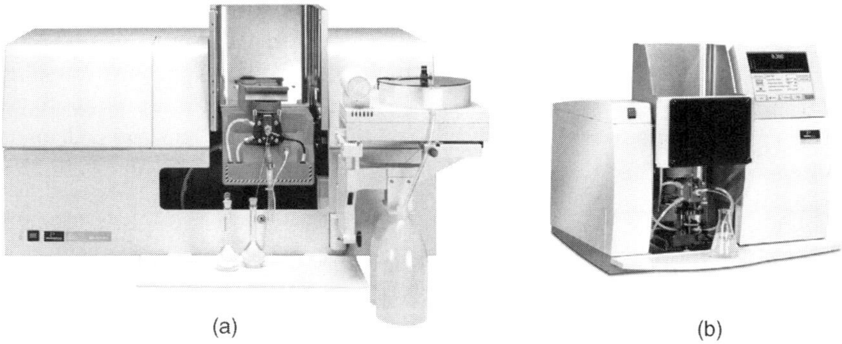

Fig. 8.1 Flame atomic absorption spectrometer models, (a) AAnalyst 800, an instrument that can be operated with either a flame or graphite furnace atomizer, with furnace autosampler, and (b) AAnalyst 200 (reproduced by kind permission of PerkinElmer Inc.).

atomization of hydrides for either AAS or AFS is usually achieved by an air–hydrogen flame of some sort.

8.3.1 Sample introduction

Despite many years of research, there is really only one efficient way of generating atoms from solids and that is (a) to harness the free energy of the dissolution process to break most of the bonds in the sample and create solvated species in aqueous solution, (b) shatter the solution into droplets by coupling the energy in a fast moving gas stream to the surface of the liquid thread, (c) fractionate the drops so that only those around 5 μm or less are entrained in the stream of fuel and oxidant gases traveling to the burner head, and (d) light the flame and adjust the fuel: oxidant ratio.

8.3.1.1 Nebulizers

The device that is almost universally applied to generate an aerosol from the sample solution is the pneumatic concentric nebulizer. The device consists of two concentric tubes of diameters a few μm and a few mm. A current generation pneumatic nebulizer is shown in Fig. 8.2. Air at a flow rate of about 8 L min^{-1} and a gauge pressure of about 40 psi is forced through the annular space between the inner and outer tubes into the spray chamber which is at a pressure slightly above atmospheric. However, the pressure at the tip of the inner tube is below atmospheric and the nebulizer draws sample solution from an external reservoir. The rapidly flowing gas stream interacts with the liquid thread that emerges from the inner tube producing instabilities in the liquid surface that rapidly result in the formation of drops. This primary spray may impact on a solid surface causing further fragmentation.

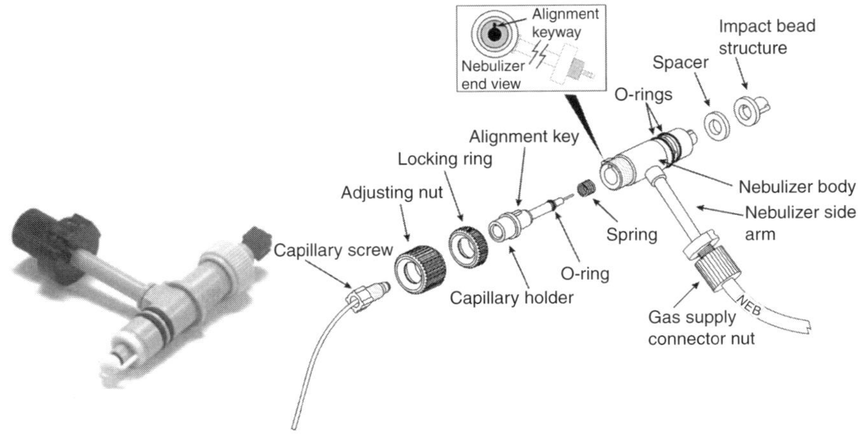

Fig. 8.2 Photograph of a complete pneumatic nebulizer with exploded view of all components including integral impact bead for PerkinElmer AAnalyst 200 and 400 spectrometers (reproduced by kind permission of PerkinElmer Inc.).

8.3.1.2 Spray chamber

The oxidant gas producing the primary aerosol is mixed with the fuel gas in a chamber which contains an outlet for the gas mixture to flow to the burner and a drain. The spray chamber may contain devices, such as baffles, to help with the gas mixing and to remove the larger aerosol droplets. The route for the gases will insure laminar flow in the mixture emerging from the burner slot. The chamber may contain a second inlet to allow the addition of more oxidant gas, which may be needed if the solution being nebulized contains a high proportion of a combustible organic solvent. A typical nebulizer, spray chamber and burner assembly is shown in Fig. 8.3.

8.3.2 Flames

The flames employed are characterized as *premixed*, so that the rate of propagation of the flame reaction is not controlled by the rate at which the fuel and oxidant are mixing, but by some fundamental features of the particular chemical reaction. Burners are designed to produce a flame that is long and thin (see Fig. 8.3). As the absorbance measured is a function of path length and number density, there is an optimum flame length if obtaining the maximum signal is the only consideration. However, it is also necessary to bear in mind the stability of the flame on the burner head. As the slot gets bigger, the gas velocity decreases and eventually falls below the critical value at which the quenching of the flame reaction at the burner slot prevents the flame reaction from passing back into the spray chamber. This could be offset by increasing the rate at

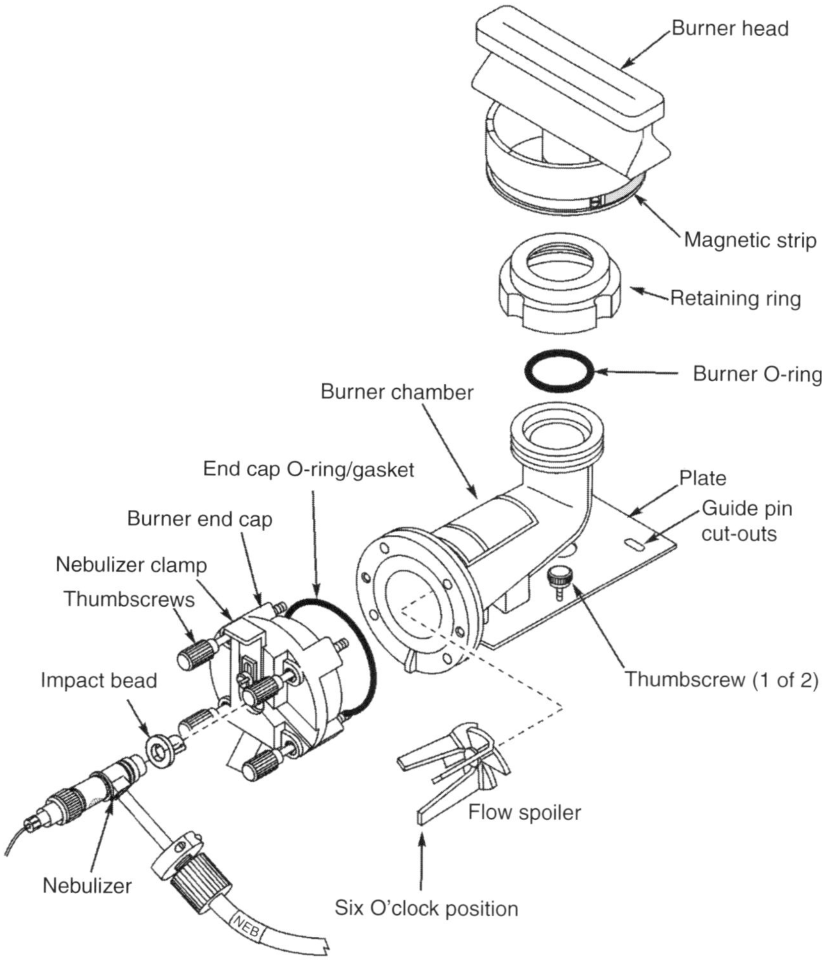

Fig. 8.3 Diagram of nebulizer, spray chamber and burner system showing a design in which secondary aerosol is produced by fragmentation of the droplets on a flow spoiler (reproduced by kind permission of PerkinElmer Inc.).

which flame gases are introduced, but this in turn reduces the number density of the atoms in the flame. Again there are factors at work which lead to a compromise in the various design and operating features in which the combination of number density and path length gives the highest absorbance. This is not quite the whole story, as the noise of the signal is an important feature of instrument performance. As will be discussed in a later section (see Section 8.7.1), flames are sources of noise for a variety of reasons one of which is *flicker*. The random fluctuations in the flame geometry caused by convection currents mean that the path length is also subject to random fluctuations.

When the droplets enter the flame, the solvent is evaporated leaving salt particles that evaporate to give gaseous molecules. Flame chemical reactions occur to produce atoms. The reactions may be simple thermolysis of the bonds in the molecule or may involve reaction with flame species such as C and CO to reduce oxide precursors to the elemental state.

This is not the complete story by any means: further reactions are possible and the whole process is dynamic, so that even atomic species that are relatively long-lived are swept out of the flame by the combustion products. A more detailed look at what goes on in the various stages of the atomization processes will be discussed later (see Section 8.7) when interferences are considered.

As elements differ in their flame chemistry, instruments will permit control over the flame chemical environment to some extent by allowing control over the fuel–oxidant ratio. Values at which a flame is not stable on the burner head will generally not be accessible through the control software. As a consequence of the various flame processes, the region in the flame in which the maximum density of atoms is found will vary from element to element and thus the instrument will allow control over the height above the burner at which the light beam passes through the flame. This is typically done by moving the entire burner–nebulizer assembly vertically while keeping the light beam fixed.

8.3.3 Light sources

The region of highest atom density in the flame is probed with a beam of light from a light source constructed, so that amongst the emitted radiation is an intense beam covering a narrow range of wavelengths around the nominal wavelength at which the target analyte atoms absorb. The way this is done is to generate atomic emission from the element in question. For the majority of elements, this may be efficiently done with an electrical discharge lamp in which the cathode is a hollow cylinder whose interior surface contains atoms of the element of interest. The lamp is filled with an inert gas (Ne or Ar) at a pressure of about 10 torr. On first striking the discharge, ions of the filled gas, formed by electron-impact ionization, bombard the cathode surface vaporizing atoms of the surface in a sputtering process. The atoms are collisionally excited and emit radiation that may be absorbed by atoms in the flame. This device is known as an HCL and is shown schematically in Fig. 8.4.

For a minority of easily volatilized elements, an alternate light source, known as the EDL, is more effective than the HCL. This device, shown schematically in Fig. 8.5, consists of a sealed quartz bulb, containing a small amount of an easily volatilized compound of the element in question and some inert gas, mounted in a coil connected to a radio-frequency power supply. Once a few electrons are seeded into the vapor, there is induction

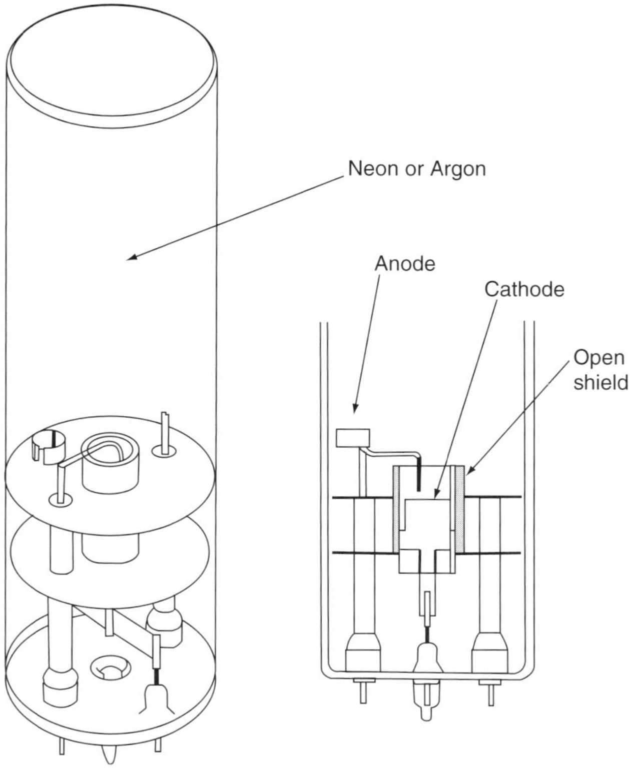

Fig. 8.4 Schematic diagram of a hollow cathode lamp (reproduced by kind permission of PerkinElmer Inc.).

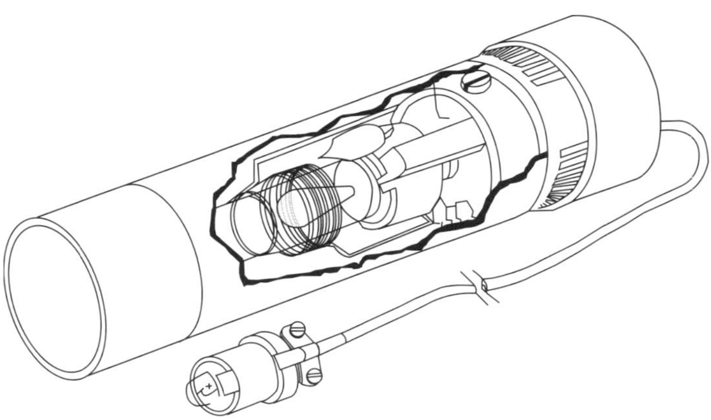

Fig. 8.5 Schematic representation of an electrodeless discharge lamp (reproduced by kind permission of PerkinElmer Inc.).

coupling with the oscillating electromagnetic field and collisional ionization and excitation reach an equilibrium state, so that a steady emission occurs. Depending on the particular instrument design, a separate power supply may be needed to run EDL light sources.

The boundary conditions referred to in Section 8.2 above are mainly related to the spectroscopic characteristics of the absorption and emission profiles (i.e. the shape of the plot of the magnitude of each phenomenon as a function of wavelength). For absorbance to be linearly related to the number density of absorbers, the absorption profile must be broader than the emission profile, so that the absorption coefficient is essentially constant for the range of wavelengths reaching the detector. The spectral profiles for emission and absorption are functions of temperature and pressure, and emission profiles also depend on the temperature gradient in the emitting region of the source. The light sources used, HCL and EDL, operate under conditions of pressure and temperature compared with those in the flame atom source such that the emission profile is narrower than the absorption profile. The light sources are also designed so that there are very few atoms in cooler regions, and so the broadening due to temperature gradients is minimized.

8.3.3.1 Atom source emission

Atoms in the flame emit radiation at the same nominal wavelength as that emitted by the light source, so a fundamental feature of any atomic absorption spectrometer is discrimination against the light emitted from the flame. This is done by modulating the light from the lamp and locking-in the frequency response of the detector electronics to the modulation frequency, so that the detector output only changes in response to changes in the a.c. light intensity. This is not to say that the light from the flame has no effect on the signal. Intense emissions, such as produced by molecular species at high number density can cause additional noise and, in severe cases, overload the detector response completely.

8.3.3.2 Stray radiation

Unlike a typical UV-vis absorption spectrometer in which the sample is placed in a light-tight compartment, an atomic absorption spectrometer collects room light as well as all of the wavelengths emitted by the flame and the light source within the spectral band pass. A substantial amount of potentially interfering stray light is blocked from reaching the detector by the monochromator, but sufficient quantities of stray light can pass through the primary optical path of the spectrometer to cause significant curvature of the calibration curve even at absorbances less than one. Calibrations can be curved for other reasons as well, and thus it is important that any computer-based calibration routine is capable of fitting an appropriately shaped curve to the calibration data. If the on-board computer is bypassed for any reason, then before a calibration function, based on a first order polynomial fitted by linear regression, is used to analyze samples, the data should be visually inspected for curvature, and/or the runs of positive and negative residuals should be examined for patterns indicating non-linearity.

8.3.4 Detectors

The role of a detector is to convert the radiant power passing through the optical system to a proportional electrical signal that can be amplified and processed to give ultimately a readout in either absorbance or, via a computer-based calibration function, in suitable concentration units. As the signal is modulated, the transduction of light intensity must be rapid. The device should also be inherently noise-free. For many years, the only practical device that fulfilled these criteria was the photomultiplier tube (PMT), but recently the performance of monolithic solid-state devices has advanced to such an extent that their performance is superior to that of most PMTs and they are now used in AA instruments. One important characteristic for which these solid-state devices have superior performance is in the response as a function of wavelength. The two devices are compared in this respect in Fig. 8.6, from which it is seen that

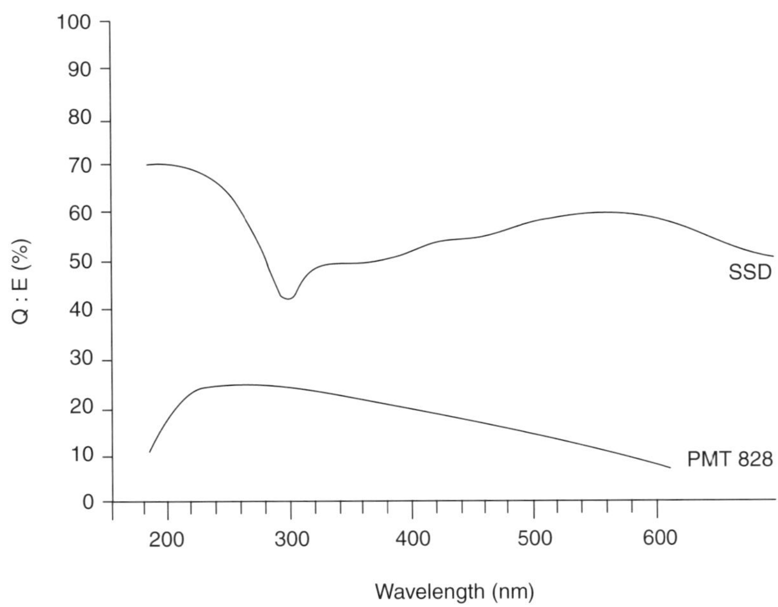

Fig. 8.6 Response curves of photodetectors for atomic absorption spectrometers, plotted as quantum efficiency as a function of wavelength. Quantum efficiency is a measure of how many photons must arrive at the detector surface per electrical event produced in the device response surface. The solid state detector (SSD) device is available in newer generations of PerkinElmer AAnalyst instruments (reproduced by kind permission of PerkinElmer Inc.).

not only does the solid-state device have a flatter response across the spectrum, but also the magnitude of the response is greater. The extent to which the electrical signal is amplified is under the control of the operator in the sense that whenever the instrument output is set to zero (often referred to as the *autozero* function), the amplification or gain of the signal processing is varied. If the light intensity arriving at the detector is low (the light source is not very bright or a significant amount of the light is absorbed by the flame gases or the air in the optical path) then the signal amplification may be such that the noise which inevitably accompanies the electrical signal manipulation may become a significant contributor to the overall noise.

8.3.5 Optical system

There are some variations in the design of optical system dependent on the manufacturer of the particular instrument. All instruments are designed so that radiation from the light source is focused on the center of the flame, and the optical path will be folded so as to make the instrument footprint smaller than if a completely in-line optical design were adopted. Most designs will avoid the use of lenses and the beam steering will be done by mirrors. The burner compartment may have windows to minimize the ingress of gases into the optical and electronic parts of the instrument. Top-of-the-range models will have features such as automated changeover of lamps, double-beam capability and background correction. The double-beam optical arrangement allows use almost immediately after switch on, as the lamp intensity drift is compensated for producing a flat baseline. It should be borne in mind that emission profile drift is not compensated for, and so there can still be a sensitivity drift as the lamp warms up. The double-beam arrangement also reduces baseline noise and can improve the detection limits. For flame determinations, background correction based on the measurement of the absorption of the light emitted from a continuum source (usually a deuterium lamp of some sort) is commonly employed. For instruments dedicated to graphite furnace atomization, Zeeman effect background correction may be employed as an alternative procedure. The optical system of a current generation spectrometer designed for use with both flame and furnace atomizers is shown in Fig. 8.7. Note that this is a double-beam-in-space design: light that passes through the reference path (part of which involves a fiber optic light guide) is imaged on the upper portion of the entrance slit of the monochromator and is thus brought to focus on a different region of the array detector surface; this allows simultaneous measurement of the intensities in the sample and reference channels. The instrument distinguishes between light from the HCL and light from the deuterium source as these are pulsed alternately (with a dark period as well).

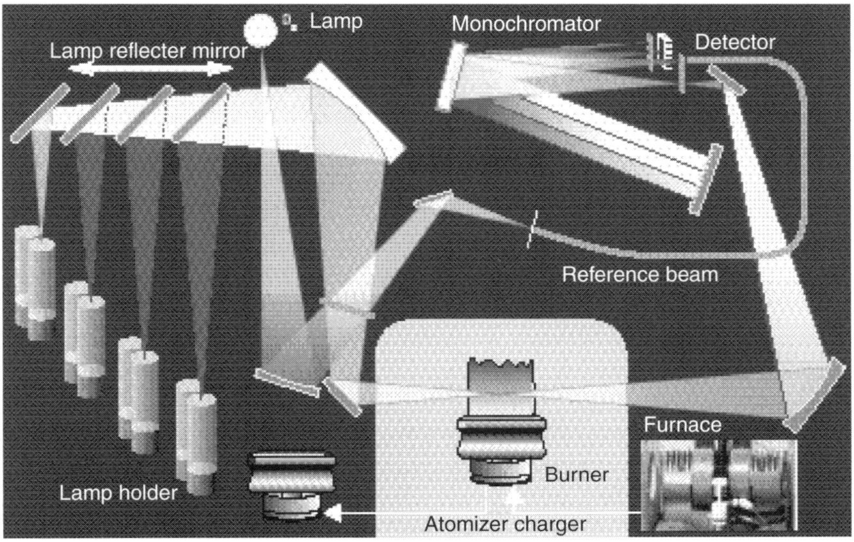

Fig. 8.7 Optical arrangement of a current generation double beam spectrometer (PerkinElmer Instruments model AAnalyst 700/800) with deuterium-source continuum background correction. Note the multiple light sources that can be accessed automatically allowing automated sequential determination of multiple analytes (reproduced by kind permission of PerkinElmer Inc.).

8.4 Instruments for hydride generation

For the determination of the hydride-forming elements by AAS, the flame atomizer is replaced by a heated quartz tube. This is mounted in the optical path of the spectrometer either in place of the burner/spray chamber unit, in which case the tube is electrically heated, or is mounted on top of the burner so that the tube may be heated by the combustion flame. The hydrides are generated either in a reaction vessel (the so-called batch mode) or in a continuous flow or flow-injection system. Whichever system is used, the gases are introduced into the quartz tube, through a side arm mounted in the tube center. Sodium borohydride solution in alkali is universally applied as the hydride-generating reagent. The hydride and hydrogen from the decomposition of the excess reagent are transferred to the quartz tube by a stream of argon. Although the procedure is considerably more complicated than the direct nebulization of sample solution, the benefits are considerable, particularly in terms of the detection limit and the concentration range that can be determined. One of the reasons for this is the considerably greater atom number density in the light beam that is obtained, compared with conventional nebulizer flame systems. The HG, stripping and transport processes have efficiencies of at least 80%, and atomization processes are much more efficient when the only precursor is a simple molecular species. Although nebulizers may transport up to 10% of the sample solution into the flame as fine droplets, the long sequence of processes that produces atoms is much less efficient. Overlaid on the consideration of the efficiencies of the various processes are considerations of mass flux. The rate at which atoms can be formed is much greater when the precursor species is the hydride, and thus the rate at which the analyte can be introduced to the quartz tube atomizer is greater than for the flame atomizer. For example, as shown in Table 8.1, the detection limit for arsenic improves from 150 to 0.03 $\mu g\ L^{-1}$ (i.e. by a factor of 5000) when the element is determined by HG (in a batch system) rather than by flame atomization. All manufacturers of AA instruments will sell an HG accessory for the instrument in question, consisting of the quartz tube atomizer and mounts, together with the reaction vessel (or flow system) and associated plumbing to connect the system. Flow-based systems can be completely automated and can be left unattended to run large numbers of samples loaded in an autosampler.

8.5 Instruments for cold vapor mercury determination

The same systems that are used for HG can also be used for the determination of mercury. In principle, the atomizer does not need to be heated as the mercury vapor is stable at room temperature; in practice, heating the tube can prevent the condensation of water, which would otherwise adversely affect the measurement. In addition to the borohydride reagent (commonly used in European laboratories), stannous chloride is a viable reducing agent for the production of mercury vapor from dissolved mercuric ions. This latter reagent is preferred by North American laboratories. In practice, it is impractical to use the same apparatus for the generation of mercury by reaction with stannous chloride and for the generation of hydrides by reaction with borohydride, as the residues of the reagents cause the production of metallic tin deposits which can severely depress the hydride formation reaction and cause spurious blank signals in the determination of mercury. For laboratories with a requirement to determine mercury in a large number of samples, the best option is to invest in an instrument dedicated to the determination of this analyte. Several manufacturers produce such an instrument. The PerkinElmer Instruments' flow injection mercury system (FIMS) is shown in Fig. 8.8. This instrument, which is a non-dispersive instrument, is designed to achieve low detection limits and has a long path length cell, a bright light source and a sensitive, low-noise detector. The detection limit is about 5 $ng\ L^{-1}$. Even lower detection limits can be achieved by the incorporation of a so-called *amalgam* accessory, a device that traps most of the mercury vapor from a relatively large volume of carrier gas (on a gold or gold/platinum metal alloy surface) and releases it rapidly, on the application of

Fig. 8.8 Flow injection mercury system (FIMS) with autosampler: an atomic absorption instrument dedicated to the determination of trace concentrations of mercury by chemical vapor generation (reproduced by kind permission of PerkinElmer Inc.).

a thermal pulse, into a relatively small volume of carrier gas. In this way, the detection limit can be lowered to about 0.2 ng L^{-1}.

8.6 Instruments for AFS

There are some theoretical reasons why AFS could give superior performance to that of AAS. The fluorescence signal is directly proportional to the light source intensity, and thus if a very intense light source is employed (such as a laser), a high signal to noise ratio (SNR) can be obtained. If the upper level involved in the spectroscopic transition is saturated, then the fluorescence signal is independent of any noise on the light source. The fluorescent radiation may be at a wavelength different from that of the exciting radiation and thus background due to scattering of the primary beam is not a problem. However, an instrument based on such design considerations is not a practical proposition, in that the market is already served by existing absorption, emission and mass spectrometry instruments. However, there is a small number of manufacturers currently producing instruments for measurements by AFS, all of which employ resonance, direct line fluorescence i.e. the emitted fluorescent light is at the same wavelength as that of the light that excites the atom to the upper state. Typically, these instruments are based on HG with atomization in an air–hydrogen diffusion flame and HCL light sources with relatively low cost optical systems, such as single lenses and interference filters. The instruments are thus limited to those elements that can be determined via HG, but also include mercury for which a flame atomizer is not needed, though a shielded atom cell is needed as molecular oxygen is quite effective in quenching the fluorescence signal. These instruments are, therefore, relatively small and low cost compared with the typical multi-purpose atomic absorption spectrometer capable of supporting flame, quartz tube and electrothermal atomizers.

8.7 Limitations to performance

The fundamental goal of instrument design is to maximize the inherent SNR under as wide a range of operating conditions as possible. However, this is offset to some extent by considerations of making a cost-effective instrument. The accuracy of the method in which the instrument is employed is also of concern as there are factors related to the presence of other components of the sample that affect the magnitude of the signal. These effects have been investigated exhaustively and it is unlikely that, in the near future at least, there will be a significant development with regard to overcoming these interferences that can be incorporated into the basic spectrometer function of the instrument beyond those which have already been developed. Interferences due to other sample components can always be overcome by incorporating chemical separation stages into the overall method, though of course this increases the complexity of the analysis and decreases the speed. Such separations are not trivial exercises as often the interferents are present in concentrations, many orders of magnitude greater than the analytes. It is also possible to envisage that such separations could be integrated with the AA instrument, but it is difficult to imagine that whatever devices were employed for the separation could be sufficiently general in application, given the large number of analyte/interferent combinations that could be encountered, so that incorporation into the instrument became a cost-effective option. Interferences can also be compensated for by the calibration strategy employed. If the standards are matched to the samples in terms of the matrix components that cause the interference(s), then (providing there is still a useful response from the instrument) accurate analyses may be made. If the matrix components are not known, the method of standard additions may be employed.

There are several categories of limitations to performance: those that are related to the way the instruments work, those that are related to analyte properties and those that are related to properties of other components of the sample. There are two basic categories of effects to be considered: those that affect the SNR and those that affect the accuracy.

8.7.1 Instrument effects

A fundamental feature of absorbance spectrometries is that the magnitude of the signal is independent of any instrument-based parameter related to the light source and detector, so essentially these components only contribute to the noise. There is a limit to the rate at which sample solution can be usefully introduced into the flame. The extent to which the processes that result in the formation of atoms can occur in the limited residence time in the flame is a function of the drop size, and so large droplets do not contribute to the atom number density. The current status of concentric nebulizer design is such that about 10% of the sample solution introduced at about 5 mL min^{-1} can be transported usefully to the flame. The spatial distribution of the appearance of atoms in the flame is a function of the drop size distribution. Very small drops will give rise to atoms lower in the flame than will larger drops. Atoms that form lower in the flame will be entrained with the expanding flame gases and will also diffuse further during residence in the flame than will atoms formed higher. The temporal distribution of atoms will exhibit fluctuations reflecting variations in the rate of the processes governing the appearance of atoms. To some extent the rate of appearance of atoms is governed by the size of the drop (which in turn governs the size of the salt particle, which in turn governs the rate at which the particle heats up). Thus a wide range of drop sizes introduced into the flame will produce more noise on the signal than would be observed with drops of uniform size. In addition, flames flicker i.e. they are dimensionally unstable and also have non-uniform and fluctuating refractive indices. Flame flicker is a significant source of noise in AAS, as both the path length and the position of the light beam, which is sensitive to the refractive index of the flame, fluctuate. It is possible to increase the number density of atoms by confining the expanding gases through the presence of a quartz tube in the flame, which has a entrance slot positioned directly over the burner slot and an exit slot at approximately 180°. The device, which is variously referred to as a *slotted tube atom trap* or *slotted tube atom retarder*, only works for a limited number of easily atomized elements. If the additional loss mechanism due to collision with the tube walls occurs at a higher rate than the accumulation of atoms in the tube interior, no gain or even a loss of sensitivity will be obtained.

8.7.2 Analyte effects (or why all elements do not have the same detection limit)

Most performance parameters such as sensitivity, detection limit, characteristic concentration are almost always given in mass/volume concentration units. It is worth bearing in mind that not all elements have the same atomic weight, and so solutions of different elements that have the same concentrations in mass/volume units do not have the same number of particles per unit volume. For example, in comparing the performance of FAAS for the determination of Li and Pb, we see from Table 8.1 that the detection limits are 0.8 and 15 $\mu g\ L^{-1}$, a ratio of about 19. Does this mean that lead does not form atoms as readily as lithium in the flame? Not necessarily: the lead (atomic weight 207) ion flux into the flame is 29 times smaller than that of lithium (atomic weight 7) for solutions of the same concentrations in mg L^{-1}. So we deduce that maybe lead forms atoms more effectively than does lithium.

The atoms of the different elements do not absorb radiation to the same extent. That is, some elements absorb more light for the same number density and path length than other elements. The reasons for this are related to the unique electronic structure of the element and of the interaction of the electrons with each other and with the nuclear charge. This is reflected in a property known as the oscillator strength for absorption, which is often listed in tables of wavelengths or other relevant properties. While this property may be useful to know, it is of less practical importance than knowing about the sensitivity or detection limit for that particular element. The second element-specific parameter that affects performance is the atomization efficiency. Although this may be influenced somewhat by the matrix and flame conditions, even when the matrix is the most favorable for atom formation, different elements will have different atomization efficiencies because each element has a different reactivity in the flame. There is a reasonable correlation between atomization efficiency and the strength of the element–oxygen bond, in the sense that the stronger this bond, the fewer atoms are formed in the flame. The extent of atom formation can be influenced by chemical reactions in the flame that reduce metal oxides. The presence of reducing agents such as carbon and carbon monoxide can increase the atom number density. The fuel to oxidant ratio that gives the maximum atom number density for one element will not necessarily be the same for another element, but for many elements the differences are small. However, for some elements the extent of atom formation in the air–acetylene flame is so small as to be useless, and a hotter flame, the nitrous oxide acetylene flame, must be used.

There is yet a third factor that governs the atom number density of a particular element under a particular set of conditions, namely ionization. The temperatures of both air and nitrous oxide acetylene flames are such that some elements are significantly ionized. As the electronic structure of the ion resulting from the removal of an electron is very different from that of the precursor atom, ions do not absorb the radiation from the HCL.

8.7.3 Sample effects

If the sample contains matrix components that interfere with the atom forming processes, then the analysis will be inaccurate unless the standards are matched to the samples in terms of the interfering components. Two sorts of effect may be discerned. In one, the sample contains compounds that react with the analyte element to form *stable* compounds that are not atomized so readily. In some cases, the signal suppression can be severe. For example, oxy-anions such as aluminate or silicate can suppress completely the signal from alkaline earth elements. This type of interference can be ascribed to the formation of particular compounds, sometimes with reasonably well-defined stoichiometry. When the concentration of phosphate is increased in solutions containing the same concentration of calcium, the resulting atomic absorption signal falls to a constant value at a composition roughly corresponding to that of tricalcium phosphate, $Ca_3(PO_4)_2$. However, the signal suppression effect can simply be the result of the presence of a high concentration of components that gives rise to relatively large salt particles in the flame which have different atomization characteristics than those of the small salt crystals formed when only the analyte is present. The second type of sample matrix effect arises from changes in solvent properties. If the sample solutions differ in viscosity, surface tension, vapor pressure or combustion characteristics compared with the standard solutions, the analysis will be inaccurate. The presence of organic solvents will typically cause these effects. Often the effects are beneficial, in the sense that a higher sensitivity is obtained when the sample contains a significant proportion of an appropriate organic solvent. This increase in sensitivity is one of the reasons for the rise in interest in methods involving transfer of the analyte from an aqueous to an organic phase, such as liquid–liquid (or solvent) extraction. If the target element was one for which significant ionization occurs and if the sample were to contain a relatively high concentration of another easily ionized element, then the analysis would be inaccurate as the degree of ionization in the sample would be suppressed compared with that in the standards.

It is also possible for sample components to affect the analysis without affecting any of the processes related to atom formation. A fairly common problem is the spectroscopic interference in which the light beam intensity is attenuated by the presence of the derivatives of sample components in the light beam. Molecular compounds, whose absorption bands overlap the wavelength of absorption of the analyte atoms, would cause the analysis to be biased high, as would the presence of salt particles big enough to scatter the light. As light scattering increases sharply as the wavelength of the light decreases for a given particle size, this effect is much more severe for wavelengths around 200 nm than it is for wavelengths well into the visible. It is also possible for molecular compounds to produce emission from the flame at the wavelength of interest of such intensity that the detector demodulator circuits are overloaded. Even if this did not occur the presence of significant emission intensity produces additional noise, degrading the detection limit.

8.8 How to get the right answer?

It might seem that for the analysis of all but the simplest sample matrices, interference effects are going to be present and that it is going to be difficult to obtain an interference-free, accurate measurement. This is true, but the real question is 'Would it matter if interferences were present?' In any method development process, the purpose to which the resulting chemical information is to be put has to be considered. In turn, this gives some guidelines as to the accuracy and precision that are needed. For example, if the pass/fail boundary for some sample solution is 10 mg L^{-1} of the target element and all the samples give signals that, from a simple calibration, correspond to concentrations around 2 mg L^{-1}, then once the most severe type of interferences have been checked for, it does not really matter that the procedure may be inaccurate or imprecise.

8.8.1 How to deal with interference effects?

Apart from separating the analyte from the interfering sample components (an approach that will be discussed later in Section 8.8.2), there are a number of procedures that are either (a) part of the sample preparation or (b) a feature of the instrument, or (c) part of the calibration procedure that can be implemented to eliminate or compensate for interference effects.

The extent of the formation of some stable compounds in the flame can be decreased considerably by the addition of a releasing agent. The most effective of these is probably lanthanum, which forms very stable compounds with many elements that interfere by the formation of stable compounds with the analyte (often those elements that form oxy anions). For example, the severe depression of phosphate on calcium may be alleviated by the addition of lanthanum in percent concentrations to the solutions for analysis. The releasing agent reacts preferentially with the interferent *releasing* the analyte for detection by atomic absorption. The effects of ionization, which cause inaccuracies in the determination of easily ionized elements, can be decreased by the addition of an ionization suppressor. Again relatively high concentrations are employed (1000–10 000 mg L^{-1}), of elements that also have a relatively low ionization potential. The electron number density in the flame is increased and the ionization of the analyte is suppressed. This can be considered to be a simple mass balance effect on a reaction ($M^* \rightarrow M^+ + e^-$) that is at equilibrium. Potassium is commonly employed as an ionization suppressor, though lanthanum will also act to some extent in this capacity. The extent of stable compound interferences (and some others) is dependent on the amount of analyte and interferent as well as the ratio of interferent to analyte. Typically, therefore, the extent of interference can be decreased by the simple expedient of diluting the sample. In principle, this can be applied in a series of steps and the interference-free signal that would be obtained by accounting for a very large dilution can be calculated. However, in practice this *infinite dilution* method of dealing with interferences is rarely applied in FAAS. Stable compound interferences can be decreased by adjusting the nebulizer to give a higher proportion of smaller droplets and by observing the atoms higher in the flame (after a longer residence time), i.e. the spectrometer can be optimized for minimum interference rather than for maximum signal. Again, this procedure is not very widely adopted.

Molecular absorption and scatter, both of which are relatively common, can be compensated for by the *background correction* feature of a suitable spectrometer. The basis for this correction is that the instrument makes two types of measurement in rapid succession: the first is of the total attenuation of the light beam by all processes (atomic and molecular absorption, and scatter) and the second is of the attenuation of the beam intensity by all processes except that of atomic absorption. The difference between these two measurements corresponds to the attenuation in light intensity due to atomic absorption alone. The first of these measurements is easy to make: it is what is seen when light from the primary source (the HCL) is monitored. The second type of measurement may be done in one of the two ways: the continuum source method or the Zeeman-effect method. In the continuum source method, the light from a continuum source is shone through exactly the same path in the flame as that traced by the light from the HCL (see Fig. 8.7). As atomic absorption has a very narrow spectral profile, the intensity removed due to this process is insignificant, so the absorption measured is that due to molecular absorption and scatter. For a laboratory with only simple sample matrices (such as drinking water) to worry about, instruments with background correction are probably not necessary. For all other work, especially, if methods are to be developed for samples not previously analysed, instruments with background correction are a worthwhile investment.

In principle, accurate analyses can always be obtained by calibration of an instrument with standards whose composition exactly matches that of the samples in terms of the interfering matrix components. In practice, this method of *matrix matched* calibration is often employed as part of the overall strategy of obtaining accurate analyses. It can make the preparation of the standard solutions a lengthy procedure, and as solutions of metals at single digit mg L^{-1} concentrations are not all that stable, a considerable portion of the working day can be taken up by the preparation of standards, especially if several different matrices are to be analyzed. It often happens that the sample matrix composition is not known, or even if it is known it may be variable. There is no way to be sure that accurate analyses are being obtained in this latter case, apart from analyzing every sample by the method of standard additions. In this procedure, analyte is added in increasing amounts to several identical portions of the sample and each portion is analyzed. An extrapolation procedure then allows the amount of analyte present in the portion of sample with no addition to be calculated, and hence the concentration can be computed by taking into account any dilution of the portions prior to measurement. This is a very time-consuming procedure, and is not very precise unless large numbers of additions (6 or 7) are made. In practice, laboratories will not use the method of standard additions for FAAS analyses apart from (a) in a method development procedure to evaluate the extent to which the slope of the calibration is different in the sample matrix from that in a simple aqueous acid matrix and (b) in the analysis of the occasional *one-off* samples. If the sample matrix is unknown, but can be considered constant for a number of samples, the first sample in the series can be analyzed by the method of standard additions and then this calibration used as a matrix matched calibration for the analyses of the remaining samples. The method of standard additions will compensate for interferences that affect the slope of the calibration, so-called *multiplicative* interferences, but will not compensate for interferences which offset the baseline, so-called additive interferences.

8.8.2 Separation and preconcentration

There are many situations for which all of the strategies for overcoming interferences are ineffective: the results are still inaccurate. And there are many situations in which the analyte concentration is too low for reliable measurement to be made by FAAS. In some of these circumstances, the use of another technique such as graphite furnace atomic absorption spectrometry (GFAAS) or ICP-MS, may solve the problem. But this is not guaranteed, as these techniques also suffer from interference effects caused by the presence of high concentration excesses of matrix components. In these situations, the approach may well be to incorporate a separation step into the sample preparation and pretreatment stages, so that the analyte is separated from the potentially interfering matrix. Often such a step can be combined with preconcentration, as many separation procedures allow the removal of analyte from a relatively large volume of sample solution with transfer to a relatively small volume.

The sample dissolution procedure offers some possibilities for the separation of analyte and matrix which will be discussed in Section 8.10. Once the sample and matrix components are completely dissolved, several possible separation procedures may be selected including: dialysis, gel-filtration, (co-) precipitation and filtration, liquid–liquid extraction, solid-phase extraction and volatilization. Clearly, properties of the analyte and matrix species are relevant in the choice of technique; for example, dialysis will only separate analyte from matrix if one of them is too large to pass through the pores of the dialysis membrane. Dialysis, typically, dilutes the analyte concentration. Many of these procedures can be implemented in a batch mode in parallel, and thus the sample throughput can be increased compared with the method in which one sample is taken throughout the procedure from start to finish before the next sample is processed. Again, such procedures may not be amenable to automated, unattended operation unless substantial capital investment is made in robotic instrumentation.

At one time, there was considerable interest in the possibilities of liquid–liquid extraction as part of the sample preparation stages in FAAS methods, as the procedures are efficient and the nebulization of most organic solvents has a beneficial effect on the performance characteristics of the determination. The degree of atom formation is typically higher than that for the same concentration of analyte in aqueous solution due to the favorable properties of the solvent in relation to nebulization (surface tension, viscosity and vapor pressure) and combustion (most solvents burn and contribute to the flame reactions, whereas water cools the flame). However, as laboratories have become more conscious of the implications of using large volumes of organic solvents in terms of safety and the costs of waste disposal, interest in these procedures has declined and has been replaced with an interest in the possibilities of solid-phase extraction. In the most convenient form, which may be difficult to achieve in practice, a large volume of the sample solution is passed rapidly through a small bed (column or disk) of a material capable of selectively retaining the analyte, followed by elution of the analyte into a small volume of non-toxic, low-cost eluent. In practice, the following parameters may need to be considered: addition of further reagents (either to adjust the pH or form a derivative with the analyte that is retained), retention of components other than the analyte, the need to wash unretained components from the bed before elution, the difficulty of redissolving the analyte species, and the need to regenerate the bed before the next sample is processed. There is a substantial, growing literature relating to the development of such procedures, but the methods may be difficult to transfer from the research laboratory into the routine analytical laboratory. One reason for this is that many of the solid-phase extraction materials described are not commercially available and have been synthesized in-house by the researcher(s) in question. Commercially available materials capable of retaining metals tend to be fairly non-specific, and thus the development of a procedure for a given analyte in a given matrix may be rather time-consuming. In addition, if the requirement is to measure more than one analyte in the presence of an interfering matrix, the method development stage may be further extended.

8.8.3 Flow injection techniques

There are considerable benefits in terms of the analytical characteristics of methods in which sample introduction is by flow injection procedures. These benefits, which have been written about in other recent texts (see the bibliography at the end of the chapter), include greater tolerance to dissolved solids, greater freedom from solvent viscosity effects, ability to work with limited sample volumes, as well as providing a methodology for the implementation of pretreatment procedures for preconcentration and matrix separation.

Although the research literature abounds with descriptions of procedures for the coupling of separation and preconcentration procedures directly with FAA instruments, the implementation of these with dedicated instrumentation provided by the manufacturers of the spectrometers is somewhat limited being confined almost exclusively to reactions for chemical vapor generation (either hydrides or atomic mercury). While it is relatively straightforward and inexpensive to assemble a continuous flow or flow injection system that is directly coupled to an FAA spectrometer for manual operation, the automation of such a system to allow unattended operation of such a system is not so easy. Thus coupling sample preparation steps directly with the instrument produces a procedure that requires constant monitoring by an operator and may be rather slow (in terms of sample throughput), depending on the complexity of the procedure(s) to be carried out, as only one sample can be processed at a time.

8.8.4 Other possibilities for improving analytical performance

In addition to the slotted tube atom retarder device described earlier, the insertion of a small bore water-cooled silica tube (or two) into the flame can provide a means to increase the sensitivity. When the tube is cold, atoms (or simple molecular forms) condense on the tube wall. Once a suitable amount has been collected, the removal of the water permits the collected analyte species to be (re-)atomized in a relatively short timescale producing a transient event whose peak height is greater than that of the steady state signal that would be obtained for conventional nebulization. In another modification of this procedure, the atomization process is aided by a transient fuel-rich event in the flame, which may be produced by the introduction of a pulse of organic solvent. These water-cooled silica tube devices are not commercially available. One or two research groups have described a high efficiency nebulization system based on forcing the sample solution through a narrow orifice with a high pressure pump; nebulization occurs due to subsequent impact of the liquid thread on a solid surface. The device is commercially available, but may require some modification of the spray chamber so as to be fitted to a particular instrument. To avoid passing the sample solution through the pump, the system is configured in the flow injection mode with a high pressure injection valve in the line connecting the pump and the orifice. There is at least one report of directing the liquid *beam* into a tube mounted in the flame.

For a limited number of sample types, it is possible to obtain useful signals from the nebulization of a slurry of the sample. For this approach to be successful, it is necessary that the sample be ground to a fine uniform particle size (the smaller the better, but 1–10 μm is a reasonable goal), and that the matrix be readily vaporized in the flame. Again, introduction via a flow injection system is useful as the amount of material might otherwise clog the nebulizer orifice, or even the burner slot, if introduced continuously. Slurry atomization is really only a viable procedure, in any general sense, for electrothermal atomization or electrothermal vaporization as the atomization efficiency can be very poor making accurate calibration impossible.

8.9 Some practical points

8.9.1 Safety

Modern instruments are fitted with numerous safety interlocks that prevent ignition of the flame gases if they are not present at appropriate pressures. The instruments will also have sensors and feedback mechanisms to turn off the gases safely should the flame be extinguished for any reason or a power failure occur while the instrument is in operation. A flashback (detonation of the pre-mixed gases in the spray chamber) is rare these days but should it occur, spray chambers are filled with rupture discs or other points of deliberate weakness to ensure that a luckless operator is not injured by mechanical failure of the nebulizer spray chamber components under the stress of such an explosion. Instruments will also be fitted with sensors to prevent attempts to light the flame if there is no back-pressure in the spray chamber drain system. If this back-pressure is caused by water in a U-tube device of some sort, it should be borne in mind that this may evaporate if the instrument is not used for a while. For an instrument with an appropriate sensor, all that will happen is that the ignition mechanism will be inoperable; however, for an older instrument without such a sensor, there is the possibility of a flashback when lighting the flame is attempted.

8.9.2 Nebulizers (sensitivity drift)

Even on the current generation instruments, the adjustment of the position of the nebulizer capillary in the venturi throat will be a manual operation. The nebulization efficiency can be surprisingly dependent on this position and it is worth checking regularly that this position is appropriate. Nebulizers (or the connecting tubing) can be blocked easily if the sample contains suspended solids, even if these are only filter paper fibers. A calibration standard should be aspirated at regular intervals to check for sensitivity changes. Such changes can be caused by changes in other operating parameters, such as wavelength setting, HCL setting, condition and position, and by changes in the flame composition or burner position. Some of these parameters may drift slowly during long operating periods.

8.9.3 Light sources

HCLs and EDLs do not last forever and will suffer a gradual deterioration in performance before catastrophic failure occurs. This often takes the form of decreased emission intensity at the required wavelength coupled with increased line broadening. The effects of these two parameters are a deterioration in SNR and a decrease in sensitivity and/or increased curvature in the calibration. The life of HCLs can be extended by decreasing the operating current, but this may be at the expense of SNR and detection limit. The cure is to purchase a new light source.

8.9.4 Calibration

As there are numerous factors which affect the magnitude of the signal obtained for a given analyte concentration, it is likely that the calibration procedure will be part of the strategy of overcoming interference effects and obtaining accurate results. For a variety of reasons (see later) the slope of the calibration and the intercept can drift over the working day and it is advisable to periodically check the sensitivity and the zero setting by aspirating a solution containing no analyte. How often a sensitivity check standard is run is a decision for the chemist in charge of the laboratory (and may sometimes be dictated by the method being run, if it is a standard or official method of some sort). Some laboratories will run such a solution between each sample and reset the baseline to zero. Clearly time spent aspirating and measuring such standards is not productive in terms of analyses of real samples, but on the other hand a significant drift in sensitivity may mean that all the results since the last drift-free measurement was made have to be discarded and the samples reanalyzed.

8.9.5 Achieving very low detection limits

Although the instrumental detection limits are very impresssive, for example values quoted for the determination of mercury by CV may be less than 1 ng L^{-1}, it may be difficult to achieve such values in a real analytical situation for a number of reasons. Even if there are no interfering matrix components, the concentration of analyte in the various reagents used for sample preparation or the generation of the chemical vapor (in the case of hydrides and/or mercury) may well be such that the baseline noise and blank signal fluctuations lead to detection limits, which are considerably higher than the values quoted in by the manufacturers. In addition, contamination by airborne particles or vapors is a very real source of high blank values and for the determination of the lowest concentrations of mercury, for example, clean sample preparation facilities will be required. The considerations of reagent purity and contamination by airborne materials are also relevant for other elements if it is necessary to deal with analyte concentrations that are close to the detection limit. The presence of matrix components needs to be carefully considered as the instrumental detection limits may not give a true indication of the method performance in the presence of high concentrations of matrix components. This is especially true if a series of chemical manipulations are needed before a solution is obtained that is suitable for introduction to the spectrometer. Some strategies for increasing SNRs (the basis for improved detection limits) were discussed in Sections 8.7 and 8.8.

8.10 Sample preparation issues

Sample pretreatment in FAAS is determined by several factors: the physical state of the sample and the availability of the target elements within that sample, and any processes that may be required by the appropriate regulatory agency.

For aqueous samples, such as beverages, little sample preparation is needed. Degassing and filtration may be all that is needed to yield a homogeneous solution. Solids, however, can be more problematic. In this case, the analyst must decide whether complete dissolution is necessary, or if the analytes can be solubilized by leaching or some form of extraction. Many samples are soluble in nitric acid or aqua regia, although more severe conditions may be needed for especially tough matrices. Microwave ovens, hot plates and hot blocks are commonly employed to aid sample dissolution. For example, the US EPA's method 3051 for the determination of trace elements in soils by FAAS, uses a combination of nitric acid and hydrochloric acid with microwave heating to leach the soluble metals into solution. Because this is a leaching process, the silica matrix of the sample is left intact. EPA method 3052, on the other hand, specifies the addition of hydrofluoric acid to the above mixture to effect total digestion. Not only solid samples can be digested: natural waters with suspended solids may require digestion, usually acidification and some form of heating. Method 3052 is for FAAS and cold vapor; both procedures are suitable for determinations by GF AAS and plasma techniques as well.

Sample pretreatment is not limited to acid digestion. Liquid–liquid extraction with an organic solvent can sometimes be a good alternative. Care must be taken, however, that appropriate sample introduction conditions for organics are used. Additionally, ultrasonic extraction can be effective. When all else fails, a fusion may be necessary to get the sample into solution. An appropriate fusion material is mixed with the pulverized sample and the mixture heated in a furnace. The resulting fusion bead is then dissolved for analysis.

When the natural concentration of analyte elements in the sample is too low to be determined directly by FAAS, preconcentration should be considered. Preconcentration can be accomplished in a variety of ways. The sample volume can be reduced through boiling off the solvent, or liquid–liquid extraction of the analyte from a large volume into a smaller one. Alternatively, the analyte can be precipitated, filtered and redissolved into a smaller amount of solution. Solid-phase extraction is a very common technique for preconcentration. In this case, the sample is passed through a column, packed with a material that retains the analyte while allowing the matrix to pass through for disposal. The adsorbed analyte is then eluted with a small volume of solvent and introduced into the spectrometer. This procedure also has the advantage of separating the analyte from a potentially interfering matrix.

8.11 Representative applications

In this section, selected examples of papers taken from the recent literature are presented to illustrate the kind of work involving FAAS that results in publications in the original primary literature. The examples are classified under the major headings of environmental, earth sciences, foods and beverages, and petroleum.

8.11.1 Earth sciences

Janusa and Beck [1] have reviewed the applications of FAAS to a wide variety of sample matrices including biota, soils and water, to environmental measurements. Their review is mainly based on papers published from 1999 to 2002.

Dundar and Altundag [2] studied heavy metal pollution in house dust samples collected after the 1999 Turkish earthquake. They collected 80 samples and controls from non-affected areas throughout Adapazari, Turkey and determined Pb, Cd, Cr, Cu, Zn and Ni by FAAS after wet digestion. The results showed that the mean concentrations of Pb, Cd, Cr, Cu, Zn and Ni were highest in houses on the streets with increased heavy traffic due to the reconstruction of city roads, sewage and water pipes, and the demolition of damaged buildings.

Roads play a major role in transporting sediment-associated, non-point source pollutants to urban stream networks via storm drains. In urban areas the relationship of erodible soil to roads may be of critical importance in controlling metal contributions to roads. Sutherland and Tolosa [3] investigated two 50-m transects (Park and School) perpendicular to roads in the Manoa basin, Oahu, Hawaii. Concentrations of nine elements were compared with those of background control soil locations and with those of five supplemental samples from nearby recreational parks. Sediment from curbside areas of roads (road deposited sediment) was collected as the starting point of each transect, and subsequently soil was sampled from two depths (0–2.5 cm and 7.5–10.0 cm) along the transects. Total and 0.5 M HCl extractable concentrations were determined for Al, Ca, Cr, Cu, Fe, Mn, Ni, Pb and Zn using ICP-AES or FAAS. They concluded that remobilization of soil stored metals in close proximity to roads can significantly prolong the environmental contamination of urban road systems and eventually stream sediments.

Ramesh *et al.* [4] developed a method for the determination of Au in geological samples such as rocks, ores, and soils by FAAS and GFAAS with Zeeman background correction. Gold was extracted into methyl iso-butyl ketone (MIBK), which was then analyzed by AAS after removal of the interfering iron from the organic phase. The procedure was suitable for the determination of gold up to 0.1 $\mu g\ g^{-1}$. The results obtained by FAAS compared favorably with those obtained by the fire assay and ICP-MS methods. They validated the methods by the analysis of international gold standard reference materials and conlcuded that reliable results, suitable for geochemical prospecting, can be obtained.

Brandvold *et al.* [5] measured base metals and Hg in stream sediments from along the Pecos River from Northeastern to southeastern New Mexico, and found elevated concentrations of Cu, Pb and Zn at two areas: below an abandoned mine waste dump and below an inactive mill. Zinc was also present at high concentrations near a fish hatchery. Partial dissolution techniques were employed on the size fractions from selected sites to estimate exchangeable, organic- and oxide-bound metal forms.

Alvarado and Jaffe [6] developed a procedure for the atomization of lead from a graphite tube, introduced by means of a graphite rod handle, into an air–acetylene flame. They compared tubes of different sizes and of different types of graphite on the basis of detection limit and useful lifetime. They found that HGA-type pyrolytically coated graphite tubes were the best with lifetimes of about 350 twenty-second heating periods for which they obtained absolute detection limits about 50 times lower than those of conventional flame atomization, which they explained as due to the extended residence time of the atoms in the absorption volume. They validated the procedure by the determination of lead in NIST Standard Reference Material 2704, Buffalo River Sediment.

Elhelu *et al.* [7] designed a study to assess the distribution and sources of lead in inner-city soils in Washington, DC. Duplicate soil samples were collected randomly from 239 unpaved front yards of homes in Washington, DC and analyzed by FAAS. They found high concentrations of lead in inner-city soils, and that the parts of the city with the highest concentrations had a large number of residents who had attained lower education levels than most of the residents in the other areas.

Liu *et al.* [8] devised a selective flow injection online preconcentration, and separation-flame atomic absorption spectrometric method was developed for the routine determination of trace amounts of Ag, Au, Pd and Pt in geological and metallurgical samples. Preconcentration on a microcolumn packed with an iso-diphenyl thiourea immobilized silica gel was followed by elution with thiourea solution.

Koscielniak and Kozak [9] developed a flow injection gradient ratio calibration method for FAAS based on recording two transient peaks, one for a sample and the other for a sample with standard addition, which are compared point by point in a defined time range. They showed that the method produced reliable results even for a non-linear calibration dependence and in the presence of interference effects by the determination of calcium in synthetic samples containing silicon, phosphate, aluminium, vanadium and titanium, and also in iron ore sample.

8.11.2 Environmental

Rumbold *et al.* [10], as part of a multi-agency study of alligator health, captured 28 American alligators (Alligator mississippiensis) along a transect through the Florida Everglades in 1999. Liver and tail muscle tissues were sampled and analyzed on a wet weight basis for total mercury by CVAAS. They found that concentrations in tissues differed significantly among locations, with animals from Everglades National Park having mean concentrations in liver and tail muscle that were twofold higher than basin-wide averages, respectively. Comparisons with a previous survey suggest that mercury concentrations have declined in some Everglades alligators since 1994.

Many of the papers published, in which the final stage in the method is quantification FAAS, describe the development and validation of a preconcentration procedure. The majority of these publications are concerned with some sort of solid-phase extraction. Wuilloud *et al.* [11] developed a method for the preconcentration and speciation of chromium in water samples by FAAS. Total chromium species were retained on an Amberlite XAD-16 resin as chromium-4-(2-thiazolylazo)-resorcinol (Cr-TAR) complexes, at pH 5.0. Cr(VI) was determined by selective formation of Cr-TAR complexes, while

Cr(III) was masked with 1,2 cyclohexane-diaminetetra acetic acid (CDTA). The detection limit, for the preconcentration of 50 mL, was 20 ng L^{-1}. Tewari and Singh [12] prepared four chelating matrices by coupling Amberlite XAD-2 with chromotropic acid, pyrocatechol, or thiosalicylic acid through an azo spacer, and impregnation of Amberlite XAD-7 with xylenol orange for the enrichment of Pb(II). They found that all four resins quantitatively sorb Pb(II) at pH 3.0–8.0 and the limit of detection was about 5 ng mL^{-1}. Saracoglu and Elci [13] developed a multi-element solid-phase extraction (SPE), preconcentration method for the determination of Bi, Cd, Co, Cu, Fe, Ni and Pb ions in drinking water, seawater and sediment samples by FAAS. The procedure was based on retention of analytes, from buffered sample solution, in the form of pyrolidine dithiocarbamate complexes on a column of Chromosorb-102 and elution with acetone.

Mendiguchia *et al.* [14] devised a procedure for the determination of Cu in seawater in which a bulk liquid membrane system was used to preconcentrate and separate the analyte prior to its analysis by FAAS. The membrane was an organic solution containing di-(2-ethylhexyl) phosphoric acid (DEHPA) in kerosene. The reliability of the new method was established by the analysis of several real samples of seawater, previously analyzed by a well-established procedure.

Giokas *et al.* [15] developed a batch preconcentration procedure for the determination of zinc by FAAS based on extraction onto a 1-(2-pyridylazo)-2-naphthol (PAN)-modified polymeric membrane. The ligand was immobilized on a water-insoluble cellulose acetate (CA) membrane and the retained zinc was eluted by rinsing with an acidified methanolic solution which was directly aspirated into the spectrometer without any other treatment. Seawater, river and lake water, wastewater as well as a reference material, were successfully analyzed.

Manzoori and Bavili-Tabrizi [16] developed a preconcentration method based on cloud point extraction for the preconcentration of Cu. The analyte in the initial aqueous solution, acidified with HCl, was complexed with O,O-diethyl dithiophosphate, and Triton X-100 surfactant was added. After phase separation at 40°C and dilution of the surfactant-rich phase with methanol, the enriched analyte was determined by FAAS with conventional nebulization. The method was successfully applied to the determination of Cu in drinking water, rainwater, serum and human hair samples.

Bujdos *et al.* [17] obtained a detection limit of 0.05 $\mu g\ L^{-1}$ for the determination of Cd by HG-AAS. They applied their procedure to the analysis of single and sequential extractions of different types of soil from contamined and uncontamined regions of Slovakia.

Van Alphen [18] tested segments of seven candles with wicks having a Pb metal core in a purpose-built combustion chamber to assess air Pb emissions. The emissions were collected on glass fiber filters, digested in concentrated

HNO_3 and analyzed by FAAS. The mean rate from the seven candles was 770 μg Pb h^{-1} and on average 20% of the Pb metal in the wick was emitted into the air with the remainder accumulating at the base of a molten wax-pool adjacent to the wick. It was concluded that the prolonged burning of candles in religious or ceremonial circumstances, or even in restaurants, may be of particular concern. Talebi [19] determined Pb in airborne particulate matter by FAAS following digestion with a mixture of nitric acid and hydrogen peroxide. The extraction procedure was checked by the analysis of an airborne particulate standard reference material (NIST SRM 1648), and it was concluded that Pb can be quantitatively (98%) extracted by the leaching process. A five-stage sequential extraction was performed to assess the potential mobility of Pb associated with airborne particulate matter.

Poykio *et al.* [20] collected woodland moss (Pleurozium schreberi) samples from 52 sampling sites around a chromium opencast mine and around ferrochrome and stainless steel works. The samples were dried, homogenized and digested with a mixture of concentrated nitric acid and perchloric acid, and the Cr, Ni and Zn contents determined by FAAS.

Jacob and Berndt [21] coupled a high temperature/high pressure flow digestion to an FAAS instrument for the rapid on-line dissolution of biological and environmental matrices. An electrically heated Pt–Ir capillary acted as a digestion tube at a temperature of 320°C. They showed that the digestion time for various matrices (wheat flour, spinach, orchard leaves, tomato leaves, bovine liver, pig kidney, sewage sludge, coal fly ash), as 2% suspensions in 2.5 M HNO_3, was about 1 min, and that the time for the whole procedure was between 3 and 8 min. As the equipment was completely glass-free, samples with a high silicate content could be digested by the addition of hydrofluoric acid.

Cuesta *et al.* [22] devised a new flow injection method for the determination of chemical oxygen demand (COD): in the first step, the sample was heated by microwave radiation, in the second step an anionic exchange resin retains the Cr(VI) that has not been reduced by the organic matter of the sample, and in the third step the Cr(VI), after being eluted, was determined by FAAS. COD values between 25 and 5000 mg L^{-1} could be determined and the limit of detection was about 7 mg L^{-1} COD.

Chang *et al.* [23] developed a rapid method for the determination of mercury in mainstream cigarette smoke in which a reusable gold–platinum gauze collected vapor-phase mercury. The particulate phase of the smoke was removed by an electrostatic precipitator or a Cambridge filter, followed by a second Cambridge filter to absorb volatile organic components. The vapor-phase mercury was determined by two-stage amalgamation CVAAS without matrix interference.

Coehlo *et al.* [24] developed a procedure for the direct determination of As(III) and As(V) in water samples by flow injection HG-AAS without pre-reduction of As(V). The flow injection system was operated in the merging zones configuration, where the sample and $NaBH_4$ are simultaneously injected (250 μL) into two carrier streams flowing at 3.6 mL min^{-1}. The detection limits were 0.3 ng mL^{-1} for As(III) and 0.5 ng mL^{-1} for As(V). A sampling rate of about 120 determinations per hour was achieved, requiring 30 mL of $NaBH_4$ and generating about 450 mL of waste.

Aucoin *et al.* [25] measured Cr, Cu, Pb and Zn in fish and sediment samples taken from Lake Boeuf, located in southeastern Louisiana. Lake Boeuf is a popular recreational fishing lake from which the following sunfish are caught and consumed: bluegill sunfish (*Lepomis macrochirus*), largemouth bass (*Micropterus salmoides*), and black crappie (*Poxomis nigromaculatus*). They extracted the metals from the fish and sediment samples with microwave digestion techniques for quantification by standard FAAS methods. The concentrations of trace metals in fish did not vary significantly between species or weight of fish; however, there was a seasonal trend. Metal concentrations in the core samples did not vary with depth, suggesting that the lake is not polluted.

8.11.3 Petroleum related

Platteau and Casabiell [26] developed a procedure for the determination of Al, Si, Fe, Ti, V, Mo, Ni and Co in catalysts used in the oil and petrochemical industries. They dissolved the samples in a mixture of nitric acid and hydrofluoric acid aided by microwave energy, followed by the addition of boric acid producing a matrix of fluoboric acid. Finally the analytes were quantified by FAAS. Recoveries better than 96% for all elements were obtained with relative standard deviations lower than 5% for major components and lower than 6% for minor components.

Mora *et al.* [27] investigated the behavior of a single-bore high-pressure pneumatic nebulizer (SBHPPN) as a tool for the analysis of lubricating oils by flow injection FAAS. They found that the SBHPPN was able, in conjunction with appropriate carrier (iso-butyl methyl ketone, IBMK), to nebulize pure lubricating oils giving higher sensitivities and lower detection limits.

Hernandez-Artiga *et al.* [28, 29] have evaluated sample treatment, based on partial microwave-assisted wet digestions in closed vessels, for the determination of wear metals in lubricating oils by FAAS. Samples, with different degree of wear, were analyzed for Fe, Cu, Pb and Cr by FAAS.

Guidroz and Sneddon [30] determined vanadium in burned and unburned Venezuelan crude oil by FAAS with a nitrous oxide–acetylene flame. They concluded, on the basis of laboratory experiments involving an open container, the surface of water, and the surface of a sandy soil, that in the burning of oil as a form of remediation, vanadium will not be removed.

Du *et al.* [31, 32] devised FAAS procedures for the determination of chromium naphthenate [31] and methylcyclopentadienyl manganese tricarbonyl

[32] in gasoline based on creating a microemulsion. They found that aqueous calibration standards were suitable.

8.11.4 Food and beverages

Dabeka *et al.* [33] developed a CVAAS method for the sub-ng g^{-1} determination of total Hg in a wide variety of foods with a CETAC Technologies M-6000A dedicated Hg analyzer. Based on a 2 g sample weight, the detection limit of the method over 12 batches averaged 0.30 ng g^{-1} wet weight. Recoveries of Hg added to 17 different foods, analyzed in a routine manner, averaged 97%, with individual recoveries ranging from 77 to 107%. Accuracy was confirmed by analysis of seven biological reference materials from the National Research Council of Canada and the National Institute of Standards and Technology.

Yaman and Dilgin [34] determined Cd in fruit samples by FAAS with a slotted tube atom trap. To identify the Cd phases most responsible for fruit-available Cd, soil near the fruit plants was also analyzed after various digestions and selective extractions. Total Cd in the soil ranged from 60 to 420 μg kg^{-1}, and they concluded that the Cd concentrations of the cherry, strawberry, apple and grape samples correlated with the Cd concentrations of the HNO_3–H_2O_2 digestion of the soil. The detection limit was 0.03 μg L^{-1}.

Solberg *et al.* [35] investigated the effect of copper coatings on net pens on farmed salmon. Samples of Atlantic salmon (*Salmo salar*), saithe (*Pollacius virens*), blue mussel (*Mytilus edulis*), brown seaweed (*Ascophyllum nodosum*) and sediment were collected from six different fish farms. Various tissue samples and seaweed were freeze dried, homogenized, wet digested and analyzed for copper by FAAS. They concluded that, from a nutritional point of view, the use of copper coatings does not affect the quality of the seafood products either within or around the net pen.

Moreno-Cid and Yebra [36] extracted Cu from solid mussel samples by a continuous ultrasound-assisted extraction system for determination by flow injection FAAS. Their method had a throughput of 11 samples per hour and a limit of detection of 0.06 μg g^{-1} (wet mass) for 30 mg of sample. They validated the procedure by the analysis of a reference standard material (TORT-1) and applied it to mussels from Galicia (Spain).

Kilic *et al.* [37] determined Ca, Fe, Mg and Zn in fresh eggs by FAAS, and Cu and Pb in chewing gum [38]. For the analysis of chewing gum, the analytes were extracted as the ammonium pyrolidine dithiocarbamate (APDC) complexes into methyl iso-buthyl ketone (MIBK). Arsenic was also determined (by HGAAS).

Frias *et al.* [39] determined Na, K, Ca, Mg, Fe and Cu by FAAS in sweet wines, bearing the denominations of origin of Lanzarote and La Palma islands (Canary Islands, Spain) from two consecutive harvests. Four analytes, Na, K,

Ca and Mg, were determined directly in dilute samples, but Fe and Cu required a microwave-assisted nitric acid digestion. They concluded it was possible to differentiate between Canarian and Malaga wines by using the content of Mg only, and separation of Lanzarote, La Palma and Malaga was achieved only with Na and Mg.

Olalla *et al.* [40] studied different methods of sample preparation for the determination of calcium in grape juice, wines, and other alcoholic beverages by FAAS, including dealcoholization, dry mineralization and wet mineralization by heating with using different acids and/or mixtures of acids. They concluded that although high-temperature mineralization with an HNO_3–HCO_4 mixture gave the best analytical results, mineralization with nitric acid at 80°C for 15 min gave the most satisfactory results in all cases, including those for wines with high levels of sugar and beverages with high alcoholic content.

Salacinski *et al.* [41] devised a method for determining the aluminium contents of water, tea leaf digest, tea and coffee infusions by means of on-line, cation-exchange flow-injection FAAS. They found that concentrations as low as 75 ng mL^{-1} could be determined with sample volumes in the range 6.0–24.0 mL.

Terres *et al.* [42] determined the Zn content of 300 food and 79 beverage samples by FAAS. They demonstrated the reliability and accuracy of their procedure by analyses of NIST and CBR-CEC reference materials. Mean zinc concentrations varied from 0.02 μg mL^{-1} in fresh water to 71.0 μg g^{-1} (fresh weight) in pork liver.

Vinas *et al.* [43] devised a slurry nebulization procedure for the rapid determination of Ca, Mg, Fe and Zn in flours obtained from several grains and legumes (wheat, wholemeal, maize, barley, rice, rye, vetch, soybean and field bean) by FAAS. Suspensions up to 3% m v^{-1} from the ground samples were directly introduced into the spectrometer with a simple flow injection manifold. Aqueous standard solutions were used for calibration.

Matusiewicz and Kopras [44] determined Ag, Cd, Cu, Fe, In, Mn, Pb, Tl and Zn in beer by FAAS with the aid of an *integrated atom trap* system mounted on a standard air–acetylene flame burner. Significant improvements in detection limits were achieved compared with those obtained with other atom trapping techniques (slotted tube and single water-cooled silica tube atom traps). For a 2-min preconcentration time, the detection limits were 3.0, 5.0, 1.2, 4.0 and 0.1 ng mL^{-1} for Cu, Fe, Mn, Pb and Zn, respectively. They also showed that the procedure could be applied to the determination of Ag, Cd, In and Tl.

Kabengera *et al.* [45] developed an HGFAAS method for the determination of As in foods. Samples were digested with HNO_3 and H_2O_2 and excess oxidants were eliminated by adding hydrochloric acid and urea. As(V) was then reduced by potassium iodide. The As contents ranged from 2.5 to 20 mg L^{-1}, and the limit of detection was 0.6 μg L^{-1}.

Plessi *et al.* [46] determined the mercury and selenium contents of fresh seafood by CVAAS and HGAAS, respectively. All the Hg values obtained were lower than the European Union's legal limit of 0.5 mg kg^{-1} fresh food, rising to 1.0 mg kg^{-1} for the edible parts of some listed species. The concentrations of Se varied between 0.073 (in perch) and 0.743 mg kg^{-1} (in tuna). In shellfish, the Hg content varied between 0.023 (in moscardino) and 0.150 mg/kg (in Mediterranean shrimp), while that of Se varied between 0.067 (in spiny lobster) and 0.605 mg kg^{-1} (in mussels). A significant difference was found between fish and shellfish for Hg, but not for Se. A large excess of Se in relation to Hg was observed (the mean molar ratio of Hg/Se is 0.23 for fish and 0.09 for shellfish), but no significant correlation was found between the two elements.

8.11.5 Clinical and forensic

Uchida *et al.* [47] determined tissue concentrations of Zn, Cu and Fe in bull terriers with lethal acrodermatitis (LAD) and tail-chasing behavior (TCB). Serum samples were taken from 29 bull terriers (9 control dogs, 6 dogs with LAD, 14 dogs with TCB), and liver and kidney specimens from 2 control dogs, 1 dog with LAD and 4 dogs with TCB. They concluded the primary cause of LAD may be copper deficiency, with zinc involved secondarily, or combined zinc and copper deficiencies. The role of ion deficiency in TCB was not clarified.

Baranguan *et al.* [48] proposed a method for the determination of Pb in hair by HGAAS based on reagent injection FIA. The detection limit was 3 ng mL^{-1} for an injected volume of 0.125 mL of 0.5 M nitric acid, 10% m/v hydrogen peroxide and 10% m/v sodium tetrahydroborate.

Chen *et al.* [49] developed a new flow injection-HG collection procedure for FAAS. They showed that interferences from concomitant elements could be eliminated by the addition of 0.2% ascorbic acid-thiourea masking reagents and that the sensitivity of their new method was one order of magnitude better than a continuous flow-HG method. The detection limit was 0.25 ng mL^{-1}. The method was validated by the accurate determination of trace Bi in standard biological reference material – human hair GBW07601.

Wang *et al.* [50] developed FAAS methods for As, Cu, Fe, Se and Zn in hair. Results from blackfoot disease patients at five clinical stages were compared with those from healthy controls. It was shown that As is a major cause of blackfoot disease, and that it antagonizes Se and Fe, which decreased in the advanced clinical stages of the disease.

Wabner and Sears [51] devised a flow injection method for FAAS that increases the analytical range for Ca, Li and Mg by use of the flow spoiler, rotation of the burner head and a flow injection dilution procedure. They applied the procedure for the determination of Ca, Mg and Li in urine, which normally has concentrations that exceed the conventional linearity limits.

The prevalence of allergic contact dermatitis (ACD) caused by nickel is increasing. The probable cause is the increased use of nickel-containing metals in intimate contact with the skin. The critical factor is the amount of nickel released from these metals (bioavailable nickel) onto the skin. Kanerva *et al.* [52] determined the amount of nickel released into synthetic sweat from metal samples by FAAS. They also found that some types of stainless steel release enough nickel to elicit allergic symptoms.

It has been suggested that selenium (Se) exhibits protective effects against mercury (Hg) toxicity in humans due to the formation of a Hg-Se complex bound to selenoprotein P in blood. Hol *et al.* [53] investigated the Se concentrations in persons who had been examined with respect to general health problems associated with dental amalgam fillings. They concluded persons with ill health related to dental amalgam might have a Se metabolism different from that of healthy people.

Fifteen metallic species, Ag, Al, Ca, Cd, Cr, Cu, Fe, K, Mg, Mn, Na, Ni, Pb, Sr and Zn, were determined in 46 cocaine samples confiscated by the Spanish police in Galicia (northwest Spain) by Bermejo-Barrera *et al.* [54]. Classification of the samples according to their geographic origin (Colombia and Venezuela) was achieved by pattern recognition techniques. Cocaine samples, around 0.5 g, were directly dissolved in 2 mL of 35.0% (v/v) HNO_3, diluted to 10 mL with ultrapure water. The metals were quantified by means of ETAAS (Ag, Al , Cd, Cr, Cu, Mn, Ni, Pb and Sr), FAAS (Ca, Fe, Mg and Zn), and FAES (K and Na). Results show that by the presence of trace and major elements the two geographic origins can be distinguished.

Zieba-Palus [55] examined two kinds of motor oil with respect to changes in their elemental composition during driving. The concentrations of Ba, Ca, Cd, Cu, Fe, Mg, Pb and Zn in oil samples were determined by FAAS and significant differences in the metal concentrations during normal car service were observed. Quantitative elemental composition of oils was considered suitable for differentiation between various oil samples for criminalistic purposes.

8.12 Bibliography

The following sources can be consulted for further and more detailed information on the topics covered in this chapter.

Atomic Absorption Spectrometry (3rd edition) by B. Welz and M. Sperling, published by Wiley-VCH in 1999 is a comprehensive text of about 950 pages. There are 6600 original publications cited. A substantial part of the book is given over to the determination of the individual elements and to their applications.

Analytical Atomic Spectrometry with Flames and Plasmas by J. Broekaert, also published by Wiley-VCH (in 2002) is a more modest text (about

370 pages) that covers the theoretical basis for almost all currently used atomic spectrometry techniques. It has minimal coverage of applications and sample preparation.

Atomic Absorption Spectrometry, Theory Design and Application edited by S. Haswell and published by Elsevier in 1991, is still a useful source of information as the bulk of the 530 pages are devoted to applications. Each of the various fields of application is covered by a well-known expert or experts.

Hydride Generation Atomic Absorption Spectrometry by J. Dedina and D. Tsalev, published in 1995 by John Wiley and Sons is still the only book devoted entirely to determinations by HG. The book is comprehensive in its coverage of the original literature up to the time of submission of the manuscript, containing about 1600 references to the original literature. The book is divided into two parts: part 1, fundamentals and general approaches occupies about 150 pages, whereas part 2, methodology and analytical applications occupies 250 pages. The material in part 2 is broken down by analyte.

Flow Analysis with Atomic Spectrometric Detectors, edited by A. Sanz-Medel and published in 1999 by Elsevier is a comprehensive coverage of the basic principles and applications of flow injection techniques for atomic spectrometry. The book is divided into three parts: the first deals with basic principles of flow injection, the second deals with flow injection techniques for sample pretreatment and the third deals with applications. The coverage is somewhat selective, whereas in Z. Fang's book, *Flow Injection Atomic Absorption Spectrometry* (published by John Wiley and Sons in 1995), the coverage of the literature up to this point is comprehensive. The book contains many excellent diagrams and figures which make visualizing the flow manifolds an easy task.

Inevitably, texts (which after all are tertiary resources) that review the recent original research literature become out of date, and to obtain an accurate picture of the activity in any area of analytical science, it is necessary to consult the primary and secondary literature. Now that many applications of flame atomic spectrometry techniques are published in a large number of applications-related journals, reading the primary literature can be a time-consuming procedure, although surveying the literature is relatively straightforward now that a number of electronic databases can be accessed from a desk-top personal computer. One way of keeping abreast of developments is to read the review literature. The most comprehensive coverage of the relevant fields is provided by the *Journal of Analytical Atomic Spectrometry* in a series of review articles, published every other month, known as the *Atomic Spectrometry Updates*. There are three application-based reviews: environmental analysis (February), clinical and biological materials, foods and beverages (April), and industrial analysis, metals, chemicals and advanced materials (December); and three technique-based reviews: atomic emission, absorption and fluorescence spectrometry and related techniques (June), atomic mass spectrometry (August), and X-ray fluorescence spectrometry (October). Each review, which covers the preceding 12 months, is written by a team of experts.

There is also much useful and relevant information available via the world wide web. Often entering a few key words into a search engine such as Google produces a wealth of information in a few minutes. Most manufacturers have websites with useful information and often there is relevant material (applications notes, *cook-books* of methods, original publications) that can be downloaded free of charge. As examples, the PerkinElmer Life and Analytical Sciences inorganic analysis site (*http://instruments.perkinelmer.com/ai/atomic/index.asp*) and the PS Analytical site (*http:/ /www.psanalytical.com/*) are well worth visiting.

References

[1] M.A. Janusa & J.N. Beck. Recent applications of flame atomic absorption spectrometry to environmental measurements, *Applied Spectroscopy Reviews*, 2002, **37**, 137–186.

[2] M.S. Dundar & H. Altundag. Heavy metal determination of house dust in Adapazari, Turkey, after earthquake, *Trace Elements and Electrolytes*, 2002, **19**, 55–58.

[3] R.A. Sutherland & C.A. Tolosa. Variation in total and extractable elements with distance from roads in an urban watershed, Honolulu, Hawaii, *Water Air and Soil Pollution*, 2001, **127**, 315–338.

[4] S.L. Ramesh, P.V.S. Raju, K.V. Anjaiah, R. Mathur, T.G. Rao, B. Dasaram, S.N. Charan, D.V.S. Rao, D.S. Sarma, M.R. Mohan & V. Balaram. Determination of gold in rocks, ores, and other geological materials by atomic absorption techniques, *Atomic Spectroscopy*, 2001, **22**, 263–269.

[5] L.A. Brandvold, V.T. McLemore & C. Oconnor. Distribution and partitioning of copper, lead and zinc in stream sediments above and below an abandoned mining and milling area near Pecos, New-Mexico, USA, *Analyst*, 1995, **120**, 1485–1495.

[6] J. Alvarado & R. Jaffe. Tube in flame atomisation: a way of enhancing detection limits in flame atomic absorption spectrometry, *Journal of Analytical Atomic Spectrometry*, 1998, **13**, 37–40.

[7] M.A. Elhelu, D.T. Caldwell & W.D. Hirpassa. Lead in inner-city soil and its possible contribution to childrens blood lead, *Archives of Environmental Health*, 1995, **50**, 165–169.

[8] P. Liu, Z.X. Su, X.Z. Wu & Q.S. Pu, *J. Anal. At. Spectrom.*, 2002, **17**, 125.

[9] P. Koscielniak & J. Kozak, *Anal. Chim. Acta*, 2002, **460**, 235.

[10] D.G. Rumbold, L.E. Fink, K.A. Laine, S.L. Niemczyk, T. Chandrasekhar, S.D. Wankel & C. Kendall. Levels of mercury in alligators (Alligator mississippiensis) collected along a transect through the Florida Everglades, *Sci. Total Environ.*, 2002, **297**, 239–252.

[11] R.G. Wuilloud, G.M. Wuilloud, J.C.A. de Wuilloud, R.A. Olsina & L.D. Martinez, *Atom. Spectrosc.*, 2002, **23**, 44.

[12] P.K. Tewari & A.K. Singh, *Talanta*, 2002, **56**, 735.

[13] S. Saracoglu & L. Elci, *Anal. Chim. Acta*, 2002, **452**, 77.

[14] C. Mendiguchia, C. Moreno & M. Garcia-Vargas, *Sep. Sci. Technol.*, 2002, **37**, 2337.

[15] D.L. Giokas, E.K. Paleologos, M.I. Prodromidis & M.I. Karayannis, *Talanta*, 2002, **56**, 491.

[16] J.L. Manzoori & A. Bavili-Tabrizi, *Microchem J.*, 2002, **72**, 1.

[17] M. Bujdos, V. Stresko, J. Medved & J. Kubova. Evaluation of cadmium determination in soil extracts by hydride generation atomic absorption spectrometry (HG AAS), *Fresenius Environmental Bulletin*, 2002, **11**, 686–690.

[18] M. van Alphen. Emission testing and inhalational exposure-based risk assessment for candles having Pb metal wick cores, *Science of the Total Environment*, 1999, **244**, 53–65.

[19] S.M. Talebi. Determination of lead associated with airborne particulate matter by flame atomic absorption and wave-length dispersive x- ray fluorescence spectrometry, *International Journal of Environmental Analytical Chemistry*, 1998, **72**, 1–9.

[20] R. Poykio, O.M. Tervaniemi, H. Torvela & P. Peramaki. Heavy metal accumulation in woodland moss (Pleurozium schreberi) in the area around a chromium opencast mine at Kemi, and in the area around the ferrochrome and stainless steel works at Tornio, Northern Finland, *International Journal of Environmental Analytical Chemistry*, 2001, **81**, 137–151.

[21] P. Jacob & H. Berndt, *J. Anal. At. Spectrom.*, 2002, **17**, 1615.

[22] A. Cuesta, J.L. Todoli, J. Mora & A. Canals, *Anal. Chim. Acta*, 1998, **372**, 399.

[23] M.J. Chang, R.L. McDaniel, J.D. Naworal & D.A. Self. A rapid method for the determination of mercury in mainstream cigarette smoke by two-stage amalgamation cold vapor atomic absorption spectrometry, *J. Anal. At. Spectrom.*, 2002, **17**, 710–715.

[24] N.M.M. Coelho, A.C. da Silva & C.M. da Silva. Determination of As (III) and total inorganic arsenic by flow injection hydride generation atomic absorption spectrometry, *Anal. Chim. Acta*, 2002, **460**, 227–233.

[25] J. Aucoin, R. Blanchard, C. Billiot, C. Partridge, D. Schultz, K. Mandhare, M.J. Beck & J.N. Beck. Trace metals in fish and sediments from Lake Boeuf, southeastern Louisiana, *Microchemical Journal*, 1999, **62**, 299–307.

[26] O. Platteau & D. Casabiell. Determination of metallic elements in catalysts by flame atomic-absorption spectrometry following microwave energy-assisted dissolution of samples, *Analyst*, 1994, **119**, 1705–1713.

[27] J. Mora, J.L. Todoli, F.J. Sempere, A. Canals & V. Hernandis. Determination of metals in lubricating oils by flame atomic absorption spectrometry using a single-bore high-pressure pneumatic nebulizer, *Analyst*, 2000, **125**, 2344–2349.

[28] D. Bellido-Milla, S.M. Ordaz-Garcia, J.L. Guerrero-Valiente & M.P. Hernandez-Artiga. Rapid procedure to determine wear metals in lubricating oils and the analysis of variance in the evaluation of sample preparation procedures, *Mikrochimica Acta*, 2002, **138**, 59–64.

[29] C. Sanz-Segundo, M.P. Hernandez-Artiga, J. de Cisneros, D. Bellido-Milla & I. Naranjo-Rodriguez. Determination of wear metals in marine lubricating oils by microwave digestion and atomic absorption spectrometry, *Mikrochimica Acta*, 1999, **132**, 89–94.

[30] J.M. Guidroz & J. Sneddon. Fate of vanadium determined by nitrous oxide-acetylene flame atomic absorption spectrometry in unburned and burned Venezuelan crude oil, *Microchemical Journal*, 2002, **73**, 363–366.

[31] B. Du, Q. Wei, S.R. Wang & W.L. Yu. Application of microemulsions in determination of chromium naphthenate in gasoline by flame atomic absorption spectroscopy, *Talanta*, 1997, **44**, 1803–1806.

[32] B. Du, Q. Wei & G.Y. Xu. Rapid determination of methylcyclopentadienyl manganese tricarbonyl in gasoline by FAAS, *Analytical Letters*, 1999, **32**, 1011–1021.

[33] R.W. Dabeka, P. Bradley & A.D. McKenzie. Routine, high-sensitivity, cold vapor atomic absorption spectrometric determination of total mercury in foods after low-temperature digestion *J. AOAC Int.*, 2002, **85**, 1136–1143.

[34] M. Yaman & Y. Dilgin. AAS determination of cadmium in fruits and soils, *Atomic Spectroscopy*, 2002, **23**, 59–64.

[35] C.B. Solberg, L. Saethre & K. Julshamn. The effect of copper-treated net pens on farmed salmon (Salmo salar) and other marine organisms and sediments, *Marine Pollution Bulletin*, 2002, **45**, 126–132.

[36] A. Moreno-Cid & M.C. Yebra. Flow injection determination of copper in mussels by flame atomic absorption spectrometry after on-line continuous ultrasound-assisted extraction, *Spectrochimica Acta Part B-Atomic Spectroscopy*, 2002, **57**, 967–974.

[37] Z. Kilic, O. Acar, M. Ulasan & M. Ilim. Determination of lead, copper, zinc, magnesium, calcium and iron in fresh eggs by atomic absorption spectrometry, *Food Chem.*, 2002, **76**, 107–116.
[38] Z. Kilic, E. Kenduzler & O. Acar. Determination of lead and copper in chewing gum samples by electrothermal-flame atomic absorption spectrometry using various chemical modifiers and arsenic by hydride generation, *Food Chem.*, 2002, **77**, 85–92.
[39] S. Frias, J.P. Trujillo, E.M. Pena & J.E. Conde. Classification and differentiation of bottled sweet wines of Canary Islands (Spain) by their metallic content, *European Food Research and Technology*, 2001, **213**, 145–149.
[40] M. Olalla, M.C. Gonzalez, C. Cabrera, R. Gimenez & M.C. Lopez. Optimized determination of calcium in grape juice, wines, and other alcoholic beverages by atomic absorption spectrometry, *Journal of Aoac International*, 2002, **85**, 960–966.
[41] H.J. Salacinski, P.G. Riby & S.J. Haswell. Coupled flow-injection analysis flame atomic-absorption spectrometry for the quantitative-determination of aluminum in beverages and water incorporating online cation-exchange, *Analytica Chimica Acta*, 1992, **269**, 1–7.
[42] C. Terres, M. Navarro, F. Martin-Lagos, R. Gimenez, H. Lopez & M.C. Lopez. Zinc levels in foods from southeastern Spain: relationship to daily dietary intake, *Food Additives and Contaminants*, 2001, **18**, 687–695.
[43] P. Vinas, N. Campillo, I.L. Garcia & M.H. Cordoba, *Food Chem.*, 1993, **46**, 307.
[44] H. Matusiewicz & M. Kopras, *J. Anal. At. Spectrom.*, 1997, **12**, 1287.
[45] C. Kabengera, P. Bodart, P. Hubert, L. Thunus & A. Noirfalise. Optimization and validation of arsenic determination in foods by hydride generation flame atomic absorption spectrometry *J. AOAC Int.*, 2002, **85**, 122–127.
[46] M. Plessi, D. Bertelli & A. Monzani. Mercury and selenium content in selected seafood *J. Food Compos. Anal.*, 2001, **14**, 461–467.
[47] Y. Uchida, A.A. Moonfanelli, N. Dodman, M.S. Clegg & C.L. Keen, Serum concentrations of zinc and copper in Bull Terriers with lethal acrodermatitis and tali-chasing behavior, *Am. J. Vet. Res.*, 1997, **58**, 808–810.
[48] M.T. Baranguan, F. Laborda & J.R. Castillo. Reagent injection FIA system for lead determination by hydride generation – quartz-tube atomic absorption spectrometry, *Analytical and Bioanalytical Chemistry*, 2002, **374**, 115–119.
[49] S.Y. Chen, Z.F. Zhang & H.M. Yu, *Anal. Bioanal. Chem.*, 2002, **374**, 126.
[50] C.T. Wang, W.T. Chang, C.W. Huang, S.S. Chou, C.T. Lin, S.J. Liau & R.T. Wang. Studies on the concentrations of arsenic, selenium, copper, zinc and iron in the hair of blackfoot disease patients in different clinical stages, *European Journal of Clinical Chemistry and Clinical Biochemistry*, 1994, **32**, 107–111.
[51] C.L. Wabner & D.C. Sears. Determination of urinary Ca, Mg and Li using flow injection with flame AAS, *Atomic Spectroscopy*, 1996, **17**, 119–121.
[52] L. Kanerva, T. Sipilainenmalm, T. Estlander, A. Zitting, R. Jolanki & K. Tarvainen. Nickel release from metals, and a case of allergic contact-dermatitis from stainless-steel, *Contact Dermatitis*, 1994, **31**, 299–303.
[53] P.J. Hol, J.S. Vamnes, N.R. Gjerdet, R. Eide & R. Isrenn. Dental amalgam and selenium in blood, *Environ. Res.*, 2001, **87**, 141–146.
[54] P. Bermejo-Barrera, A. Moreda-Pineiro, J. Moreda-Pineiro, A. Bermejo-Barrera & A.M. Bermejo-Barrera. A study of illicit cocaine seizure classification by pattern recognition techniques applied to metal data, *Journal of Forensic Sciences*, 1999, **44**, 270–274.
[55] J. Zieba-Palus. Examination of used motor oils by flame AAS for criminalistic purposes: a diagnostic study, *Forensic Science International*, 1998, **91**, 171–179.

9 Chemometrics in elemental analysis

Michael J. Adams and Melissa J. Romeo

9.1 Introduction

It is common for analytical spectroscopists to be involved in studies and investigations requiring the measurement of elemental composition. In many cases, selection of an appropriate technique and the actual analysis can appear to be the simplest and least controversial aspect of the study. Since the outcome of the investigation can depend almost exclusively on the quality of the analytical data, the analysis is worthy of considerable attention; from perhaps reformulating the questions to be answered by the study, identifying the information required, as well as then making the measurements and interpreting these data. Chemometrics is the study and application of mathematical and statistical techniques to the acquisition, manipulation and processing, and interpretation of analytical chemistry data. It lies at the very heart of analytical methodology and, as such, represents a key area of study for all analysts.

With such a wide brief and limited space, it is appropriate and necessary for us to limit our discussion of chemometrics in elemental spectrometry to those techniques in common use and those that illustrate important features. The interested reader is encouraged to refer to the original research papers for detailed experimental discussions, and to the many more general texts on chemometrics for further information and background reading.

Chemometric techniques are applied generally to multi-variate and multi-analyte (multi-element) studies. To this end, we will examine here examples and applications in experimental design and optimization, data processing (including calibration) and data interpretation involving pattern recognition methods.

9.2 Experimental design and optimization

The formal, statistical design of an experiment should provide an efficient and effective procedure for executing the experiment and obtaining information regarding the variables contributing to the data. The first step in experimental design is the recognition and then selection of variables for further study. A well-chosen experimental design will maximize information gained for a given effort, and will provide estimation on the influence of the different variables on the response variable or function. Once the relative importance of each variable has been assessed then, usually, an optimization procedure is undertaken in order to identify the best combination of a reduced number of variables. An excellent account of experimental design for chemists is presented by Davies [1], and Deming and Morgan [2] provide a thorough and comprehensive chemometric approach to the subject.

Applications of a wide range of experimental design techniques are to be found throughout the analytical literature, and the methods are applied in designing effective sampling regimes, developing sample-processing treatments, and investigating and optimizing instrument and experimental conditions. In experimental design terminology, the variables selected for study are referred to as *factors* and the value that each might take is its *level*. Thus, denoting the number of factors as k, the number of levels to be examined for each factor as m, and assuming no interaction between factors, the complete response profile would be obtained from $k \times m$ experiments or *trials*. Unfortunately life is rarely this simple and generally we can expect at least some of the factors to *interact*, i.e. our chosen factors are not statistically independent variables. In the presence of interaction *factorial-based* experimental designs are generally employed. The extreme design is represented by the *full factorial design* in which all k factors are varied simultaneously over all m levels. This leads to m^k trials of different factor combinations. For eight factors, each at two levels, then 256 experiments are required and at four levels the number of trials increases to over 65 000. Because of the exponential increase in number of trials required with increasing k, popular design strategies are the 2^k procedures. All combinations of all factors are studied, but each factor is allowed only two levels, designated *high* and *low* and a factor effect is defined as the difference in average response between the experiments conducted using the high level of the factor and the experiments carried out at the low level of the factor. Factor combinations for full 2^2 and 2^3 experimental designs are illustrated in Figs 9.1(a) and (b).

A simple example serves to illustrate the application and analysis of 2^k designs. Figure 9.2 presents a graph of a two-factor model described by a linear equation of form,

$$y_i = b_0 + b_1 x_{1,i} + b_2 x_{2,i} + b_{12} x_{1,i} x_{2,i} \tag{9.1}$$

where y_i represents the measured response for trial i conducted with factors x_1 and x_2 at the high or low levels. The regression coefficient b_0 represents an intercept, b_1 and b_2 are linear single-factor terms and b_{12} is an *interaction* term.

If x_1 and x_2 can both take high and low levels respectively of, say, 2 and 5 then the experimental design, along with recorded response values, is shown in Table 9.1.

In matrix form, Equation 9.1 can be written as

$$\mathrm{y} = \mathrm{X} \cdot \mathrm{b} \tag{9.2}$$

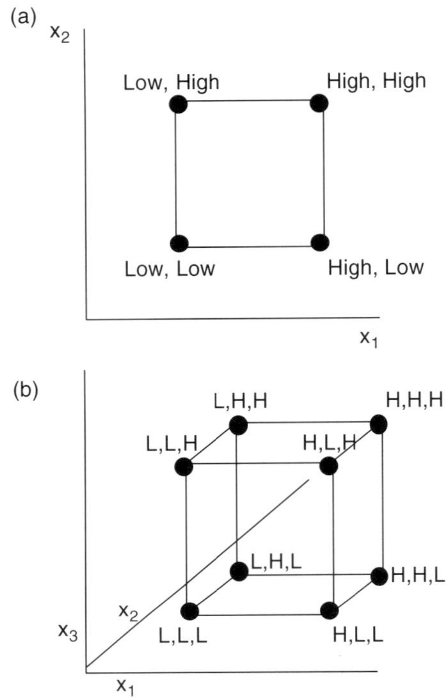

Fig. 9.1 Factor combinations for (a) 2^2 design and (b) 2^3 design.

where

$$y^T = [6.20\ 6.50\ 8.00\ 8.75]$$

and

$$X = \begin{bmatrix} 1 & 2 & 2 & 4 \\ 1 & 2 & 5 & 10 \\ 1 & 5 & 2 & 10 \\ 1 & 5 & 5 & 25 \end{bmatrix}$$

Equation 9.2 can be solved for coefficients b by the least squares method,

$$b = (X^T \cdot X)^{-1} \cdot X^T \cdot y \tag{9.3}$$

where X^T is the transpose of matrix X.

Substituting for X, X^T and y, then

$$b^T = [5.0\ 0.50\ 0\ 0.05] \tag{9.4}$$

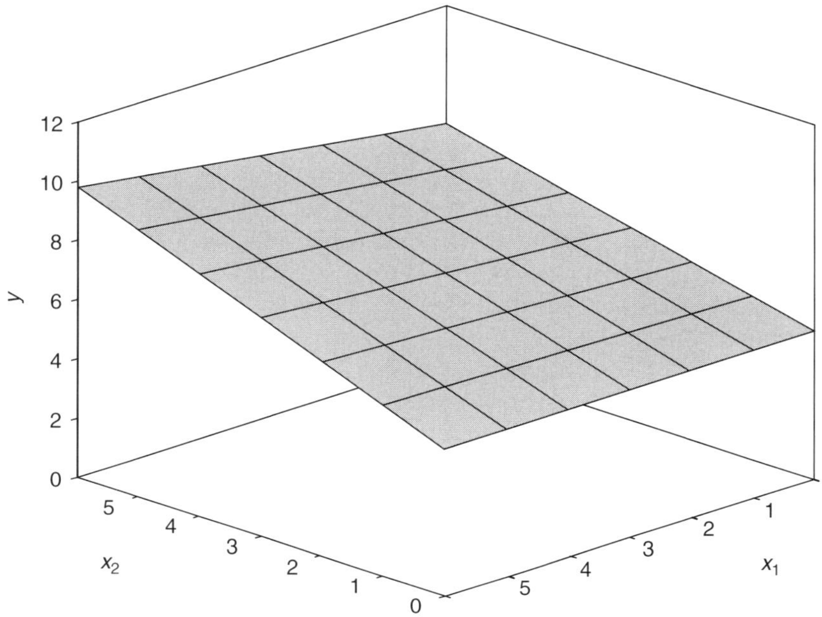

Fig. 9.2 Graph of the model, $y = 5.0 + 0.5\,x_1 + 0.05\,x_1x_2$.

Table 9.1 A 2^2 experimental design with recorded response values

Experiment			Response
	x_1	x_2	y
1	2	2	6.20
2	2	5	6.50
3	5	2	8.00
4	5	5	8.75

These values agree with the model data used to construct Fig. 9.2.

One important feature of this result is that the coefficient b_2 is zero, but this does not imply that factor x_2 is not important. Although the b_2x_2 term does not influence the model, x_2 is an important factor because of its role in the interaction term $b_{12}x_1x_2$.

The relative magnitude of the estimated regression coefficients can provide information on the importance of variables in the model, but care must be taken since these values will be highly dependent on the scale of variable, i.e. its units of measurement. As with regression analysis in modeling and calibration, the use of standardized or normalized variables can help to avoid misinterpretation.

The use of 2^k designs and their interpretation is frequently undertaken, classically, by applying such tools as the Yates' algorithm: a formal procedure for estimating the b coefficients for full two-level factorial designs described in many common statistics texts [3, 4]. Deming and Morgan provide a useful comparison of classical and regression factor modeling [2].

As the number of factors to be examined increases, the number of experiments required to produce a full 2^k factorial design grows exponentially. In many research projects the number of factors for initial study can be large, and a full 2^k design and analysis is not warranted. What may be required is a so-called *screening method* that helps in identifying a few important factors on which a more detailed study can be undertaken.

To investigate seven variables, a 2^k design requires 128 experiments. In most investigations it is often assumed that the influence of interactions of third order or higher is extremely small and can be excluded from the model. Therefore, to estimate the mean value (the constant term in the model), main effects and second order interaction effects of seven variables, performing 128 experiments may not be necessary as only 29 parameters need to be investigated (i.e. mean value, seven main effects and 21 second order interactions). If only the main effects are of interest, only $k + 1$ experiments are required and the response function is given by

$$y = b_o + \Sigma b_i x_i + \epsilon \tag{9.5}$$

It is possible to add experiments in order to separate or estimate interaction effects and the response (second order) function is given by

$$y = b_o + \Sigma b_i x_i + \Sigma\Sigma b_{ij} x_i x_j + \epsilon \tag{9.6}$$

It is not always necessary to investigate interactions between all the variables. Estimating the main effects is recommended using a linear model and after this evaluation the variables that have the largest influence on the result can be selected for further study. Therefore, large numbers of experimental variables can be investigated without the need to perform too many experiments.

Fractional factorial designs are selected fractions of full factorial designs. The size of the fraction will determine the number of effects to estimate and the number of experiments required. Fractions (1/2, 1/4, 1/8, 1/16 . . . $1/2^p$) of a factorial design require 2^{k-p} experiments, where k is the number of variables and p the size of the fraction.

To illustrate this in practice, seven variables can be studied in a 2^{7-4} fractional factorial design, where the design is defined by the model matrix for a 2^3 factorial design. A full factorial design with 7 variables requires 128 experiments, whereas the 2^{7-4} fractional factorial design requires eight experiments (1/16 of the full design). Although fractional factorial designs have the advantage that large numbers of variables can be investigated whilst requiring fewer experiments to be undertaken, less information is gained as a result. Fewer experiments cause contamination of the main effects by interaction effects, and the main effects are said to be *confounded.*

The application of fractional factorial designs to elemental analysis by atomic spectrometry is well represented in the research literature. Recent examples include examination of sequential extraction methods for metal partitioning in soils and sediments [6], a study of cold vapor and hydride generation procedures for electrothermal atomic absorption spectrometry [7], and sample pretreatment for ultrasound-assisted acid leaching of metals from human hair [8].

A specific fractional design procedure that has proved popular is the Plackett–Burman model [2, 7, 8]. Its advantage is that it provides an efficient design for screening a large number of factors in order to identify those of most significance, with a small number of experiments. The design is referred to as a *saturation* design and all interactions between factors are confounded with new main effects. The number of experiments conducted is equal to the number of factors considered. Once the most important factors have been identified by the experimental design, the next step for the experimenter is usually to optimize the factor's levels; i.e. determine the combination of main factors providing the best result.

One commonly referenced optimization technique is the *simplex method* [5, 9]. A simplex is a geometric figure defined by a number of coordinates, the number of points being equal to one more than the number of factors examined. Thus, for two factors the simplex is a triangle, and in three dimensions (i.e. three factors) it is defined by a tetrahedron. The simplex method is described by a sequential series of experiments that aim to drive the simplex toward the region of optimum (usually maximum) response. At each step, the new corner of the new simplex defines the next set of experimental conditions to be examined. The *walk* through factor space is illustrated in Fig. 9.3. The simplex method has several disadvantages. The number of experiments required to attain the optimum region is not known in advance, the method can be slow to converge on a complex response surface, and the technique is difficult to implement if the number of factors exceeds four or five. In addition, the simplex method does not provide information on factor interaction effects, and there is always the danger of the simplex procedure converging to a local rather than a global optimum.

An often preferred optimization technique that incorporates elements of experimental design is the *central composite design* procedure [2, 5]. Central composite design methods belong to the class of techniques referred to as *response surface methodology*, by which the shape of the response surface as a function of the factors studied is modeled and an optimum determined from the model. To be able to determine an optimum (a turning point in the surface graph) it is necessary for the polynomial model to contain quadratic terms,

$$y = b_o + \sum_{i=1}^{k} b_i x_i + \sum_{i=1}^{k} b_{ii} x_i^2 + \sum_{j=1}^{i-1} \sum_{i=1}^{k} b_{ij} x_i x_j + \epsilon \qquad (9.7)$$

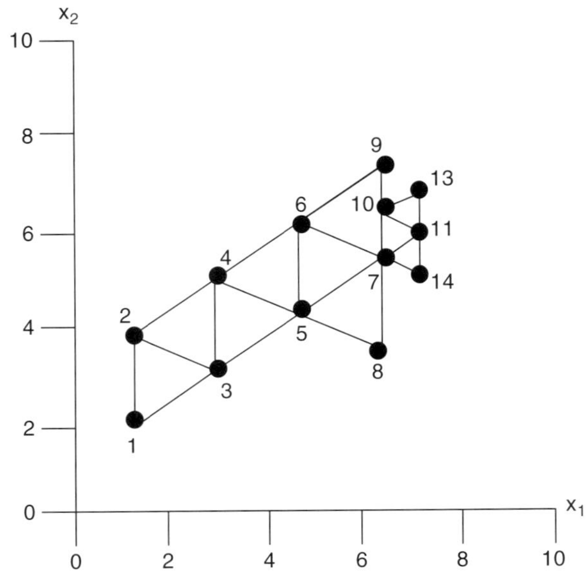

Fig. 9.3 A two-factor simplex and typical movement through the factor space to some optimum region.

A full central composite design includes the following [5]:

(a) A full factorial or fractional factorial design.
(b) Experiments at the center of the factorial region.
(c) Axial points situated on the axes away from the origin.

The designs for two and three variable systems are illustrated in Fig. 9.4 and these have proved popular with experimental spectroscopists. For further details the reader is referred to the tutorial by Lundstedt *et al.* [5] and the recent papers [6, 8].

9.3 Multi-variate statistics

Before continuing with our review of applications of chemometrics in elemental analysis and atomic spectrometry, it is worthwhile considering and introducing some of the basic terms and concepts that are fundamental to this study. In particular, we need to examine the statistics associated with variable interactions. This issue has already arisen in our discussions on experimental design and optimization, where it is evident that simple *one-variable at a time* studies will not provide the information required. In data processing and analysis, similar considerations must be examined.

Commonly in analytical chemistry, we assume our recorded data to have arisen from a parent population with a *normal distribution*. For a single variable

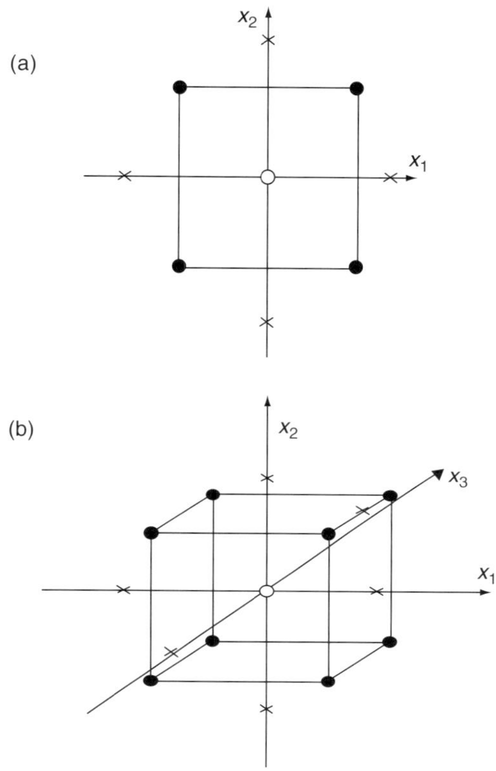

Fig. 9.4 Central composite designs for (a) two variables and (b) three variables. The different points refer to those from the factorial design (●), the center point (o) and the axial points (×) situated away from the center.

the univariate data can be summarized by its *mean* value, providing a location of maximum likelihood along the variable's range axis, and its spread as given by the *variance* or *standard deviation*. We can describe multi-variate data in a similar manner. Each variable will have its mean and variance, but in addition there is a measure of interaction referred to as *covariance*. This is defined as

$$\mathrm{Cov}_{x_1,x_2} = \frac{\sum_{i=1}^{n}(x_{1,i} - \bar{x}_1)(x_{2,i} - \bar{x}_2)}{n - 1} \tag{9.8}$$

where x_1 and x_2 are the variables under study and n is the number of measurements. Note that if there is only one variable, then $x_1 = x_2$ and Equation 9.8 is the well-known definition of variance.

It is usually convenient in multi-variate data analysis to employ matrix representation. If X is the matrix of variate-centered data then,

$$\mathrm{Cov} = \frac{X^{T} \cdot X}{n - 1} \tag{9.9}$$

It is apparent from Equation 9.9 that the covariance matrix is both square and symmetric, with the diagonal elements being the variances of the individual variances. Just as a variance value is sensitive to the units of measurement, so is covariance, and to estimate the degree of interaction between variables without the influence of units, the correlation coefficient, r, can be employed.

$$r_{x_1,x_2} = \frac{\mathrm{Cov}_{x_1,x_2}}{\sigma_{x_1} \cdot \sigma_{x_2}} \tag{9.10}$$

where σ_x is the standard deviation of variable x. The multivariate curves in Fig. 9.5 illustrate graphically the role of covariance and correlation in describing interactions or dependence between variables.

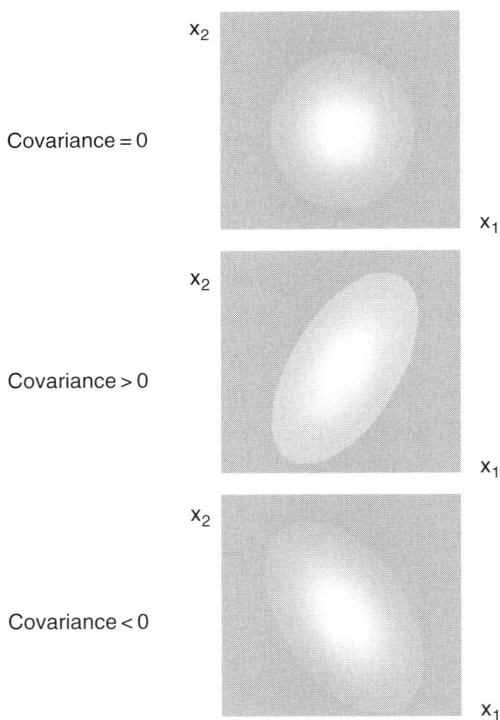

Fig. 9.5 The bivariate normal distribution for variables having negative, zero and positive correlation.

Consider the data presented in Table 9.2 representing common metal ions determined in 15 samples of spring water from three different supplies. The mean, variance and standard deviation for each variable are also provided. Following mean-centering of this data the between-analyte covariance and correlation matrices can be calculated, and these are shown in Table 9.3(a, b) respectively. Of course, we can also study relationships between samples, and Table 9.4

Table 9.2 Concentration (expressed as mg kg^{-1}) of common trace elements in spring waters. The sample codes (A . . . C) indicate the origin of the samples

Sample No.	Ca	Mg	K	Na
A1	3.0	5.1	0.9	13.8
A2	3.5	4.8	0.8	13.6
A3	3.4	5.3	1.0	13.7
A4	2.6	5.6	1.1	14.2
A5	2.1	4.9	1.0	14.4
B1	9.7	2.8	0.8	16.3
B2	10.5	3.4	0.7	16.5
B3	10.8	3.2	1.2	15.9
B4	9.6	2.7	1.1	15.7
B5	9.8	3.5	1.3	15.9
C1	8.5	4.2	1.0	11.8
C2	7.6	4.0	1.1	12.3
C3	8.1	3.9	1.2	12.1
C4	8.0	4.5	0.8	12.4
C5	7.9	4.1	0.7	11.9
Mean	7.0	4.1	0.98	14.0
Variance	9.91	0.81	0.03	2.89
Standard deviation	3.15	0.90	0.19	1.70

Table 9.3 The between-analyte covariance matrix (a) and correlation matrix (b) for the data from Table 9.2

		Ca	Mg	K	Na
(a)	Ca	9.91	−2.55	0.05	1.60
	Mg	−2.55	0.81	−0.02	−0.08
	K	0.05	−0.02	0.04	0.03
	Na	1.60	−0.75	0.03	2.89
(b)	Ca	1	−0.90	0.09	0.30
	Mg	−0.90	1	−0.10	−0.49
	K	0.09	−0.10	1	0.10
	Na	0.30	−0.49	0.10	1

Table 9.4 The correlation matrix between the samples from Table 9.2

	A1	A2	A3	A4	A5	B1	B2	B3	B4	B5	C1	C2	C3	C4	C5
A1	1	0.9982	0.9996	0.9981	0.9969	0.8369	0.8237	0.7938	0.8261	0.8393	0.8097	0.8699	0.8368	0.8592	0.8431
A2	0.9982	1	0.999	0.9926	0.9915	0.8679	0.8562	0.8288	0.858	0.8703	0.8429	0.8979	0.8681	0.8876	0.8733
A3	0.9996	0.999	1	0.9967	0.9945	0.8455	0.8337	0.8042	0.8348	0.8483	0.8224	0.8798	0.8478	0.8705	0.8548
A4	0.9981	0.9926	0.9967	1	0.9985	0.8017	0.7872	0.7549	0.79	0.8042	0.7731	0.8383	0.8019	0.8274	0.8096
A5	0.9969	0.9915	0.9945	0.9985	1	0.8057	0.7887	0.7576	0.7948	0.8071	0.7673	0.8357	0.7993	0.8211	0.8038
B1	0.8369	0.8679	0.8455	0.8017	0.8057	1	0.9985	0.9967	0.9997	0.9998	0.9825	0.991	0.9926	0.9812	0.9841
B2	0.8237	0.8562	0.8337	0.7872	0.7887	0.9985	1	0.9985	0.998	0.9992	0.9909	0.9936	0.997	0.9874	0.9907
B3	0.7938	0.8288	0.8042	0.7549	0.7576	0.9967	0.9985	1	0.9974	0.9968	0.9873	0.9861	0.9927	0.9791	0.9843
B4	0.8261	0.858	0.8348	0.79	0.7948	0.9997	0.998	0.9974	1	0.9992	0.9804	0.9879	0.9906	0.9774	0.981
B5	0.8393	0.8703	0.8483	0.8042	0.8071	0.9998	0.9992	0.9968	0.9992	1	0.9859	0.9934	0.9949	0.985	0.9876
C1	0.8097	0.8429	0.8224	0.7731	0.7673	0.9825	0.9909	0.9873	0.9804	0.9859	1	0.9925	0.9973	0.9957	0.9982
C2	0.8699	0.8979	0.8798	0.8383	0.8357	0.991	0.9936	0.9861	0.9879	0.9934	0.9925	1	0.998	0.9976	0.9973
C3	0.8368	0.8681	0.8478	0.8019	0.7993	0.9926	0.997	0.9927	0.9906	0.9949	0.9973	0.998	1	0.9965	0.9983
C4	0.8592	0.8876	0.8705	0.8274	0.8211	0.9812	0.9874	0.9791	0.9774	0.985	0.9957	0.9976	0.9965	1	0.9994
C5	0.8431	0.8733	0.8548	0.8096	0.8038	0.9841	0.9907	0.9843	0.981	0.9876	0.9982	0.9973	0.9983	0.9994	1

presents the between-samples correlation matrix. Since the covariance matrix provides information about the distribution of individual variables as well as the degree of interaction, or dependence, between all pairs of variables, its calculation is usually the starting point for most multivariate analysis techniques.

9.4 Exploratory data analysis

Once multivariate (multi-element) data has been obtained from many, possibly hundreds, of the samples, then a common starting point in its interpretation is to search for patterns in the data. This process may be undertaken at the same time as a reduction in dimensionality of the data. One of the most widely used methods of pattern recognition is that of *cluster analysis*.

9.4.1 Cluster analysis

Cluster analysis can aid in the determination and identification of underlying patterns and structure from tables of data. The first step in cluster analysis generally involves converting the data into a measure of similarity or dissimilarity between samples. This is to ensure that similar samples are grouped together in a cluster, whilst maximizing the differences between the clusters. The number of clusters observed and the class membership of samples may be dependent on the similarity measure chosen.

The correlation coefficient is often used as a measure of similarity in cluster analysis. The correlation coefficient estimates interrelation, free from the effects of measurement units, between variables. Once the correlation coefficient matrix has been formed, samples are grouped together based on the largest correlation values, and a *dendrogram* is drawn which reflects the correlation of samples, and forms clusters accordingly. The process is hierarchical and agglomerative in that at each stage of the process more samples are included in clusters until, eventually, all samples are combined in a single cluster.

Examination of the data in Table 9.4 indicates that samples B1 and B5 have the highest mutual correlation. In hierarchical cluster analysis these two samples are *combined* to form a *new* sample, or cluster, B1B5. The correlation between this *sample* and all others is taken as the average value of that between B1 and others and B5 and others. The resulting correlation matrix is thus reduced in size by one row and one column and the process repeated until only one cluster remains, containing all samples. Further operation details and worked examples are presented by Adams [10]. At each stage the correlation between connected pairs of samples or clusters is noted and used to produce the final result, a dendrogram. For the spring water samples, this is presented in Fig. 9.6.

In general, the correlation coefficient is not a good measure of similarity with which to perform cluster analysis on data based on elemental composition. The correlation coefficient is only a measure of colinearity between data

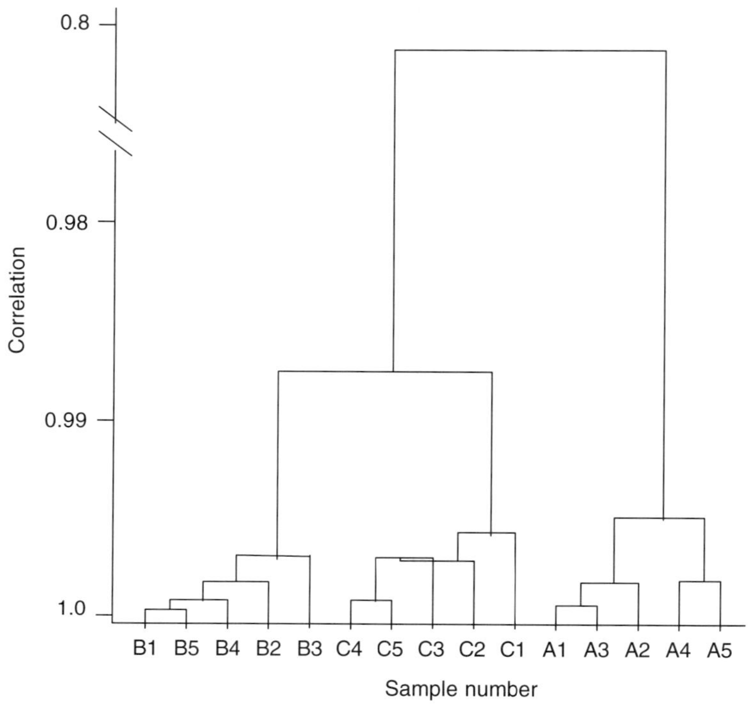

Fig. 9.6 Dendrogram of the spring water samples from Table 9.2 using inter-sample correlation values (Table 9.4) as the measure of similarity.

and does not take account of non-linear relationships or the absolute magnitude of variates. More common measures of similarity (or more correctly, dissimilarity) are offered by distance metrics. Using a distance measure, it is assumed that samples lying close together in the multidimensional space defined by the recorded variables will be more similar than those far apart. A wide range of between-sample distance metrics are available for cluster analysis [10], but the most common is the Euclidean distance, d_{AB}, defined as the distance between samples A and B according to

$$d_{AB} = \sum_{i=1}^{k} (x_{A,i} - x_{B,i})^{1/2} \tag{9.11}$$

where $x_{A,i}$ and $x_{B,i}$ are the values for variable i for samples A and B. For two variables Equation 9.11 reduces to the well-known Pythagoras equation.

The popularity of cluster analysis as an efficient tool for exploratory data analysis resides in its simplicity of use (most statistical software packages generally include a range of clustering algorithms) and its ability via the dendrogram to reduce complex data sets to a single, two-dimensional plot. Examples of the application of cluster analysis to elemental analysis and atomic spectrometry are numerous and wide-ranging. Marcos *et al.* using X-ray fluorescence (XRF) have employed the technique to relate trace element concentrations in tea to geographical origin [11], Custs *et al.* have employed cluster analysis to classify soils [12], and Haswell and Walmsley have applied the technique to a study of coffee and wine samples [13].

The disadvantages of cluster analysis lie in the severe distortion of relationships between samples brought about by compression to two dimensions, and the lack of information to be gained as to the origin and cause for the clusters. Thus, from Fig. 9.6, cluster analysis correctly identifies three distinct groups of water sample and correctly classifies each sample to its correct parent group. Which analytes are responsible, or play a significant role in this classification are not evident. With increasing numbers of samples and analytes, the situation becomes even more difficult to resolve.

To help overcome this problem, pre-processing of the data can be undertaken to reduce the dimensionality of the data. One powerful technique employed to achieve this is the *principal components analysis* (PCA). Not only does PCA enable efficient dimensionality reduction, it also provides information on structure within data.

9.4.2 Principal components analysis

When modeling experimental data the recorded response is termed the *dependent* variable and the set factors are the *independent* variables. This terminology is somewhat unfortunate since, as we have seen in the examples above, these experimental factors are rarely independent in a statistical sense. From Table 9.3(b), the concentrations of magnesium and calcium are highly (negatively) correlated. Given this high colinearity between these two analytes, it is fair to say that together they are not as valuable in describing differences, or similarities, between samples as two elements of low or zero correlation. The information content of a variable can be considered as being related to its variance, and its value as a single source of information is given by its covariance, with other variables. A variable or a factor of importance in modeling will exhibit a high variance and have low covariance values, i.e. it will have low or zero correlation with other variables.

PCA is a mathematical manipulation of the analytical data (or more accurately its covariance matrix) in order to form a new set of factors that are independent (i.e. have zero correlation) and are ranked according to relative variance of each new factor. These new factors, called *principal components*, are linear combinations of the original variables and represent the result of a geometric transformation of original variables to new, artificial axes. A principal component, PC, can be defined by

$$\mathrm{PC} = \mathrm{X} \cdot \mathrm{A} \tag{9.12}$$

where PC is the resultant matrix with each column representing a sample expressed in the new factor space and X is the original data. The transformation matrix A is referred to as the loadings matrix and each column of A is an *eigenvector* derived from the covariance matrix of X.

In the case of two variables, say the magnesium and calcium concentrations in the spring water samples, then Equation 9.12 can be expressed as

$$\begin{aligned} \mathrm{PC1} &= a_{1,1} \cdot [\mathrm{Ca}] + a_{1,2} \cdot [\mathrm{Mg}] \\ \mathrm{PC2} &= a_{2,1} \cdot [\mathrm{Ca}] + a_{2,2} \cdot [\mathrm{Mg}] \end{aligned} \tag{9.13}$$

where $a_{i,j}$ are the elements of the 2×2 matrix of eigenvector.

The calculation of these coefficients, the eigenvector matrix, is discussed and illustrated in numerous chemometrics texts [10, 14, 15]. Eigen-analysis functions are standard in most commercial mathematical and statistical software packages and the algorithms employed need not concern us here.

For the calcium and magnesium data from Table 9.2 the eigenvectors are

$$\mathrm{A} = \begin{bmatrix} -0.97 & -0.25 \\ 0.25 & -0.97 \end{bmatrix} \tag{9.14}$$

and substituting into Equation 9.13 provides the two new variables or factors, PC1 and PC2. The calcium and magnesium mean-centered original concentration data are shown in the *scatter-plot* of Fig. 9.7(a) and the 15 samples in the new factor space in Fig. 9.7(b).

Associated with each computed eigenvector is another value, the *eigenvalue* that represents the variance of the new factor, PC1 or PC2. For this bivariate data, the first eigenvalue is 10.58 and the second is 0.14. The sum of these eigenvalues is the same as the sum of the variances of the original [Ca] and [Mg] variables, but the new factors PC1 and PC2 are now uncorrelated ($r = 0$) and $\sigma\mathrm{PC1}^2 > \sigma\mathrm{PC2}^2$. Thus, the first PC is oriented along the axis of greatest variance of the original variables, the next PC is that vector orthogonal to the first and explains the next largest proportion of variance. Subsequent PCs can be calculated (if the number of original variables is greater than two) that will be orthogonal to all preceding axes and that explain decreasing amount of variance.

For the full 15×4 matrix of concentration data from Table 9.2, eigenanalysis produces the results shown in Table 9.5. These results provide considerable information about the nature and structure inherent in the original data that is not immediately obvious from inspection of Table 9.2.

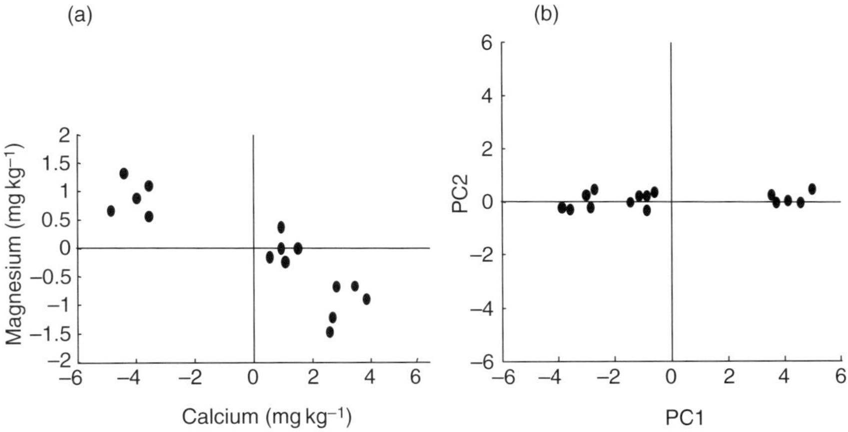

Fig. 9.7 The 15 samples from Table 9.2 displayed in (a) the mean-centered calcium and magnesium concentration space and (b) in the pattern space defined by the derived two principal components, PC1 and PC2.

Table 9.5 Eigenvectors and eigenvalues extracted from the covariance matrix in Table 9.3(a)

	Factor 1	Factor 2	Factor 3	Factor 4
(a) Eigenvectors				
Ca	0.944	0.236	−0.230	0.002
Mg	−0.253	0.072	−0.965	−0.007
K	0.006	−0.008	0.005	−0.999
Na	0.211	−0.970	−0.128	0.008
(b) Eigenvalues				
Eigenvalue	10.96	2.55	0.10	0.04
% Variance	80.29	18.68	0.73	0.30
% Cumulative contribution	80.29	98.97	99.7	100

Firstly, the eigenvalues as they decrease in magnitude from PC1 to PC4 indicate that the first two PCs contain nearly 99% of the variance associated with the data. Thus the 15 samples can be displayed in a two-dimensional space defined by PC1 and PC2 with only a 1% loss of information. This plot is shown in Fig. 9.8, and the three groups of samples, each containing five samples are evident and clearly separated.

Secondly, we can investigate the structure of these principal components to discover the underlying reasons for the group separations. The eigenvectors associated with PC1 and PC2 are presented in Table 9.5. For PC1, the eigenvector is dominated by a large, positive value for calcium concentration. The PC1 axis of Fig. 9.8 will, therefore, serve to separate samples according to the relative concentration of this analyte. The second eigenvector, associated with PC2 and having the next largest eigenvalue, is characterized by a large negative

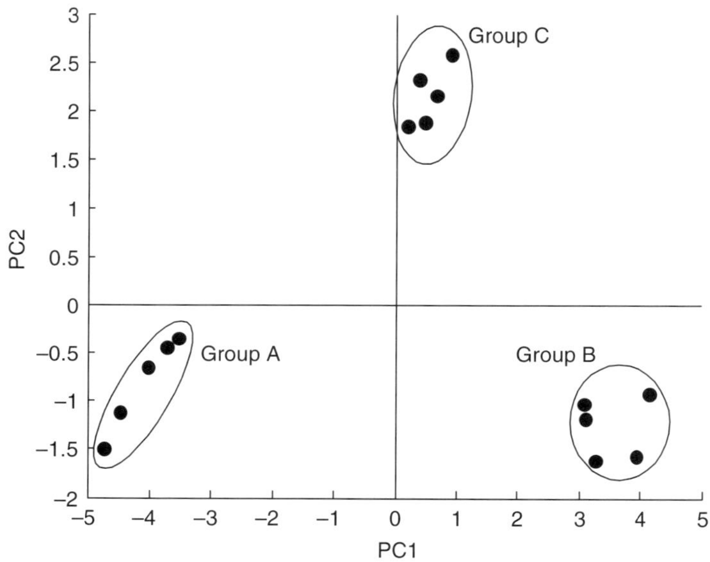

Fig. 9.8 Projection of spring water elemental data on to PC1 and PC2 axes.

coefficient for sodium. The PC2 axis will separate samples according to this analyte's concentrations.

This simple PCA investigation has highlighted several important features. The original data can be efficiently and effectively represented in a much lower dimensional space than the original four variables. The reduced dimensionality permits direct visual identification of clusters or groups within the data. Finally, we can identify the underlying causes for the patterns and groups within the data. Given that in addition to these features, the PCs represent new, uncorrelated factors it is not surprising that PCA represents one of the most common and important chemometric techniques employed in data analysis and modeling. Selected PCs can be used as variables in further processing techniques, such as cluster analysis or discriminant analysis, or used to great advantage in regression analysis when developing calibration models.

Benninghoff *et al.* have used PCA and cluster analysis to reveal potential classification of human tissue samples for diagnostic cancer recognition according to trace element analysis by total reflection XRF spectroscopy [16]. Klemene *et al.* [17] using inductively coupled plasma-atomic emission spectrometry (ICP-AES) data, have examined the classification of archeological copper ingots with the aid of PCA, and Zhuang *et al.* [18] have reported the study of coastal water pollution using ICP-AES and flame emission spectrometry with PCA treatment of the elemental data.

9.5 Discriminant analysis

Discriminant analysis represents a class of pattern recognition techniques. However, unlike cluster analysis, which requires no *a priori* assumptions regarding structure or likely groups within our data, discriminant analysis aims to assign an unclassified sample to one of the two (no more) known and characterized groups. This technique is sometimes referred to as *supervised pattern recognition*. Parametric and non-parametric methods are available and widely used. One of the most common non-parametric technique is that referred to as *k-nearest neighbors* (*k-NN*).

With reference to Fig. 9.9, two groups of samples are described by two variables, and our task is to assign the new unknown sample to group A or B. Application of the *k-NN* algorithm first involves computing the distance between the new sample and all samples in the known groups; the Euclidean distance is the most widely used metric. For the single nearest-neighbor procedure, the new sample is then assigned to that group containing the sample's nearest neighbor. A *3-NN* procedure can also be followed, and here the sample is assigned to the group containing the majority of nearest neighbors. The *k-NN* algorithm is easy to implement but can be time consuming if thousands of objects make up the known groups, as the unknown sample distance to each known object has to be calculated and compared.

Parametric methods rely on knowing, or assuming, the form of distribution of the parent population of the objects in the known and classified groups. For

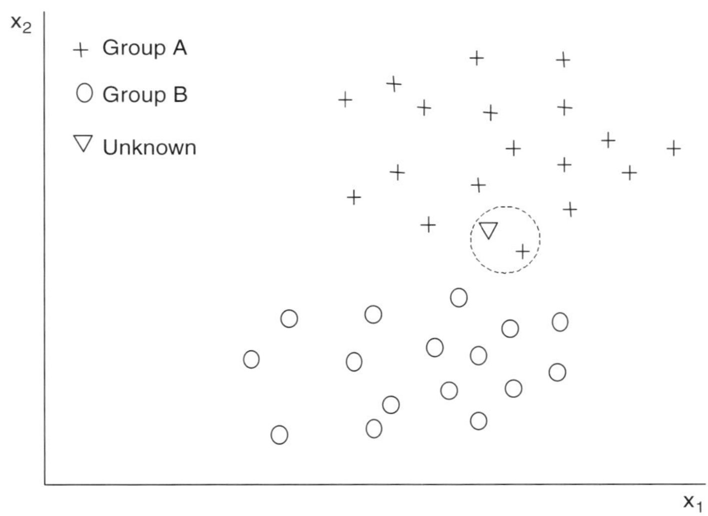

Fig. 9.9 Two characterized groups, A and B, of objects defined by variables x_1 and x_2, and an unknown sample assigned to group A according to the 1-*NN* rule.

example, if we have two groups of objects and can reasonably assume that both are from normally distributed parent populations of similar covariance, *linear discriminant analysis* can be employed. This situation is illustrated in Fig. 9.10(a). A straight line can be fitted between the two profiles that efficiently separates the groups A and B.

A sample is assigned to group A if

$$((\mu_A{}^T \cdot Cov^{-1} \cdot x) - 0.5(\mu_A{}^T \cdot Cov^{-1} \mu_A)) > ((\mu_B{}^T \cdot Cov^{-1} \cdot x) - 0.5(\mu_B{}^T \cdot Cov^{-1} \cdot \mu_B)) \quad (9.15)$$

where μ_A is the vector of variable means for group A, Cov is the covariance matrix of all the objects, and x is the vector of variables defining the new, unclassified sample.

If the two reference groups have dissimilar covariance matrices, then a linear discriminant function may no longer provide effective group separation and a quadratic function should be derived, Fig. 9.10(b) [10]. Rodriguez *et al.* studied the mineral composition of milk samples by AAS, and using discriminant analysis achieved a 98% correct classification of samples including infant formula, human and pasteurized milks [19].

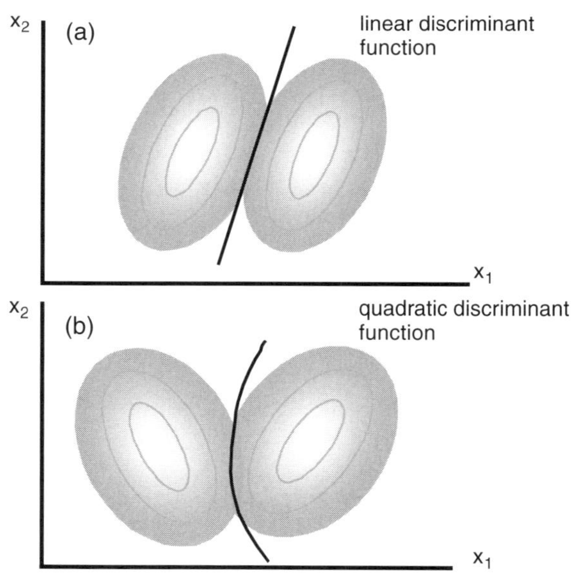

Fig. 9.10 Bivariate normal distribution for two groups of objects with (a) similar covariance and (b) widely differing covariance. Efficient differentiation between the groups can be achieved with a linear function in the former case, whilst the latter requires a quadratic discriminating function.

9.6 Calibration models

The design development, application and validation of appropriate calibration models for the prediction of unknown analyte concentrations from suitable standards have led to a remarkable number of publications. The situation in elemental, atomic spectroscopy appears far less settled than in molecular spectrometry in the optical spectral regions where orthogonal polynomial regression analysis, usually via principal components regression (PCR) or partial least squares regression (PLSR), has become a standard procedure offered by most instrument manufacturers. The space available here only permits us to review the basic features of multivariate regression analysis and to direct the reader to specific published applications and developments. The simplest situation for quantitative analysis is when the system contains a single active component, and a simple linear model can be employed. This model can be used for multi-component mixtures if spectra of individual components are sufficiently distinct and do not overlap.

The single-component linear model assumes the recorded spectroscopic signal y, which is a linear function of component concentration, x:

$$y = k_0 + k_1 x \tag{9.16}$$

where k_0 is a constant term corresponding to some background signal and k_1 is a sensitivity value. For M standard samples, M pairs of (y, x) are obtained, and the best estimates of the calibration parameters k_0 and k_1 are obtained by the least squares method. Application of the ordinary least squares (OLS) model serves to minimize the square of the differences, E, between predicted, $\hat{y}$, and measured, y, response values.

$$E = \sum_{i=1}^{M} (\hat{y}_i - y_i)^2 \tag{9.17}$$

By substitution,

$$E = \sum_{i=1}^{M} (k_0 + k_1 x_i - y_i)^2 \tag{9.18}$$

and differentiating E with respect to k_0 and k_1 provides the so-called normal equations:

$$\begin{aligned} k_0 M + k_1 \Sigma x_i &= \Sigma y_i \\ k_0 \Sigma x_i + k_1 \Sigma x_i^2 &= \Sigma (x_i y_i) \end{aligned} \tag{9.19}$$

Solving the simultaneous equations for k_0 and k_1 provides the OLS model for the best straight line describing the regression of y on x.

Application of the OLS model requires three assumptions to be satisfied.

- Experimental error is associated only with the dependent variable, y.
- Repeated measurements of y_i are normally distributed.
- Variance of y is independent of sample concentration, x, i.e. the data are homoscedastic.

A visual representation of the data and prediction equation is provided by the calibration graph. Visual pots of data are always important since they allow the user to gauge the fit of the equation and identify outliers or rogue values.

The multi-component linear model is a generalization of the single-component model. Application of the multi-component model assumes the following conditions:

- Spectral responses of each component are independent of other components.
- Spectral response of zero corresponds to zero concentration of each component.
- At each wavelength measured, the spectral response is linearly dependent on the component concentration in the whole concentration range studied.

These conditions imply that the response, y_i, at the ith coordinate (wavelength, wavenumber, etc.) in a system of N_c components, is given by the sum:

$$y_i = \sum_{j=1}^{N_c} x_j \cdot k_{ij} \tag{9.20}$$

where x_j is the concentration of component j, and k_{ij} is the sensitivity coefficient of component j at wavelength i. If a spectrum is recorded at N_w wavelengths, it can be represented by the N_w dimensional vector y, and

$$\mathrm{y} = \mathrm{K} \cdot \mathrm{x} \tag{9.21}$$

where x is the N_c dimensional column vector of component concentrations, and K is the matrix of sensitivity coefficients (the calibration matrix of pure component spectra).

If there are fewer wavelengths than components ($N_w < N_c$), then a solution cannot be obtained. If $N_w > N_c$ then the system is over-determined and a solution is obtained by the method of least squares.

$$\begin{aligned} y &= K \cdot x \\ K^T \cdot y &= K^T \cdot K \cdot x \\ (K^T \cdot K)^{-1} \cdot K^T \cdot y &= x \end{aligned} \tag{9.22}$$

The effect of increasing the number of wavelengths, in an attempt to improve the results, can lead to greater errors by inclusion of noise. This method can be used for moderately complex mixtures. However, its application does rely on the knowledge of the complete composition (concentration of every component) of the calibration mixture.

The *inverse least squares* (ILS) method is sometimes referred to as the P-matrix method. The calibration model is transformed so that component concentrations are defined as functions of the recorded spectral responses,

$$X = P \cdot Y \tag{9.23}$$

Each column of the matrix P, of dimensions $N_c \times N_w$, contains calibration coefficients for each component in a mixture, at each wavelength.

For a series of calibration mixtures, the concentration of each component in each mixture is put in the matrix X ($N_c \times M$) and the individual spectrum in columns of Y ($N_w \times M$).

Equation 9.23 is solved for P by the least squares method,

$$P = X \cdot Y^T (Y \cdot Y^T)^{-1} \tag{9.24}$$

This can be solved if $M > N_w$, and then for the unknown mixture sample, with its spectrum in the column vector, y,

$$X = P \cdot y \tag{9.25}$$

As ILS requires more standard calibration mixtures than the number of wavelengths, then in many applications satisfying the inequality, $M \geq N_w$, can be a problem. This can limit the spectral range or resolution of the data set used in the analysis. This restriction does not cause a problem for applications requiring a large set of calibration standards, and for applications in which the relevant spectral information is contained in relatively few wavelengths or factors (e.g. PCR, PLS). It is often the case with multiple regression analysis involving large numbers of independent variables that collinearity or high correlation exists between these variables. Thus, wavelength selection becomes an important issue. Correlation adds redundancy to the regression model, as more variables may be included than is necessary for optimal performance. The following rules-of-thumb provide a useful guide to variable selection:

- Select the smallest number of variables possible.
- Use the maximum information contained in the independent variables.
- Choose independent variables that are not highly correlated with each other.

Employing data compression and extracting the eigenvectors to form new, orthogonal variables address the apparent conflict between these rules. Two methods are in common use, PCR and PLSR.

PCR uses PCA. The principal components of the independent variable data are calculated and a selected few are used in the regression equation. The most frequent application of PCA is in cases where the independent variables are expected to be collinear – hence its popularity in spectral analysis. This collinearity means that the independent variables will contain some dominating types of variability that carry most of the available information. Redundancy and smaller noise variabilities can then be removed.

The purpose of PCA is to express the main information in the independent variables by a lower number of derived variables, the so-called principal components.

Equation 9.23 gives the general ILS regression model. The aim in PCR is to replace the original Y matrix by a smaller set of variables that are orthogonal, i.e. principal components.

$$X = p \cdot Z \tag{9.26}$$

Rearranging this equation gives p,

$$p = x \cdot Z^T \cdot (Z \cdot Z^T)^{-1} \tag{9.27}$$

Concentrations for unknown solutions are then obtained by mapping their spectral data to the selected principal components and solving for x.

In order to reduce the number of variables in the regression equation we need to reduce the number of principal components. Analogous to multiple linear regressions we can select principal components according to their correlation with the dependent variable concentration. Thus,

$$X = p \cdot Z^* \tag{9.28}$$

where Z* represents the reduced matrix of the selected principal components.

The regression equation expresses the relationship between analyte concentration and principal components, and the principal components are combinations of the original spectral variables.

Therefore,

$$z_1 = v_1{}^T \cdot Y_o$$

and
$$z_2 = v_2{}^T \cdot Y_o \tag{9.29}$$

where v_i is the eigenvector associated with the ith principal component.

The regression coefficients related to the recorded variables are therefore given by

$$B = p \cdot V^{*T} \tag{9.30}$$

with V* here representing the reduced eigenvector matrix.

The non-linear iterative partial least squares (NIPALS) algorithm [15] extracts one factor (principal component) at a time. Each factor is obtained iteratively by repeated regression of response data on principal components to obtain improved eigenvectors.

In employing principal components as regression factors, we have succeeded in fully utilizing all the measured variables and developed new uncorrelated variables. In selecting which eigenvectors to use, the first employed is that corresponding to the largest eigenvalue, the second corresponding to the next largest eigenvalue, and so on. This strategy assumes that the major eigenvectors correspond to phenomena in the data matrix of relevance in modeling the dependent variable. Although this is generally accepted as being the case for most analytical applications, another data compression method can be employed if variables having high variance but little relevance to *y* are thought to be present. This method is *partial least squares regression* (PLSR).

With collinear data, PCR can give substantial improvement over classical regression methods since it defines projections of the measured data as the dominant phenomena. The choice of eigenvectors for prediction generally tends to use those with the largest eigenvalues, i.e. the greatest variance. However, it may be the case that some major eigenvectors have no relevance to modeling or predicting concentration.

PLSR provides an alternative calibration technique. It differs from PCR by actively using the response variable in the decomposition of the data matrix. By balancing the X- and y-information the method reduces the impact of large, but irrelevant X-variations in the calibration modeling. Probably the most common algorithms for PLS regression with a single y-variable are *PLS1*, as provided by Martens and Naes [15], and SIMplied partial least squares (SIMPLS), from de Jong [20]. The SIMPLS algorithm was proposed and developed by de Jong as a novel scheme for PLSR. The algorithm calculates the PLS factors directly as linear combinations of the original variables. It is simple to implement and has been demonstrated to be faster than the conventional PLS1 procedure. SIMPLS is entirely equivalent to PLS1 for univariate analyte determination.

Adams and Allen have investigated the application of PCR and PLSR for quantitative elemental analysis by XRF spectrometry, as alternatives to the more traditional polynomial models requiring prior knowledge of sample composition [21, 22]. Facchin *et al.* have applied multivariate calibration models to energy-dispersive X-ray spectrometry data for the determination of lead and sulfur [23], and Schramm has examined the use of PCR in energy-dispersive XRF [24]. The application and comparison of chemometric modeling techniques for calibration using electrochemical/ICP-MS data have been reported by Donachie *et al.* [25], and treatment of ICP-MS data using PLS has also been reported by Moberg *et al.* [26].

9.7 Conclusion

As both instrumental analytical techniques and their applications become more sophisticated, and ever more information is demanded from the analyst, so the need grows for sophisticated tools to aid interpretation and evaluate data. Chemometrics is developing to respond to these needs and is providing fruitful areas for research and application in elemental analysis. Here we have been able to give only an overview and summary of some of the basic tools available. The text references provide more detailed background and discussion, and the original research papers will serve to illustrate and guide users on typical trends and applications. Many software packages that reduce or eliminate the problems associated with coding mathematical and data analysis functions are commercially available and the interested reader is encouraged to experiment with their data, to seek and investigate what structures and information can be extracted.

References

[1] L. Davies, *Efficiency in Research, Development and Production: The Statistical Design and Analysis of Chemical Experiments*, The Royal Society of Chemistry, Cambridge, UK, 1993.

[2] S.N. Deming & S.L. Morgan, *Experimental Design: A Chemometric Approach*, 2nd edn, Elsevier Science Pub., Amsterdam, The Netherlands, 1993.

[3] C. Chatfield, *Statistics for Technology*, Chapman and Hall, London, UK, 1976.

[4] J.C. Miller & J.N. Miller, *Statistics for Analytical Chemistry*, Ellis Horwood, Chichester, UK, 1993.

[5] T. Lundstedt, E. Seifert, L. Abrams, B. Thelin, A. Nyatrom, J. Pettersen & R. Bergman, *Chemomet. and Intell. Lab. Systems*, 1998, **42**, 3–40.

[6] R. Santamaria-Fernandez, A. Moreda-Pineiro & S.J. Hill, *J. Environ. Monit.*, 2002, **4**, 330–336.

[7] P. Bermejo-Barrera, A. Moreda-Pineiro & A. Bermejo-Berrera, *J. Anal. At. Spectrom.*, 2000, **15**, 121–130.

[8] J. Moreda-Pineiro, P. Lopez-Mahia, S. Muniatcgui-Lorenzo, E. Fernandez-Fernandez & D. Prada-Rodriguez, *Spectrochim. Acta* (Part B), 2002, **57**, 883–895.

[9] M.J. Adams in S.J. Haswell (ed.), *Practical Guide to Chemometrics*, Marcel Dekker Inc., New York, USA, 1992.
[10] M.J. Adams, *Chemometrics in Analytical Spectroscopy*, The Royal Society of Chemistry, Cambridge, UK, 1995.
[11] A. Marcos, A. Fisher, G. Rea & S.J. Hill, *J. Anal. At. Spectrom.*, 1998, **13**, 521–525.
[12] G. Custs, S. Boeykens, D. Cicerone & C. Vazquez, *X-Ray Spectrom.*, 2002, **31**, 132–135.
[13] S.J. Haswell & A.D. Walmsley, *J. Anal. At. Spectrom.*, 1998, **13**, 131–134.
[14] E.R. Malinowski, *Factor Analysis in Chemistry* (2nd edn), J. Wiley & Sons, New York, USA, 1991.
[15] H. Martens & T. Naes, *Multivariate Calibration*, J. Wiley & Sons, Chichester, UK, 1991.
[16] L. Benninghoff, D. von Czarnowski, E. Denkhaus & K. Lemke, *Spectrochim. Acta* (Part B), 1997, **52**, 1039–1046.
[17] S. Klemene, B. Budic & J. Zupan, *Anal. Chim. Acta*, 1999, **389**, 141–150.
[18] Z. Zhuang, X. Wang, C. Yang, E. Zhu & P. Yang, *Microchim J.*, 1995, **51**, 138–144.
[19] E.M. Rodriguez, M. Sanz Alaejos & C. Diaz Romero, *J. Agri. Food Chem.*, 1999, **47**, 1520–1524.
[20] S. de Jong Chemomet, *And Intell. Lab. Systems*, 1993, **18**, 251–257.
[21] M.J. Adams & J.R. Allen, *J. Anal. Atom. Spectrom.*, 1998, **13**, 119–124.
[22] M.J. Adams & J.R. Allen, *Analyst*, 1998, **123**, 537–541.
[23] I. Facchin, C. Mello, M.I.M.S. Bueno & R.J. Poppi, *X-Ray Spectrom.*, 1999, **28**, 173–177.
[24] R. Schramm, *Anal. Chim. Acta*, 2000, **420**, 197–203.
[25] A. Donachie, A.D. Walmsley & S.J. Haswell, *Anal. Chim. Acta*, 1999, **378**, 235–243.
[26] L. Moberg, K. Petterson, I. Gustavsson & B. Karlberg, *J. Anal. Atom. Spectrom.*, 1999, **14**, 1055–1059.

Index